RADAR SCATTERING AND IMAGING OF ROUGH SURFACES

SAR Remote Sensing
A SERIES

Series Editor
Jong-Sen Lee

Polarimetric SAR Imaging
Theory and Applications
Yoshio Yamaguchi

Imaging from Spaceborne and Airborne SARs,
Calibration and Applications
Masanobu Shimada

For more information about this series, please visit: https://www.routledge.com/SAR-Remote-Sensing/book-series/CRCSRS

RADAR SCATTERING AND IMAGING OF ROUGH SURFACES

Modeling and Applications with MATLAB®

Kun-Shan Chen

CRC Press is an imprint of the
Taylor & Francis Group, an **informa** business

First edition published 2021
by CRC Press
6000 Broken Sound Parkway NW, Suite 300, Boca Raton, FL 33487-2742

and by CRC Press
2 Park Square, Milton Park, Abingdon, Oxon, OX14 4RN

CRC Press is an imprint of Taylor & Francis Group, LLC

ISBN: 978-1-138-54126-9 (hbk)
ISBN: 978-1-351-01157-0 (ebk)

Typeset in Times
by MPS Limited, Dehradun

Visit the Support Materials: https://www.routledge.com/9781138541269.

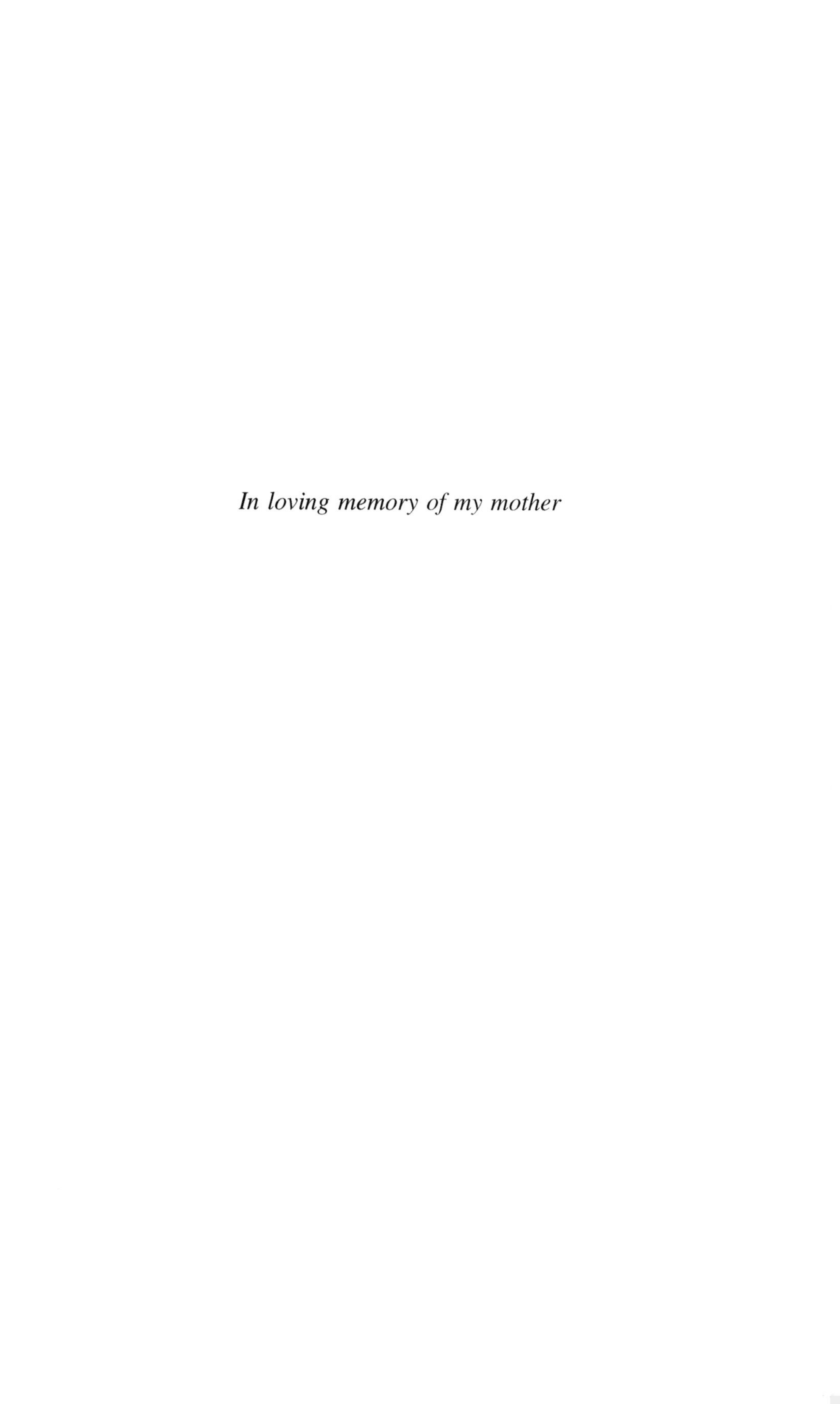

In loving memory of my mother

This book is dedicated to
Professor Adrian K. Fung for his mentorship
my students for their patience and inspiring feedback
my beloved wife, Jolan, for her lifetime of support and encouragement
my beloved children—Annette, Vincent, and Lorenz—for
bringing me joy

Contents

Preface

The rough surface is everywhere in our physical world. The radar scattering from a rough surface is a difficult boundary value problem. However, it is of particular interest to the remote sensing society since all the natural surfaces are rough in the full spectrum of electromagnetic waves. Due to the complexity of the problem, mathematically tractable solutions or computationally manageable solutions are still in high demand. Research on wave scattering from random media has a very long history, back to the early 1960s. Great advances have been observed in the past 30 years since then.

This book attempts to systematically give the physical mechanisms and numerical computation aspects on radar scattering and imaging of the rough surface. Understanding of electromagnetic wave scattering from a dielectric rough surface is of interest and is vital in lines of geophysical parameters retrieval from microwave scattering and emission measurements, surface finish, roughness-caused phased error of a reflective surface, etc. Radar imaging of such rough surfaces is also found in many application domains. This book covers both subjects and their physical connection and offers useful models for improving understanding and for practical applications. Electromagnetic waves scattering from a randomly rough surface is of palpable importance in many disciplines and bears itself in various applications, spanning from surface treatment to remote sensing of terrain and ocean. For example, in microwave remote sensing of terrain, it has been a common practice to retrieve, by analyzing the sensitivity of the scattering behavior and mechanisms, the geophysical parameters of interest from the scattering and/or emission measurements. Another example is that by knowing the backscattering patterns, one may be able to detect the presence of the undesired random roughness of a reflective surface such as an antenna reflector, and thus accordingly devise a means to correct or to compensate the phase errors with respect to the phase center. Therefore, it has been both theoretically and practically motivated to study the electromagnetic wave scattering from the random surfaces. To tackle the complex and yet intricate mathematical derivations, while to retain a high level of accuracy, models beyond conventional have been developed under certain physical-justified assumptions. Among the assumptions, one was to use a simplified Green's function by dropping off the phase term associated with the random surface height. Doing so might be more profoundly critical among all assumptions but greatly alleviating the burden of mathematical derivations, and yet unavoidably degrading the model accuracy, to certain extents, depending on the surface property and observation geometry.

This book differs in that it does not cover these topics from pure electromagnetic wave theory in the framework of Maxwell's equations, but treats the subjects from an application-oriented manner while still maintaining minimum theoretical foundations. In line with these points, the statistical description of the randomly rough surface is introduced. The uncertainties of the surface parameters that are inputs to the scattering model are addressed to facilitate the tracking of model predictions. In terms of parameter retrieval from radar measurements, global sensitivity analysis is regarded as a premier issue to warrant a rich information content of which the radar scattering signal contains in particular configurations. Following the sensitivity analysis is the approach to estimating the surface parameters by least-square fitting and a neural network. Examples of inversion from various radar satellites are provided.

The second part of the book, closely related to the scattering theory, covers the radar imaging—in the context of image formation of the scattering process. The coverage is motivated by the fact that in recent years there has been an explosive growth of interest in radar imaging of random media, from remote sensing of terrain to medical imaging. The advance of theoretical research in stochastic electromagnetic fields and modern processing techniques has been progressively made at a swift pace like never before.

The electromagnetic wave-media interactions and the image formation are naturally connected, so that some critical issues regarding image sensitivity and information content, uniqueness, accuracy, and resolution can be properly addressed. The subject covers a broad treatment of subjects closely related to the imaging of random surface and media through theory modeling, numerical simulation, and laboratory measurements. Computer codes in MATLAB® for selected model algorithms, accompanied by users' applications, are available as Support Materials at the CRC website: https://www.routledge.com/9781138541269.

Acknowledgments

Many pioneers in the research and development of wave scattering and imaging inspired me to explore this fascinating field. In many ways, it is not possible to include them all. Professor Adrian K. Fung, my thesis advisor, taught me the rough surface scattering. He has been a rich source of consultation in my career as a teacher.

I am indebted to countless individuals for their original contributions. I am grateful to Dr. Jong-Sen Lee, a pioneer in SAR polarimetry and a life-long mentor and friend, who introduced me to the Lee filter and taught me how to filter out the speckle in life. Inspiration from Professor George W. Pang has been a valuable asset to grip the essence of wave scattering theory. The book is a collective material from my teaching and research in the past 30 years. Many colleagues and students contributed, in one way or the other, to the idea fermentation and implementation presented in this book. Drs. Peng Xu, Yu Liu, Jiangyuan Zeng, Ming Jin, and Xiaofeng Yang participated in research projects that generated the outcomes given in this book. My former student Dr. Ying Yang contributed to advance the IEM model, originally developed by A. K. Fung. Dr. Chiang-Yen Chiang designed and implemented computer simulations of radar images. Drs. Chih-Yuan Chu, Johnson Ku, and Jhih-Syuan Huang conducted measurements in both field and laboratory. Dr. Rui Jiang has offered feedback on the subjects. Dr. Dengfeng Xie developed the codes to explore the ocean scattering. Mr. Tie-Yan Yin contributed to the TR imaging through random media. Ms. Suyun Wang developed the radar calibration technique. Of the many current students, Zhen Xu did a great job in simulating the Moon-based SAR system, which is presented in the last chapter. Tingting Li has provided a thorough bistatic SAR imaging simulation. I am also thankful for my students, Xiuyi Zhao and Mingde Li, for implementing the MATLAB® codes, and Tian Ma, Zhihua Song, Jinlin Zhou, and Tingyu Meng for their contributions.

The microwave anechoic chamber provided by Xuchang University in Henan has been an indispensable facility to conduct many experimental measurements. Thanks are due to Deans Dr. Genyuan Du and Zhili Zhang, and many individuals, for their support. Guilin University of Technology in Guilin has generously provided resources in many aspects to support my writing of this book. Drs. Wen-Yen Chang and Amy Liu have been friendly critics and sources of information in many ways. I am thankful to a great friend, Dr. Jinxue Wang, for his permanent enthusiasm in manuscript reading and advising. Special thanks go to Ying Yang, who keenly helped to compile and edit all the scattered material and has offered timely technical assistance during this book project. My long-time research secretary, Peiling Chen, provided both administrative and technical support. I owe her all my sincere thanks.

I must thank my book series editor, Dr. Jong-Sen Lee, who invited and encouraged me to write this book. His input and advice were invaluable for this project. I would also like to extend my gratitude to my editor, Ms. Irma Britton, and her team members at CRC Press. Her patience and assistance has made this book possible.

Author

Kun-Shan Chen earned a PhD in electrical engineering at the University of Texas at Arlington in 1990. From 1992 to 2014, he was a professor at the National Central University, Taiwan. From 2014 to 2019, he was with the Institute of Remote Sensing and Digital Earth, Chinese Academy of Sciences, China. Since 2019, he has been a professor at Guilin University of Technology, where his research interests include microwave remote sensing theory, modeling, system, and measurement, and intelligent signal processing and data analytics for radar.

He has authored or co-authored over 160 refereed journal papers, contributed 10 book chapters, co-authored (with A. K. Fung) *Microwave Scattering and Emission Models for Users* (Artech House, 2010), authored *Principles of Synthetic Aperture Radar: A System Simulation Approach* (CRC Press, 2015), and co-edited (with X. Li, H. Guo, X. Yang) *Advances in SAR Remote Sensing of Ocean* (CRC Press, 2018).

His academic activities include being a guest editor for a special issue on Remote Sensing for Major Disaster Prevention, Monitoring and Assessment (2007) in *IEEE Transactions on Geoscience and Remote Sensing*, a guest editor for the special issue on Remote Sensing for Natural Disaster (2012) in *Proceedings of the IEEE*, IEEE GRSS Adcom member (2010–2014), a founding chair of the GRSS Taipei Chapter, an associate editor of the IEEE *Transactions on Geoscience and Remote Sensing* since 2000, founding deputy editor-in-chief of IEEE *Journal of Selected Topics in Applied Earth Observations and Remote Sensing* (2008–2010). He served as guest editor of the special issue of Data Restoration and Denoising of Remote Sensing Data and special issues of Radar Imaging Theory, Techniques, and Applications, both for *Remote Sensing*, and was co-chair of the Technical Committee for IGARSS 2016 and IGARSS 2017. He served as a member of the editorial board of the *Proceedings of the IEEE* (2014–2019), and has been a member of the editorial board of the *IEEE Access* since 2020. He is a Fellow of IEEE.

1 Introduction

1.1 SURFACE SCATTERING AS A RANDOM PROCESS

Surface scattering occurs because of the presence of an irregular surface which separates two electromagnetic media through which the wave propagates. Here the surface, being a dielectric or perfect conducting boundary, is a random process, so is the scattering process itself. By random process, we mean the surface height is statistically random in time or in space, or in both time and space. Understanding the radar image of the scene requires a firm grip of the multiphysics mechanisms which participate in the formation of the image that is the result of the scattering process in view of the receiving of the returned signal off from the target scene of interest. By multiphysics, it also implies multiscale, particularly keen in the world of electromagnetic waves. Imaging of a rough sea is a good example, where the sea surface contains many scales with varying wavelengths (Figure 1.1). When there are objects, usually complex in geometry, riding on the rough sea, physical, mathematical, and numerical modeling of the scattering process becomes very complicated.

A photo of a rough surface shown in Figure 1.1 with random height $z(x, y; t)$ can be (1) spatially isotropic or anisotropic; (2) homogeneous or heterogeneous depending upon the lower medium properties; (3) stationary or nonstationary; (4) ergodic or non-ergodic [1]. In general, the surface has spatial wavenumber and spectral wavenumber—they are related by a dispersion relationship. The surface height and its spectral density are given by

$$Z(K_x, K_y; \omega) = \Im[z(x, y; t)] = \int_{-\infty}^{\infty}\int_{-\infty}^{\infty}\int_{-\infty}^{\infty} z(x, y; t)\exp[+j(K_x x + K_y y + \omega t]dxdydt \tag{1.1}$$

$$z(x, y; t) = \Im^{-1}[Z(K_x, K_y; \omega)] = \frac{1}{(2\pi)^3}\int_{-\infty}^{\infty}\int_{-\infty}^{\infty}\int_{-\infty}^{\infty} Z(K_x, K_y; \omega)\exp[-j(K_x x + K_y y + \omega t]dK_x dK_y d\omega \tag{1.2}$$

where

$$\vec{K} = (K_x, K_y) = (K, \varphi) \tag{1.3}$$

FIGURE 1.1 A surface with random height in space, time, or both (@KSC).

is the spatial wavenumber vector, and ω is temporal frequency. If the spatial wavenumber vector is a function of both magnitude of wavenumber K and direction, we say the surface is spatially anisotropic; otherwise, it is isotropic. We shall discuss this more in the next chapter. Equation 1.2 states that an arbitrary random surface can be decomposed into various scales via Fourier expansion—each scale has its own amplitude and periodicity. By the theory of resonance, each scale may be responsible for the scattering within the probing wavelength and observation geometry.

From Equation 1.1, let's freeze the surface in time, we then obtain the power density function:

$$W(k_x, k_y) = |Z(k_x, k_y)|^2 \tag{1.4}$$

which describes the distribution of energy in K-space.

By Wiener–Khinchin theorem, the surface covariance function is [1]

$$C(\tau_x, \tau_y) = \frac{1}{(2\pi)^2} \int_{-\infty}^{\infty} \int_{-\infty}^{\infty} W(k_x, k_y) \exp[-(k_x\tau_x + k_y\tau_y)]dxdy \tag{1.5}$$

which describes the distribution of variance in spatial domain.

As a random variable, let the probability density function of $z(x, y; t)$ be $p_z(z(x, y; t))$. For a certain time t_τ, if

$$p_z(z(x, y; t)) = p_z(z(x, y; t + \tau_t)), \tag{1.6}$$

then we say the surface is strictly sense stationary. The following two conditions

$$\bar{z}_t(t) = \bar{z}_t \tag{1.7}$$

$$C(t, t+\tau) = C(\tau) \tag{1.8}$$

state that the surface mean height over time and the covariance function are both invariant to a time shift. Then the surface is said to be wide sense stationary (WSS). If conditions of Equations 1.7 and 1.8 are not met, the surface is said to be nonstationary.

Now define the time average over a period T_0 is

$$\langle z(t) \rangle = \lim_{T_0 \to \infty} \frac{1}{2T_0} \int_{-T_0}^{T_0} z(t)dt \tag{1.9}$$

If z is WSS, and if

$$\bar{z}_{xy} = \bar{z}_t \tag{1.10}$$

$$\mathrm{C}_{xy} = \mathrm{C}_t(\tau) = \langle z(t)z(t+\tau) \rangle, \tag{1.11}$$

then we have surface ergodic in mean and correlation function. In other words, the surface has a mean value in time equal to its counterpart in space.

Two distribution models of surface heights or heights differences are commonly used in remote sensing of rough soil surface. For zero mean and standard deviation,

$$\textit{Gaussian distribution } p_z(z) = \frac{1}{\sqrt{2\pi}\,\sigma} e^{-z^2/2\sigma^2} \tag{1.12}$$

$$\textit{Modified exponential distribution } p_z(z) = \left(\frac{1}{2\sigma} + \frac{|z|}{\sigma^2}\right)\exp\left(-2\frac{|z|}{\sigma}\right). \tag{1.13}$$

The modified exponential distribution is formed by adding a constant to the exponential function.

Previous studies show that the surface height distribution is the dominant factor for coherent scattering, compared to surface correlation function. A no-Gaussian height distribution generally causes stronger coherent scattering [2]. Other than the surface height distribution, in surface scattering, the difference of the surface heights, two independent random variables, is of importance. For example, in high-frequency limits, the scattering strength is proportional to the surface slope distribution [3–6]. The following two models are commonly applied: Cauchy distribution and Laplace distribution.

$$\textit{Cauchy distribution } p_s(z) = \frac{1}{\pi\gamma}\left[\frac{\gamma^2}{z^2+\gamma^2}\right] \tag{1.14}$$

where γ is the scale parameter which specifies the half-width at half-maximum [7–10]. Since the Cauchy distribution has no moment generating function, there are no defined mean and standard deviation. The Cauchy distribution is used to model the ratio of the surface heights whose distributions follow the Gaussian model. It was proposed to model the surface elevation differences when there is a lack of dependence on wavelength of the roughness parameter.

$$\textit{Laplace Distribution } p_s(z) = \frac{1}{2\beta}\exp\left(-\frac{|s-\alpha|}{\beta}\right) \tag{1.15}$$

where α is a real scalar and β a positive real parameter. For Laplace distribution, mean is $\bar{z} = \alpha$ and the standard deviation is $\sqrt{2}\beta$. The Laplace distribution is used to model the difference of the surface heights whose distributions follow the exponential model. In other words, for exponentially distributed height, the surface slope becomes Laplace distribution. For the ocean surface, the height distribution is more complicated due to wind-blown and the interactions of short waves and long waves in the process of energy transfer. Treatments of non-Gaussian height distribution are to as references [2,10–15], and the references cited therein.

1.2 NATURE OF WAVE SCATTERING FROM ROUGH SURFACE

Electromagnetic waves scattering from a randomly rough surface is of palpable importance in many disciplines and bears itself in various applications spanning from surface treatment to remote sensing of terrain and sea [16–20]. For example, in microwave remote sensing of terrain, it has been a common practice to retrieve, by analyzing the sensitivity of the scattering behavior and mechanisms, the geophysical parameters of interest from the scattering and/or emission measurements. Another example is that by knowing the scattering patterns, one may be able to detect the presence of the undesired random roughness of a reflective surface such as an antenna reflector, and thus accordingly devise a means to correct or to compensate the phase errors with respect to the phase center. Therefore, it has been both theoretically and practically motivated to study the electromagnetic wave scattering from the random surfaces.

As shown in Figure 1.2, two rays, s_1,s_2, incident on a rough surface at angle of incidence θ_i, the path difference is $\Delta r = 2h\cos\theta$, and the phase difference is

$$\Delta\phi = \frac{2\pi}{\lambda}\Delta r = \frac{4\pi h}{\lambda}\cos\theta \tag{1.16}$$

when $\Delta\psi = 0$, the surface is said electromagnetically flat; otherwise, it is electromagnetically rough. If the surface is flat, the refection law applies everywhere on the surface. When it is rough, the specular reflection changes into diffuse scattering. Hence, the phase difference is a good measure of surface roughness. The question is how rough is rough? However, it is difficult to quantitatively determine the value of $\Delta\psi$ to divide the surface from "rough" to "smooth." One possible choice is Rayleigh criterion: if $\Delta\psi < \frac{\pi}{2} \rightarrow h < \frac{\lambda}{8\cos\theta}$, then surface is said smooth. Rayleigh criterion is

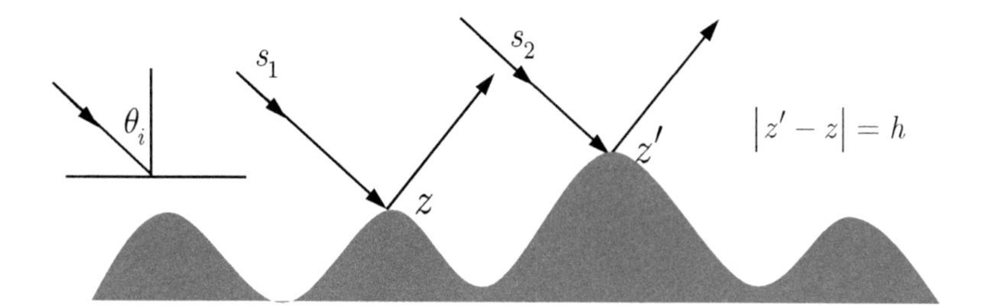

FIGURE 1.2 Simple measure of surface roughness.

somewhat loose. Another choice is Fraunhofer criterion: $\Delta\psi < \frac{\pi}{8} \rightarrow h < \frac{\lambda}{32 \cos\theta}$. Generally, the Fraunhofer criterion is more consistent with experimental observations. Fraunhofer criterion is also applied to determine the far-field range in measuring the radar cross-section, as will be given in the next chapter. In the preceding discussion, only surface height difference h is concerned. We should keep in mind that to be more reasonable and more realistic, the roughness criterion must consider the surface correlation as well. Nevertheless, the degree of roughness is dependent on the probing wavelength; everything about the physical size must be in terms of wavelength. The total scattering from a rough surface is viewed a sum of coherent scattering and incoherent scattering. When the surface is perfectly flat, there exists only the coherent scattering which obeys the reflection law, to be discussed in the next chapter. When there is presence of roughness, the coherent scattering reduces, and the incoherent scattering occurs and becomes stronger as roughness increases, and is eventually completely dominant in total scattering. As is conceptually illustrated in Figure 1.3, the coherent scattering is specular, while the incoherent scattering is diffuse, and is more so when the surface becomes rougher. In general, the scattering pattern in space is bistatic, and the pattern's beam-width is controlled by the roughness.

1.3 RADAR IMAGING MECHANISMS AND COMPUTATION

Radar images record the scattering process that is strongly correlated with surface geometric and dielectric properties. Imaging through random medium requires extensive interplay between the scattering and the signal processing. Because of the physical size limitation of the radar antenna, the spatial resolution in the azimuthal direction is finite. A synthetic aperture radar (SAR) has been indispensable for Earth sensing [20–25]. SAR is an aperture-limited system with azimuthal resolution proportional to the aperture size and is independent of radar wavelength. This is quite different from the diffraction-limited optical system, where angular resolution is inverse proportional to the aperture size and is proportional to the wavelength. Figure 1.4 illustrates commonly used SAR observation modes: strip map mode, spotlight mode, and circle mode (ISAR). SAR is moving along the azimuthal direction with the antenna pointing to the right, left, or alternating. Figure 1.4 only shows the right-looking.

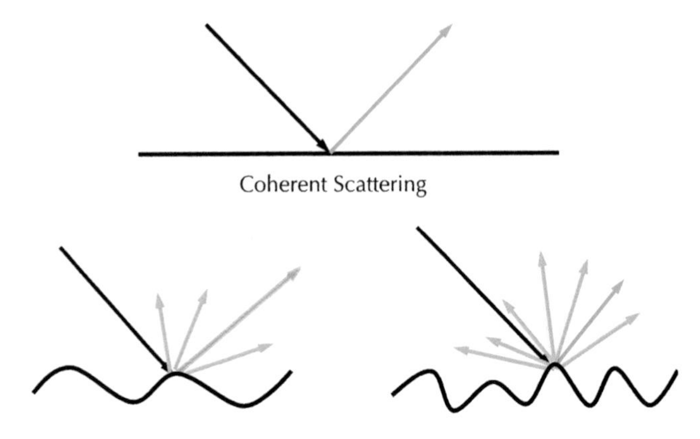

FIGURE 1.3 Schematics of wave scattering from flat to rough surfaces.

When an electromagnetic radar wave is incident upon targets, the scattering and propagation take place at the boundary between the two media. Referring to the scattering geometry in Figure 1.4, the total electric field is

$$\mathbf{E}(\vec{r}) = \mathbf{E}^i(\vec{r}) + \mathbf{E}^s(\vec{r}) \tag{1.17}$$

where $\mathbf{E}^i(\vec{r})$ is the incident field and $\mathbf{E}^s(\vec{r})$ is the scattered field. The volume integral equation (VIE) [26–27] governing the total field for a three-dimensional dielectric object is

$$\mathbf{E}(\vec{r}) = \mathbf{E}^i(\vec{r}) + k_0^2 \iiint_V \bar{\bar{\mathbf{G}}}(\vec{r}, \vec{r}')\cdot[\varepsilon_r(\vec{r}) - 1]\mathbf{E}(\vec{r}')d\vec{r}' \tag{1.18}$$

where k_0 is the free space wavenumber, ε_r is the relative permittivity of the dielectric object, and $\mathbf{G}(\mathbf{r}, \mathbf{r}')$ is the electric dyadic Green's function in free space:

$$\mathbf{G}(\mathbf{r}, \mathbf{r}') = \left(\mathbf{I} + \frac{\nabla\nabla}{k_0^2}\right)\frac{e^{ik_0|\mathbf{r} - \mathbf{r}'|}}{4\pi|\mathbf{r} - \mathbf{r}'|} \tag{1.19}$$

Numerical methods for solving the VIE in Equation 1.19 have been well documented and references cited therein [28–30]. These include the full-wave moment method, finite difference time domain (FDTD), Fast multiple pole, wavelet transform. As for approximate solution, the iterative approaches such as Born approximation, Rytov approximation, and their high order versions can be found in [16, 31–32]. According to Maxwell's equations, the surface currents are related to the total electric field as

$$\mathbf{J} = j\omega\frac{\varepsilon - \varepsilon_0}{\varepsilon}\mathbf{E} \tag{1.20}$$

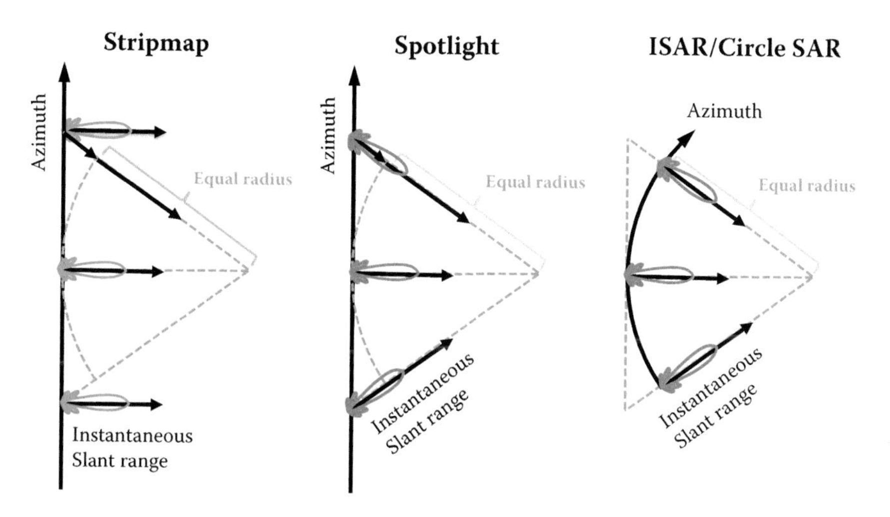

FIGURE 1.4 SAR observation modes.

where ε is the permittivity of the dielectric object and ε_0 is the permittivity in free space.

In synthetic aperture processing, the received signal from the scattered field is obtained by integrating the surface fields over an illumination area on the targets to gather two-dimensional data in the range (fast time τ) and azimuth directions (slow time η). Suppose that the SAR system moves along the y direction with velocity u, we obtain the slant range R varying with slow time:

$$R(\eta) = \sqrt{R_0^2 + u^2(\eta - \eta_c)^2} \tag{1.21}$$

Then, the time-harmonic scattered field received by SAR traveling along the y direction, at far-field range R, can be expressed as

$$\mathbf{E}_s(R, t, \eta) = \frac{e^{i\omega t} e^{-ikR}}{4\pi R} \int \mathbf{J}(\mathbf{r}') e^{i\vec{k}\cdot\vec{r}'} e^{-\frac{(x'-x_c)^2\cos^2\theta_s}{R_0^2\beta^2}} e^{-\frac{(y'-y_c)^2}{R_0^2\beta^2}} dS' \tag{1.22}$$

where $\mathbf{J}(\mathbf{r}')$ are surface fields given in Equation 1.20, with a two-dimensional Gaussian antenna gain pattern with full beam-width β, centered at a resolution cell (x_c, y_c), are considered (see Figure 1.5).

In SAR [22, 25], the received signal, called the echo signal, is a coherent sum of all scattered fields received at R

$$\mathbf{E}^s(t, \eta) = \int_0^{L_{sa}} \mathbf{E}^s(t, R(\eta))dy \tag{1.23}$$

where L_{sa} is the synthetic length (see Figure 1.5).

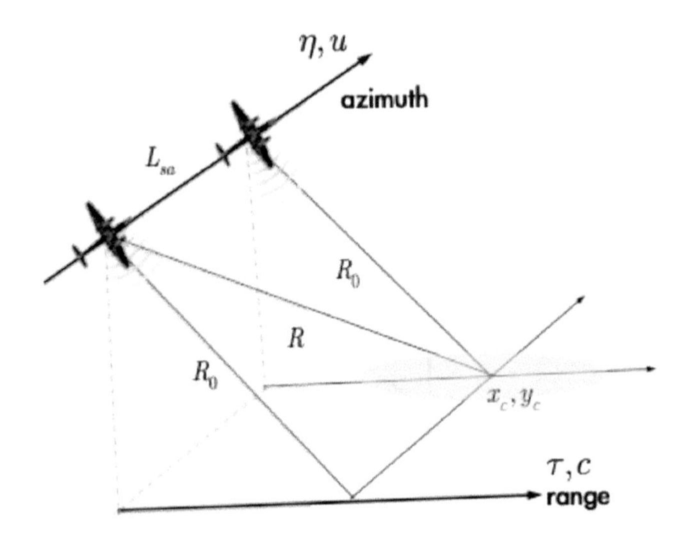

FIGURE 1.5 SAR imaging geometry.

Note that for a linear frequency modulated (LFM) transmitting signal called chirp, a factor $e^{\pi a_r (t - \frac{R(\eta)}{c})^2}$ should be included, where a_r is the chirp rate. The echo signal at this point remains a vector field. In receiving, if the antenna is $\hat{\mathbf{p}}$ -polarized, then we have

$$E_{sp} = \hat{\mathbf{p}} \cdot \mathbf{E}_s \tag{1.24}$$

The SAR echo signal after demodulation [25] is

$$s_0(\eta, t) = E_{sp} \otimes w_r\left(t - \frac{2R}{c}, T_r\right) w_a(\eta - \eta_c) \exp\left(-j\frac{4\pi f_0 R(\eta)}{c} + j\pi a_r\left(t - \frac{2R(\eta)}{c}\right)^2\right) \tag{1.25}$$

where $\otimes$ is convolution operator, T_r is the pulse duration, f_0 is the carrier frequency, t is the analog to digital conversion (ADC) sampling time or fast time, and $w_r(\cdot)$ is a rectangular function along the range direction:

$$w_r(t, T_r) = rect\left(\frac{t}{T_t}\right) = \begin{cases} 1, & \left|\frac{t}{T_t}\right| \le 0.5 \\ 0, & \text{otherwise} \end{cases} \tag{1.26}$$

where $w_a(\eta - \eta_c)$ is the antenna gain pattern, a function of slow time. Taking the Fourier transform to the echo signal both in range and azimuth directions, we obtain

$$S_{2df}(f_\eta, f_t) = \tilde{E}_{sp} W_r(f_t) W_a(f_\eta - f_{\eta_c}) \exp(j\theta_a(f_\eta, f_t)) \tag{1.27}$$

where $\tilde{E}_{sp}$ is Fourier transform of Equation 1.24, and the phase term is

$$\theta_a(f_\eta, f_t) = -\frac{4\pi R_0}{c}\sqrt{(f_0 + f_t)^2 - \frac{c^2 f_\eta^2}{4v_r^2}} - \frac{\pi f_t^2}{a_r} \tag{1.28}$$

Now that the matching filter is designed according to Equation 1.28, several focusing algorithms have been proposed, including range-Doppler (RD), Omega-K, and Chirp-Scaling, along with improved versions. The refined Omega-K method [23] is a fast-focusing algorithm while maintaining fine spatial resolution and small defocusing error. While the form varies between implementations, the Omega-K algorithm requires a change of variables in the frequency domain referred to as Stolt mapping. The phase term in Equation 1.28 includes the target phase due to the range encoding, range cell migration (RCM), range-azimuth coupling, and azimuth encoding.

The first step in focusing is a reference function multiplication (RFM). The reference function is calculated for a selected range. A target at the reference range is correctly focused by the RFM, but targets away from that range are only partially focused. Generally, the RFM is considered bulk compression.

$$\theta_R(f_\eta, f_t) = \frac{4\pi R_{ref}}{c}\sqrt{(f_0 + f_t)^2 - \frac{c^2 f_\eta^2}{4v_r^2}} + \frac{\pi f_t^2}{a_r} \tag{1.29}$$

After filtered by RFM, the phase remaining in the two-dimensional signal spectrum is approximately

$$\theta_{RFM}(f_\eta, f_t) = -\frac{4\pi(R_0 - R_{ref})}{c}\sqrt{(f_0 + f_t)^2 - \frac{c^2 f_\eta^2}{4v_r^2}} \tag{1.30}$$

Stolt mapping is the second focusing step to process the residual phase term. This process completes the focusing of targets away from the reference range by remapping the range frequency axis according to

$$\theta_{focused}(f_\eta, f_t) = -\frac{4\pi(R_0 - R_{ref})}{c}(f_0 + f_t') \tag{1.31}$$

To achieve perfect range compression and registration, the mapping transforms the original range frequency variable f_t into a new range frequency variable f_t', such that the phase is now linear in f_t'

$$\theta_{focused}(f_\eta, f_t') = -\frac{4\pi(R_0 - R_{ref})}{c}(f_0 + f_t') \tag{1.32}$$

Theoretically, the range inverse Fourier transform (IFT) results in perfect range compression and registration. The mapping has effectively removed all the phase terms higher than the linear term and implements residual azimuth phase and range-azimuth coupling. For this reason, Stolt mapping can be viewed as differential compression. Finally, a two-dimensional IFT is performed to transform the data back to the time domain, i.e., the SAR image domain.

Besides the popular SAR imaging, Time Inverse (TR) imaging [33–35] is a commonly used imaging technique which is based on the time-invariant of the fluctuation equation, and the specific implementation process is that the time-reversal array receives a reflected signal from the target, reverses it in time and resends it to the imaging targets, thus enabling the target to be imaged. The time-reversal signal achieves simultaneous focus on time and space at the target position, equivalent to a two-dimensional space-time matching filter. Spatial focus refers to the time-reversal signal's ability to effectively utilize the multipath effect through the random medium, adaptive focusing at the target by multiple reflections, scattering, and re-fraction, and time focus refers to the time-reversal signal of multiple time-reversal arrays that eventually reaches the target position at the same time, thus enabling the fusion detection of the target. This property is important for the imaging of targets in complex environments.

Referring to Figure 1.6, suppose that the target position is at $\mathbf{X}_m$, N is the number of time-reversal array antennas, each positioned at $\mathbf{R}_i$, $i = 1, 2, \ldots N$, from whereby it transmits a signal $s_t(t)$ toward the time-reversal array antenna. Then the receiving signal of ith by time reversal array antenna at $\mathbf{R}_i$ is

$$s_{ri}(t) = s_t(t) \otimes_t h_{\mathbf{X}_t\mathbf{R}_i}(t) \tag{1.33}$$

where $\otimes_t$ is the time domain convolution, $h_{\mathbf{X}_t\mathbf{R}_i}$ is the impulse response between the source point at $\mathbf{X}_m$ and the receiving antenna at $\mathbf{R}_i$, with $h_{\mathbf{X}_t\mathbf{R}_i}(t) = h_{\mathbf{R}_i\mathbf{X}_t}(t)$ from

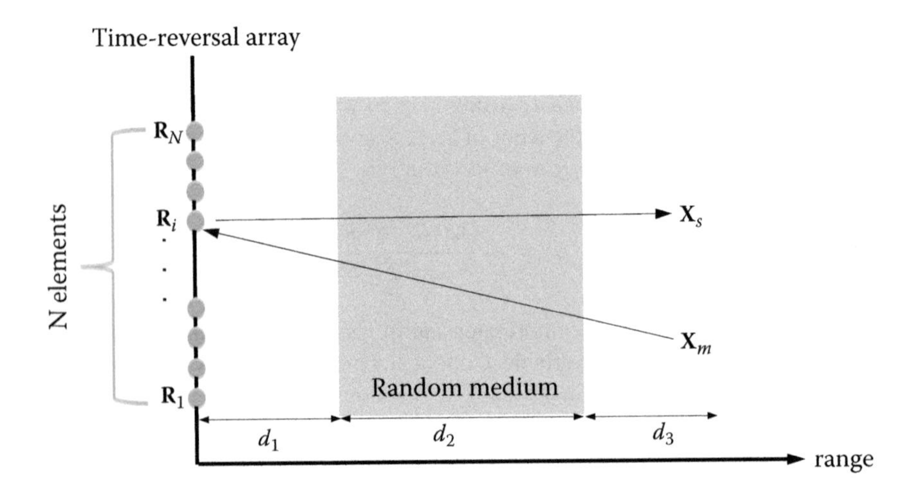

FIGURE 1.6 Schematic geometry of time reversal imaging.

duality of the Green's function. The signal of Equation 1.33 then is redirected to target area and is received at $\mathbf{X}_s$.

By recording the received signal from each array antenna and reversing its time, we obtain the time-reversal signal for i^{th} array

$$f_i(-t) = s_t(-t) \otimes_t h_{\mathbf{r}_t\mathbf{r}_i}(-t) \tag{1.34}$$

Now resending the time-reversal signal to the probing space, and the time-reversal signal of the first antenna received at the original source point is

$$\begin{aligned} p_i(t) &= s_{ri}(-t) \otimes_t h_{\mathbf{r}_i\mathbf{r}_t}(t) \\ &= s_t(-t) \otimes_t h_{\mathbf{r}_t\mathbf{r}_i}(-t) \otimes_t h_{\mathbf{r}_i\mathbf{r}_t}(t) \end{aligned} \tag{1.35}$$

where $h_{\mathbf{r}_t\mathbf{r}_i}(-t) \otimes_t h_{\mathbf{r}_i\mathbf{r}_t}(t)$ is a time-domain matching filter.

For an N-elements array antenna, multi-baseline interferometric processing will improve the time-reversal signal response. The received signal at any point in the target area is the sum from every antenna:

$$p(\mathbf{X}, t) = \sum_{i=1}^{N} s_t(-t) \otimes_t h_{\mathbf{XR}_i}(-t) \otimes_t h_{\mathbf{R}_i\mathbf{X}}(t) \tag{1.36}$$

where it is understood that the time-inverted signal waveform matches exactly the source point, and the unrelated items cancel out each other as the receiving antenna at $\mathbf{X}$ which is positioned away from the source point at $\mathbf{X}_t$. For random media with ample multipaths, the correlation process will get better focus in time and therefore in space as well. The time-reversal process can be viewed as a two-dimensional space-time matching filter, and the time-reversal signal can be used to focus the target simultaneously in time and space.

We make a quick comparison of the imaging performance of three algorithms: TR, SAR, TR-MUSIC for target obscured by a random medium using the simulation parameters given in Table 1.1. The results are shown in Figures 1.7 and 1.8, using a Gaussian beam of the form [36–37]

$$P_g(\omega) = A_0\left(\frac{2\sqrt{\pi}}{B}\right)\exp\left(-\frac{(\omega - \omega_c)^2}{B^2}\right) \tag{1.37}$$

where A_0 is the amplitude, ω_c is carrier angular frequency, and B is the bandwidth. For illustration, we may simply set $A_0 = 1$.

From Figures 1.7 and 1.8, it is clear that, for a fixed optical thickness τ, the image resolution from all three methods degrades quickly as albedo increases, and is more so for TR and SAR. When the optical thickness increases from 1.0 to 5.0, poor results delivered by TR and SAR are evident, while acceptable results are seen from TR-MUSIC, except for high albedo of 1.0 [38]. Overall, the TR-MUSIC imaging methods performed the best, followed by SAR imaging and the least TR

TABLE 1.1
Simulation Parameters for Comparison of TR, SAR, TR-MUSIC Imaging Performance

Parameter	Value	Parameter	Value
d1	12λ	Number of array elements	41
d2	13λ	Element spacing	0.5λ
d3	25λ	Pixel size	$0.5\lambda \times 0.5\lambda$
Target location	$(0\lambda, 50\lambda)$	Transmitted waveform	Guassian pulse
Optical thickness	1 & 5	Carrier frequency	0.3 GHz
albedo	0.1, 0.5, 1.0	Bandwith	0.01*0.3 GHz

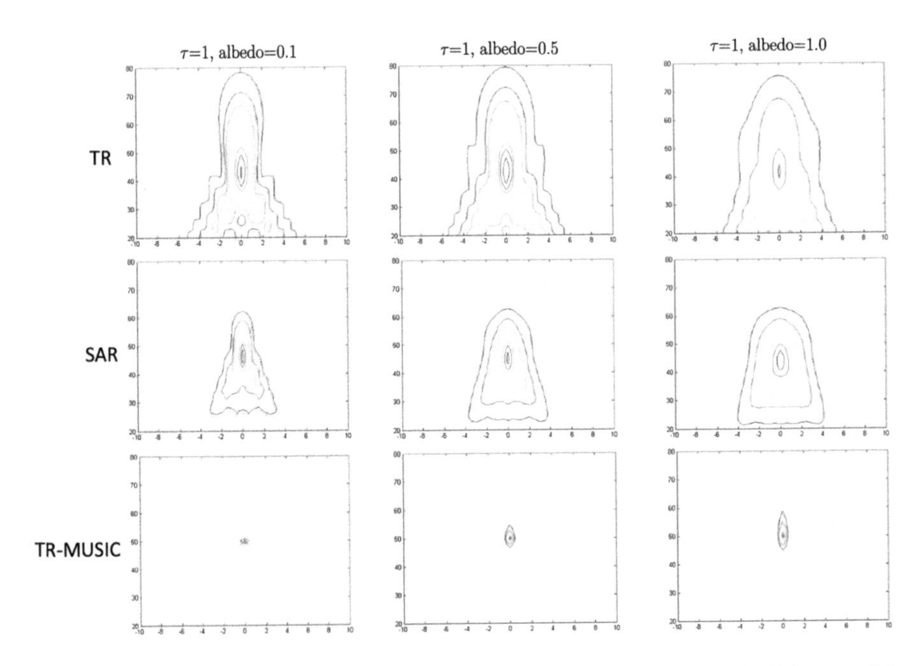

FIGURE 1.7 Image performance of TR, SAR, TR-MUSIC with optical thickness of 1 wavelength, and albedo of 0.1, 0.5, and 1.0.

imaging. It is not sufficient only to consider the optical thickness of the random medium when evaluating the imaging performance. The influence of albedo is equally important, and the scattering thickness, which is closely related to optical thickness and albedo, is the main cause of degradation about imaging performance. Though the results of the three methods are all degraded by the presence of random media, it could be clearly seen that TR-MUSIC performs well.

Although the algorithm of the time-reversal imaging is conceptually simple, the spatial resolution it produces is normally poor. When there are multiple targets

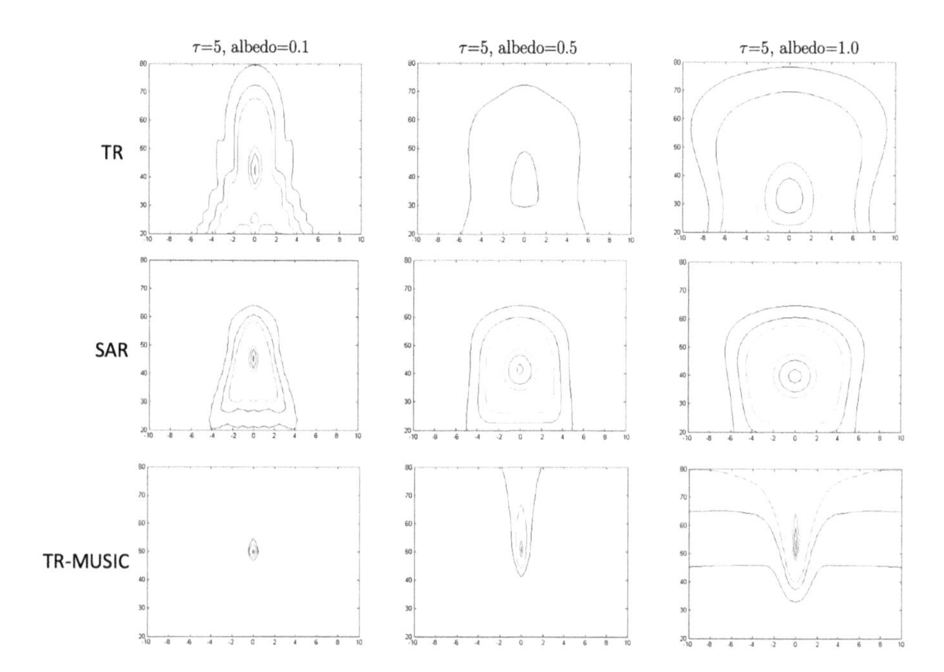

FIGURE 1.8 Same as Figure 1.7 except optical thickness of 5 wavelengths.

presented in the probing space or target area, only the strongest scattering target perhaps can be focused distinctly, and the targets with weaker scattering tend to be blurred, and perhaps in some cases, are simply vanished from the imaged scene. More details and demonstrations will be given in Chapter 8.

1.4 PROGRESS OF THE SUBJECT

Surface scattering is not new but has a long history of research due to its theoretical and application interests [20, 39]. Bistatic scattering is a subject of increasing interest in applications from surface treatment to remote sensing of terrain and sea. Extensive studies have shown that some bistatic configurations may provide desirable performance in remote sensing of soil moisture and in distinguishing the influence of various geophysical parameters on observed returns. The global navigation satellite signal reflectometry (GNSS-R) technique is an emerging bistatic radar technique for the remote sensing of soil moisture [40–41]. Both the GNSS-R simulations and measurements have shown great potential for sensing soil moisture. Most GNSS-R bistatic radar studies focus on the specular region due to received power considerations. Limited studies have looked into the potential out-of-plane bistatic radar configurations for soil moisture estimation. In the simulation studies [42–44], they found that using bistatic geometries away from the specular region can provide more useful information on soil moisture retrieval than the traditional monostatic observations. However, previous studies have largely centered on the in-plane configuration.

The scattering behavior in azimuth direction was rarely analyzed despite the importance of the azimuthal angular pattern in bistatic scattering.

Recent studies have revealed that there exists a local minimum or dip in the normalized radar cross-section as a function of the azimuthal scattering angles, and it has a close relationship with surface parameters. Johnson and Ouellette [45] found that the minima locations are distinct for HH and VV polarized returns in the case of slight roughness, and that the minimum locations for VV polarization are dependent on the permittivity of the surface. In [46–47], it was pointed out that the forward region is preferable for soil moisture sensing regardless of the correlation functions, and in out-of-plane, the bistatic scattering has low returned signal, especially within the null regions. It has been illustrated that the depth and location of the dip in the azimuthal plane is closely related to the permittivity of media [48]. Those studies strongly suggested that the angular patterns of bistatic scattering from rough surface have the potential for soil moisture remote sensing. By knowing, at depth and breadth levels, the dip features, we are able to avoid the undesired, and useless, dips in configuring a radar bistatic observation geometry, and thus, accordingly, devise more effective means to retrieve surface parameters of interest, e.g., moisture content. As for Brewster angles, there are numerous works on their behaviors, and it is found that they generally shift in polar scattering direction in the presence of surface roughness and lossy medium [49]. For Brewster's scattering induced dip with the incident angle equal to, or near the ordinary Brewster angle, the behaviors could be quite different.

Generally, we have three types of modeling approaches, i.e., the empirical models, the semi-empirical models, and the theoretical models [20]. The empirical and semi-empirical models often involve several fitted parameters derived from limited field observations. Although often working well on the same or similar dataset, the models are site-dependent, and, as a result, are limited, to a certain extent, in the modeling of surface scattering and emission for a variety of surface profiles. Therefore, the site-independent theoretical models rooted in strictly physical laws have become an important and popular tool to help us achieve a good understanding of the interaction between electromagnetic fields and natural surfaces. The theoretical models can be categorized into two principal classes, i.e., the numerical models and the analytical models. The former is based on the numerical solution of Maxwell's equations for the electromagnetic field. Though the numerical models can yield very high accurate simulations, they are extremely computationally expensive, which makes their direct application for analysis of the microwave satellite data rather difficult. On the other hand, the analytical models give a good balance between simulation accuracy and computational efficiency. The allure of the analytical models is that they greatly improve the computational efficiency compared with the numerical methods while still ensuring a fairly high accuracy in their application domain, which makes them easily and efficiently employed in numerous microwave remote sensing applications [20].

Among the analytical models, the integral equation method (IEM) was developed by Fung et al. [18–19, 50] to bridge the gap between the Kirchhoff approximation (KA) and small perturbation method (SPM). An IEM model was later evolved to an AIEM by removing some of the assumptions, resulting in more accurate scattering prediction, particularly for bistatic scattering and microwave emissivity simulations [51]. The AIEM model has been used as a forward model to simulate the scattering coefficients and emissivity of bare soil surfaces with various ground conditions, and also as an inverse model to surface parameters retrieval. Together with model development, the analysis and examination of the model validity offer insights into the potential and limitations of the model, which can be very beneficial for model refinements as well as model applications. Many efforts have been devoted to the evaluation of the IEM/AIEM through inter-comparison with numerical simulations, experimental measurements, or both. However, most works still focus on the backscattering comparisons, and limited studies have looked into the capability of AIEM in reproducing the bistatic scattering and microwave emissivity. To the best of our knowledge, currently not much work has yet been carried out for a comprehensive analysis and assessment of the validity and limitations of rough soil surface scattering and emission predicted by the AIEM. Quantitative and systematical understanding of the AIEM performance should be implemented to explore potential model applications for microwave remote sensing. Great progress of computing power and the increasing abundance of measurement data propitiously offer an opportunity to attempt a detailed evaluation of the AIEM predictions by using both the numerical simulations and experimental measurements with a wide range of radar and ground surface parameters.

The properties of bistatic scattering from an inhomogeneous rough surface, in this study, is modeled by the transitional layer as a function of depth. The lower medium of the rough surface is horizontally uniform but vertically inhomogeneous. Both linear and circular polarizations are investigated in light of the dependences of transition rate, background dielectric constant, and surface roughness. The presence of dielectric inhomogeneity generally leads to several features that do not appear in the homogeneous surface [52].

Polarimetric bistatic radar is a subject of increasing interest in applications to remote sensing [21]. The use of bistatic polarimetry for radar remote sensing may be desirable in distinguishing the influence of various geophysical parameters on observed returns for providing greater capability to extract, interpret, identify, and classify targets' physical features. To fully benefit from the potential of polarimetry in bistatic configurations, it is critical to finely analyze the characteristics of bistatic scattering from targets as was performed in monostatic systems. As for bistatic polarimetry, some pioneering analysis attempts have been made [53–55]. They gave some understanding of the bistatic polarimetry theory. In practice, the electromagnetic problem is much more complex in the natural world. More complex targets, such as terrestrial rough surfaces should be attempted to investigate the validity of bistatic polarimetry theory on complex targets.

For a randomly rough surface, it generally concurs that backscattering returns are dominated by incoherent scattering for a moderate to large incident angle, and backscattering enhancement is produced for a very rough surface where multiple scattering is greatly increased [39, 56]. For the bistatic scattering of a rough surface, the scenario could be quite different. One of the objectives that is of interest is to explore the scattering mechanism pertinent to the theory of bistatic radar polarimetry for effective retrieval of surface parameters. Natural snow-covered surfaces, a typical randomly rough surface, exhibit multilayer structure and may behave, driven by wind force, similar to microstructures with a roughness spectrum that contains a substantial amount of high-frequency components. For example, Sastrugi snow surface in vast areas of Greenland, Arctic, and Antarctic regions is inherently a corrugated surface consisting of ridges separated by grooves such that these ridges and grooves are perpendicular to the direction of the driving force, e.g., prevailing wind, and one-dimensional roughness could be the force direction. Such a structure, being azimuthal asymmetrical, is of both theoretical and practical interest because it generates both coherent and incoherent scattering with strongly directional patterns in the scattering plane. Understanding the characteristics of bistatic polarimetric scattering of the snow layer is vital to the effective retrieval of snow parameters that are influential in the study of climate change.

To acquire the scattering matrix from which the polarimetric parameters are derived, a physical-based full-wave numerical simulation from Maxwell's equations is an effective approach. Because solving Maxwell equations, which are governing the scattered field, can provide many ensembles (realizations) that satisfy the statistical requirements and consider speckle or clutter. Thus, in this context, an important objective is to extract physical information from the scattering matrix, or equivalent coherency matrix or Kennaugh matrix, from full-wave simulation, thus addressing a set of methods known collectively as target decomposition theories in the bistatic case.

In the context of radar imagery, understanding the speckle properties is imperative for both image de-noising and applications such as land-cover classification and parameter retrieval [21]. Radar speckle arises due to the coherent sum of randomly distributed scattering contributions. A general assumption is the "fully developed" speckle model, which is only applicable when the resolution cell size is much larger than the correlation length of the ground surface [57]. This requirement may be met in low-to-medium image resolution. The fully developed speckle model is based on the following two hypotheses: (1) total scattering is contributed by independent scatters; (2) there is a sufficiently large number of scatters contributing to a resolution cell so that a complex Gaussian distribution is followed. For very high resolution radar images, these assumptions are no longer valid. As a result, the K-distribution is often applied to model the speckles [21].

Backscattering and its speckle statistic variation with the resolution cell size or antenna footprint has been a subject of interest in rough surface scattering. The dependence of polarimetric backscattering characteristics from Gaussian correlated rough surfaces on the SAR resolution cell size was studied by modified theoretical models and indoor SAR experiments in EMSL [58–59]. The SAR images of rough soil surfaces were measured and analyzed at different resolution cell sizes [60]. As the radar image resolution reaches from tens of meters to approximately one meter or even a smaller size, this topic becomes more important currently for land observations. The topic of VHR radar speckle statistics using indoor SAR experiments and supporting full-wave simulations has been addressed.

Several algorithms in the field of time-reversal (TR) imaging include time-reversal mirror algorithm, space-space DORT (Decomposition of the Time Reversal Operator) algorithm, and space-space TR-MUSIC (Multiple Signal Classification) algorithm

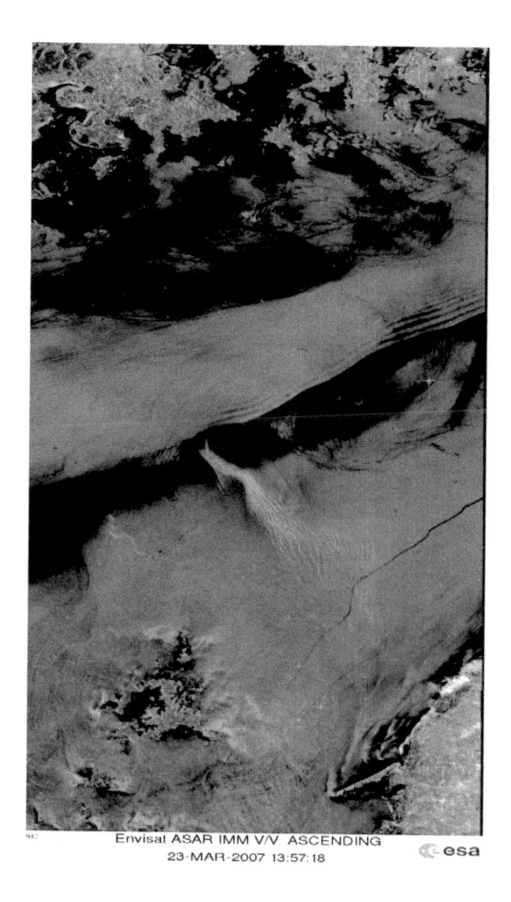

FIGURE 1.9 A SAR image acquired by ESA's Envisat over ocean surface.

[61] have been widely used. Time-reversal mirror is the most primitive time reversal algorithm, which can only focus on the strongest scattering target, but loses the ability to focus on the weaker target in the same detection area, and, also, the algorithm offers a lower imaging resolution. The acquisition of space-space multi-static data matrix is complicated. Each antenna of the time-reversal array needs to emit electromagnetic waves to the detection area in turn. All antennas receive and store the echo signals of the target at the same time, which makes a real-time focusing impossible. The space-frequency multi-static data matrix is an antenna transmitting signal, all antennas receive echo signals so that the real-time imaging becomes feasible. For this reason, a space-frequency multi-static data matrix is preferred. Both space-space DORT and space-space TR-MUSIC are based on the space-space polymorphic data matrix. It has been realized that space-space DORT has better anti-noise performance, but comes with poorer resolution. On the contrary, the space-space TR-MUSIC algorithm has better imaging resolution, but with inferior anti-noise performance. There is room to continue to improve the TR-based imaging algorithm in lines of imaging performance and computational efficiency.

For vast area mapping, SAR is still the dominant imaging method for its effective and efficient imaging performance. Many useful imaging modes, such as stripmap, spotlight, scan-mode, and sliding-mode, have been put into operating in airborne and spaceborne platforms. The advances of hardware and software also push the SAR technique into a new era of digital beamforming [62–63], MIMO (multiple input–multiple output) [64–65], compressive sensing [66–67], just to name some. However, from a wave scattering point of view, SAR image focusing based on a point target model loses information of multiple scattering within an image resolution cell. How to preserve such interactions among targets within that resolution cell while retaining high focusing efficiency is still desired.

Theoretical modeling and numerical computation for the radar imaging targets over a rough surface are still challenging. In this context, it involves multi-physics mechanisms in the process of radar scattering into and from such a complex scene. For illustration, a SAR image over a complex ocean scene is displayed in Figure 1.9 [68], where strong air-sea interactions associated with bottom topography effects were seen. Several ships moved downward and up-bound could be identified from both strong ships' body scattering and associated wake behind them. Two internal wave systems were observed—one wide bright band across the image scene, and the other, much smaller in scale with multiple tails downward. Sometimes from SAR images we only see the wake of the ship but not the ship itself.

It may be because the ship is specially designed to make the radar backscattering very weak, or it may be caused by the passage of an underwater ship. This SAR image contains rich information about the scene it imaged but also profoundly suggests that surface roughness—no matter how it is caused—is the dominant factor to determine the radar scattering strength, but not the only one. SAR System simulation of the SAR images over a rough surface is only possible by fully exploiting the physical mechanisms and empowering computation capability; as far as we may concern, such work is challenging but promising progress is optimally foreseen.

REFERENCES

1. Papoulis, A., and Pillai, S. U., *Probability, Random Variables and Stochastic Processes*, McGraw-Hill, New York, 2002.
2. Wu, S. C., Chen, M. F., and Fung, A. K., Scattering from non-Gaussian randomly rough surfaces-cylindrical case, *IEEE Transactions on Geoscience and Remote Sensing*, 26(6), 790–798, November 1988.
3. Beckmann, P., and Spizzichino, A., *The Scattering of Electromagnetic Waves from Rough Surfaces*, Macmillan, New York, 1963.
4. Bass, F. G., and Fuks, I. M., *Wave Scattering from Statistically Rough Surfaces*, Pergamon Press, Oxford, UK, 1979.
5. Ulaby, F. T., Moore, R. K., and Fung, A. K., *Microwave Remote Sensing: Active and Passive, Volume II, Radar Remote Sensing and Surface Scattering and Emission Theory*, Artech House, Norwood, MA, 1982.
6. Voronovich, G., *Wave Scattering from Rough Surfaces*, 2nd Edition, Springer, Berlin, Germany, 1999.

7. Marcus, A. H., Application of a statistical model to planetary astronomy, *Journal of Geophysical Research*, 74(20), 4958–5082, 1969.
8. Marcus, A. H., Distribution of slopes on a cratered planetary surface: Theory and preliminary applications, *Journal of Geophysical Research*, 74(22), 5253–5267, 2012.
9. Simpson, R. A., Surface roughness estimation at three points on the lunar surface using 23-CM monostatic radar, *Journal of Geophysical Research*, 81(23), 4407–4416, 2012.
10. Beckmann, P., Scattering by non-Gaussian surfaces, *IEEE Transactions on Antennas and Propagation*, AP-21, 169–175, 1973.
11. Brown, G. S., Scattering from a class of randomly rough surfaces, *Radio Sciences*, 17, 1274–1280, 1982.
12. Eom, H. J., and Fung, A. K., A comparison between backscattering coefficients using Gaussian and non-Gaussian surface statistics, *IEEE Transactions on Antennas and Propagation*, AP-31, 635–638, July 1983.
13. Fung, A. K., and Chen, K. S., Kirchhoff model for a skewed random surface, *Journal of Electromagnetic Waves and Applications*, 5(2), 205–216, 1991.
14. Chen, K. S., Fung, A. K., and Amar, F., An empirical bispectrum model for sea surface scattering, *IEEE Transactions on Geoscience and Remote Sensing*, 31(4), 830–835, July 1993.
15. McDaniel, S. T., Microwave backscatter from non-Gaussian seas, *IEEE Transactions on Geoscience Remote Sensing*, 41(1), 52–58, January 2003.
16. Ishimaru, A., *Wave Propagation and Scattering in Random Media*, Academic Press, New York, 1978.
17. Tsang, L., Kong, J. A., and Shin, R. T., *Theory of Microwave Remote Sensing*, Wiley, New York, 1985.
18. Fung, A. K., *Microwave Scattering and Emission Models and Their Applications*, Artech House, Norwood, MA, 1994.
19. Fung, A. K., and Chen, K. S., *Microwave Scattering and Emission Models for Users*, Artech House, Norwood, MA, 2010.
20. Ulaby, F. T., and Long, D. G., *Microwave Radar and Radiometric Remote Sensing*, University of Michigan Press, Ann Arbor, MI, 2014.
21. Lee, J. S., and Pottier, E., *Polarimetric Radar Imaging: From Basics to Applications*; CRC Press, Boca Raton, FL, 2009.
22. Franceschetti, G., and Lanari, R., *Synthetic Aperture Radar Processing*, CRC Press, Boca Raton, FL, 1999.
23. Cumming, I., and Wong, F., *Digital Signal Processing of Synthetic Aperture Radar Data: Algorithms and Implementation*, Artech House, Norwood, MA, 2004.
24. Elachi, C., and van Zyl, J., *Introduction to the Physics and Techniques of Remote Sensing*, John Wiley & Sons, Hoboken, NJ, 2006.
25. Chen, K. S., *Principles of Synthetic Aperture Radar: A System Simulation Approach,* CRC Press, Boca Raton, FL, 2015.
26. Stratton, J. A., *Electromagnetic Theory*, Wiley-IEEE Press, New York, 2007.
27. Kong, J. A., *Electromagnetic Wave Theory*, Wiley, New York, 1986.
28. Chew, W. C., Jin, J. M., Michielssen, E., and Song, J., *Fast and Efficient Algorithms in Computational Electromagnetics*, Artech House, Norwood, MA, 2000.
29. Jing, J. M., Theory and Computation of Electromagnetic Fields, 2nd Edition, Wiley-IEEE Press, New York, 2015.
30. Graglia, R. D., Peterson, A. F., *Higher-Order Techniques in Computational Electromagnetics*, SciTech Publishing, Raleigh, NC, 2016.
31. Chew, W. C., *Waves and Fields in Inhomogeneous Media*, IEEE Press, Piscataway, NJ, 1995.
32. Marks, Daniel L., A family of approximations spanning the Born and Rytov scattering series, *Optics Express*, 14(19), 8837, 2006.

33. Devaney, A. J., Time reversal imaging of obscured targets from multistatic data, *IEEE Transactions on Antennas and Propagation*, 53(5), 1600–1610, 2005.
34. Devaney, A. J., *Mathematical Foundations of Imaging, Tomography and Wavefield Inversion*, Cambridge University Press, Cambridge, UK, 2012.
35. Ishimura, A., *Electromagnetic Wave Propagation, Radiation, and Scattering: From Fundamentals to Applications*, 2nd Edition, Wiley-IEEE Press, Hoboken, NJ, 2017.
36. Ishimaru, A., Jaruwatanadilok, S., and Kuga, Y., Imaging through random multiple scattering media using integration of propagation and array signal processing, *Waves in Random & Complex Media*, 22(1), 24–39, 2012.
37. Ishimaru, A., Jaruwatanadilok, S., and Kuga, Y., Time reversal effects in random scattering media on super resolution, shower curtain effects, and backscattering enhancement, *Radio Science*, 42, RS6S28, 2007. doi:10.1029/2007RS003645.
38. Yi, T. Y., and Chen, K. S., A comparative study of radar imaging of the target obscured by random media, *Proceedings of the 2019 IEEE International Geoscience and Remote Sensing Symposium*, 2019. doi:10.1109/IGARSS.2019.8898984.
39. Tsang, L., Ding, K. H., Huang, S. W., and Xu, X. L., Electromagnetic computation in scattering of electromagnetic waves by random rough surface and dense media in microwave remote sensing of land surfaces, *Proceedings of the IEEE*, 101(2), 255–279, 2013.
40. Zavorotny, V. U., Gleason, S., Cardellach, E., and Camps, A., Tutorial on remote sensing using GNSS bistatic radar of opportunity, *IEEE Geoscience and Remote Sensing Magazine*, 2(4), 8–45, 2014.
41. Schiavulli, D., Nunziata, F., Pugliano, G., and Migliaccio, M., Reconstruction of the normalized radar cross section field from GNSS-R delay-doppler map. *IEEE Journal of Selected Topics in Applied Earth Observations and Remote Sensing*, 7, 1573–1583, 2014.
42. Xu, P., Chen, K. S., Liu, Y., and Li, Z. W., Multimode coherent pattern in bistatic scattering from randomly corrugated surfaces with irregular grooves at L-Band, *IEEE Transactions on Geoscience and Remote Sensing*, 547, 4143–4152, 2016.
43. Zeng, J., Chen, K. S., Bi, H., Chen, Q., and Yang, X. F., Radar response of bistatic scattering to soil moisture and surface roughness at L-band, *IEEE Geoscience and Remote Sensing Letters*, 13(12), 1945–1949, 2016.
44. Xu, P., Chen, K. S., Liu, Y., Shi, J., Peng, C., Jiang, R., and Zeng, J., Full-Wave simulation and analysis of bistatic scattering and polarimetric emissions from double-layered Sastrugi Surfaces, *IEEE Transactions on Geoscience and Remote Sensing*, 55(1), 292–307, 2017.
45. Johnson, J. T., and Ouellette, J. D., Polarization features in bistatic scattering from rough surfaces, *IEEE Transactions on Geoscience and Remote Sensing*, 52(3), 1616–1626, 2014.
46. Liu, Y., Chen, K. S., Liu, Y., Zeng J. Y., Xu, P., and Li, Z. L., On angular features of radar bistatic scattering from rough surface. *IEEE Transactions on Geoscience and Remote Sensing*, 55(6), 3223–3235, 2017.
47. Liu, Y., Chen, K. S., Xu, P., and Li, Z. L., Bistatic coherent polarimetric scattering of randomly corrugated layered snow surfaces. *IEEE Journal of Selected Topics in Applied Earth Observations and Remote Sensing*, 10(11), 4721–4739, 2017.
48. Chen, K. L., Chen, K. S., Li, Z. L., and Liu, Y., Extension and validation of an advanced integral equation model for bistatic scattering from rough surfaces., *Progress in Electromagnetics Research*, 152, 59–76, 2015.
49. Yang, Y., Chen, K. S., and Li, Z. L., A note on Brewster effect for lossy inhomogeneous rough surfaces. *IEEE Transactions on Geoscience and Remote Sensing*, 99, 1–9, 2020.
50. Fung, A. K., Li, Z., and Chen, K. S., Backscattering from a randomly rough dielectric surface, *IEEE Transactions on Geoscience Remote Sensing*, 30(2), 356–369, 1992.

51. Chen, K. S., Wu, T. D., Tsang, L., Li, Q., Shi, J. C., and Fung, A. K., Emission of rough surfaces calculated by the integral equation method with comparison to three-dimensional moment method simulations, *IEEE Transactions on Geoscience and Remote Sensing*, 41(1), 90–101, 2003.
52. Yang, Y., and Chen, K. S., Full-polarization bistatic scattering from an inhomogeneous rough surface, *IEEE Transactions on Geoscience and Remote Sensing*, 57(9), 6434–6446, 2019.
53. Germond, A. L., Pottier, E., and Saillard, J., Nine polarimetric bistatic target equations, *Electronics Letters*, 33(17), 1494–1495, 1997.
54. Czyz, Z. H., On theoretical foundation of coherent bistatic radar polarimetry, *Proceedings of SPIE*, 3120, 69–105, 1997.
55. Titin-Schnaider, C., Polarimetric characterization of bistatic coherent mechanisms, *IEEE Transactions on Geoscience and Remote Sensing*, 46(5), 1535–1546, 2008.
56. Yang, Y., Chen, K. S., Tsang, L., and Liu, Y., Depolarized backscattering of rough surface by AIEM model, *IEEE Journal of Selected Topics in Applied Earth Observations and Remote Sensing*, 10(11), 4740–4752, 2017.
57. Chen, K. S., Tsang, L., Chen, K. L., Liao, T. H., and Lee, J. S., Polarimetric simulations of SAR at L-Band over bare soil using scattering matrices of random rough surfaces from numerical 3d solutions of Maxwell Equations, *IEEE Transactions on Geoscience and Remote Sensing*, 52(1), 7048–7058, 2014.
58. Allain, S., Ferro-Famil, L., Pottier, E., and Fortuny, J., Influence of resolution cell size for surface parameter retrieval from polarimetric SAR data. In *Proceedings of IEEE IGARSS*, Toulouse, France, 2003, pp. 440–442.
59. Park, S. E., Ferro-Famil, L., Allain, S., and Pottier, E., Surface roughness and microwave surface scattering of high-resolution imaging radar, *IEEE Geoscience and Remote Sensing Letters*, 12(4), 756–760, 2015.
60. Nesti, G., Fortuny, J., and Sieber, A. J., Comparison of backscattered signal statistics as derived from indoor scatterometric and SAR experiments, *IEEE Transactions on Geoscience and Remote Sensing*, 34(5), 1074–1083, 1996.
61. Mu, T., and Song, Y., Time reversal imaging based on joint space–frequency and frequency–frequency data. *International Journal of Microwave and Wireless Technologies*, 11(3), 207–214, 2019.
62. Younis, M., Fischer, C., and Wiesbeck, W., Digital beamforming in SAR systems, *IEEE Transactions on Geoscience and Remote Sensing*, 41(7), 1735–1739, July 2003.
63. Mao, C., et al., X/Ka-Band dual-polarized digital beamforming synthetic aperture radar, *IEEE Transactions on Microwave Theory and Techniques*, 65(11), 4400–4407, 2017.
64. Li., J., and Stoica, P., *MIMO Radar Signal Processing*, Wiley, Hoboken, NJ, 2009.
65. Krieger, G., MIMO-SAR: Opportunities and pitfalls, *IEEE Transactions on Geoscience and Remote Sensing*, 52(5), 2628–2645, 2014.
66. Chen, C. H., ed., *Compressive Sensing of Earth Observations*, CRC Press, Boca Raton, FL, 2019.
67. Yang, J., Jin, T., Xiao, C., and Huang, X., Compressed sensing radar imaging: Fundamentals, challenges, and advances, *Sensors*, 19(14), 3100, 2019.
68. Chen, K. S., Chiang, C. Y., and Guo, H., *Introduction to Synthetic Aperture Radar*, in *Advances in SAR Remote Sensing of Ocean*, X. F. Li, H. Guo, K. S. Chen, and X. F. Yang, eds., CRC Press, Boca Raton, FL, 2018.

2 Statistical Description of Rough Surfaces

2.1 TYPES OF ROUGH SURFACE

The electromagnetic waves reflected from a plane interface between two media is well known via reflection law and Snell's law. However, the reflection from a rough surface is much more complicated in the sense that the reflected energy is statistically distributed in all directions, though some favor on certain directions than on another; its directional pattern is determined by the properties of the rough surface. The rough surface can be periodically, randomly, or quasi-periodically randomly rough. For natural surfaces, it is rarely being purely periodical—mostly being quasi-periodical.

2.1.1 Quasi-Periodic and Random Rough Surfaces

A good example of a quasi-periodically randomly rough surface is the surface formed on account of the wind erosion, saltation of snow particles, and deposition surface, known as the Sastrugi surface [1], frequently presented in the polar region. The Sastrugi surface inherently has periodic-like microstructures and may be mathematically modeled as a corrugated surface with a random number of ridges formed by natural forces, e.g., wind sculpture. The corrugated surface consists of ridges separated by grooves such that these ridges and grooves are perpendicular to the direction of the driving force, e.g., prevailing wind, and 1-D roughness is modeled along the forcing direction. Wind-blown grubby sand dunes, which cover the vast area on Earth, present yet another interesting surface feature in view of electromagnetic wave scattering. Mathematically, the Sastrugi surface may be given as a superposition of "ridges and grooves":

$$z = z(x, y) = \sum_{m=-\infty}^{\infty} \sum_{n=-\infty}^{\infty} a_m \zeta(x - x_m, y - y_n) \tag{2.1}$$

where the amplitude coefficient, a_m, is assumed to be uniformly random distribution in $[h_0(1 \pm 5\%)]$ with mean height h_0. The position x_m, y_n may be assumed to be Poisson random distribution and is determined by the ridge density (number of ridges per unit length). Notice in particular that for the purpose of simulation, the total surface length L is finite in both directions. The "ridge" function $\zeta(x, y)$ can be of the form:

$$\zeta(x, y) = s(x)\text{rect}(x/\delta_x)\text{rect}(y/\delta_y) \tag{2.2}$$

where $s(\cdot)$ is the shape function that specifies the ridges' shape, $\text{rect}(\cdot)$ is the rectangular function, and δ_x, δ_y define the width and length of the ridge, respectively.

Another class of naturally occurring surfaces such as terrain and sea is best, and, perhaps, only can be modeled by a random process. Two classical approaches are commonly treated in the next section: fractal and statistical approaches.

2.1.2 Isotropic and Anisotropic Surfaces

The spatial anisotropy can be statistically described by directional correlation function [2–4]. The degree of anisotropy is shown in Figure 2.1 with the correlation length varying with the azimuthal angle ψ, that is

$$L(\psi) = L_x\cos^2\psi + L_y\sin^2\psi \tag{2.3}$$

where L_x and L_y are the correlation lengths along two orthogonal directions, and their difference determines the degree of spatial anisotropy. For illustration, L_y was selected as 20 cm, and L_x varied from 1 to 19 cm in steps of 3 cm, so that the degree of anisotropy migrates from weak to strong. The equivalent correlation length of the anisotropic rough surface may be defined by integrating out the directional dependence as

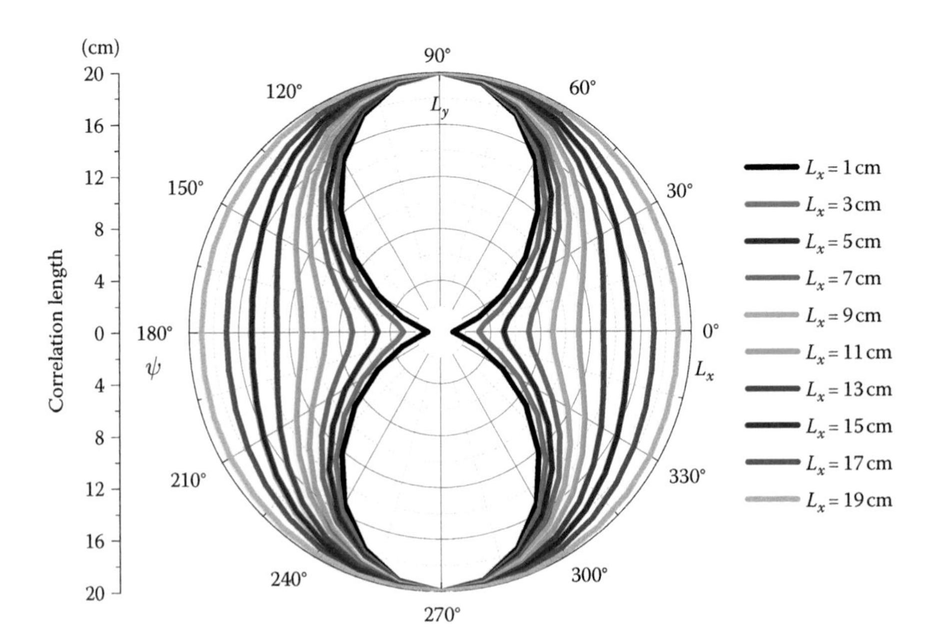

FIGURE 2.1 Correlation length angular function with different degrees of anisotropy.

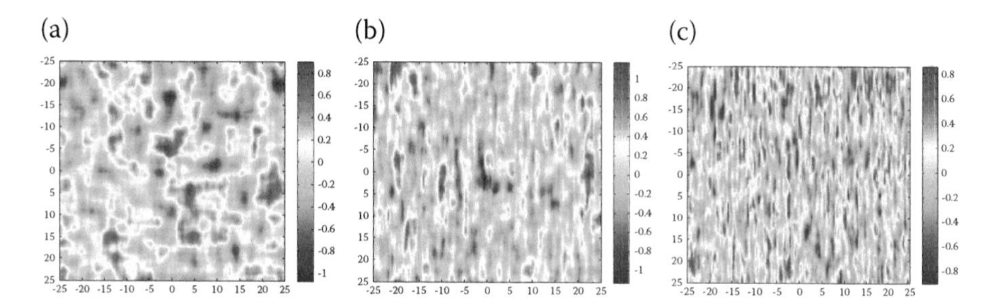

FIGURE 2.2 Numerically generated realization of the rough surfaces with three degree of anisotropy, (a) weak anisotropy, (b) moderate anisotropy, (c) strong anisotropy.

$$L_e = 1/2\pi \int_0^{2\pi} L(\psi)d\psi \tag{2.4}$$

Figure 2.2 illustrates the numerically generated realization for three types of rough surfaces with the degree of anisotropy varying from weak, moderate, to strong. By comparison, the strongly anisotropic rough surface shows stronger orientation dependence.

2.2 STATISTICS OF RANDOMLY ROUGH SURFACE

In the problem of wave scattering from a randomly rough surface, the surface is usually described in terms of its high probability distribution and correlation function. Much of the literature assumes that the high distribution is Gaussian for simplicity although it is very dependent on the surface that is considered. In what follows, we will assume the surface heights obey the Gaussian distribution. The non-Gaussian distribution will be treated in the context of sea surface scattering in the following chapters.

2.2.1 Fractal Approach

It has been proved that the fractal geometry provides a sound and reliable description of natural surfaces [5–11]. However, to apply the fractal geometry to remote sensing, it is mandatory to devise a fractal model for the surface that allows a possibly approximate evaluation of the scattered electromagnetic field. A viable fractal model for the surface description is the fractional Brownian motion (fBm). By definition, an isotropic surface $z(x, y)$ belongs to the class of fBm if [11]:

$$P\{z(x, y) - z(x', y') < \bar{\xi}\} = \frac{1}{\sqrt{2\pi}T^{(1-H)}\tau^{H}} \int_{-\infty}^{\bar{\xi}} \exp\left\{ - \frac{\xi^2}{2T^{2(1-H)}\tau^{2H}} d\xi \right. \tag{2.5}$$

where H is the Hurst coefficient, T is the topothesy and the lag distance ζ:

$$\tau = \sqrt{(x - x')^2 + (y - y')^2} \tag{2.6}$$

Hence, an fBm surface is characterized by stationary Gaussian distributed height increments, while the surface is not stationary. Its description relies on only the two parameters H and T. We note that different choices can be made for the set of parameters representing the fBm surface [5–9]. For instance, the fractal dimension D and the standard deviation of the height increments at unitary displacement can be used without any restriction on the following analysis: in fact, D and s are related to H and T:

$$\begin{aligned} D &= 3 - H \\ s &= T^{(1 - H)} \end{aligned} \tag{2.7}$$

Even the surface is described through an fBm random process; it is not straightforward to obtain the corresponding evaluation of the scattered field, because fulfilling of model hypotheses, mathematical developments [12–14], limits of validity [13,14], and related issues need to be addressed. In particular, we stress that a surface satisfying Equation 2.5 for any small value of the distance t would be not differentiable at any point, and its root mean square (RMS) slope would be infinite. On the other hand, a surface satisfying Equation 2.5 for any large value of the lag distance τ would suffer from the infinite variance problem (infrared catastrophe). Fortunately, real natural surfaces satisfy Equation 2.5 only in a limited (although wide) range of lag distance (or, equivalently, of spatial frequencies); in addition, whenever a scattering problem is in order, the range of distances t to be considered is limited on one side by the illuminated patch size (pixel size), and on the other by the incident electromagnetic wavelength l (values of t much smaller than l need not be considered). Therefore, it is argued that the band-limited fBm surfaces are adopted [9,11–14], and hence lower and upper cut-off frequencies must be introduced. Limits of validity of scattering models obviously depend on such cut-off frequencies [13,14]. However, we here want to stress again that these latter parameters are often not related to surface properties, but rather to its illumination (illuminated patch size, wavelength), and therefore they are a priori known.

An example of fractal random sea surface is taken from [14] and is briefly illustrated here:

$$z(x, y) = \sigma C \sum_{n=0}^{N_f - 1} b^{(s - 2)} \sin [K_0 b^n (x \cos \beta_n + y \sin \beta_n) + \alpha_n]; \quad b > 1, 1 < s < 2 \tag{2.8}$$

where $z(x, y)$ is a band-limited generalized Weierstrass surface with fractal dimension $D = s + 1$; K_0 is fundamental wavenumber; α_n is random phase uniformly distributed over $[-\pi, \pi]$ and independent on β_n; β_n is random phase with mean $\bar{\beta}_n$ and some kind of distribution depending on what type of rough surfaces; σ is the surface *RMS* height. The normalized coefficient such that z has *RMS* height:

$$C = \left[\frac{2(1 - b^{2(s-2)})}{1 - b^{2N_f(s-2)}} \right]^{1/2} \quad (2.9)$$

The autocovariance function of Equation 2.8 is

$$\mathfrak{C}_z(\xi, \zeta) = \mathrm{E}\left\{ \frac{\sigma^2 C^2}{2} \sum_{n=0}^{N_f - 1} b^{2(s-2)n} \cos K_n(\xi \cos \beta_n + \zeta \sin \beta_n) \right\} \quad (2.10)$$

where E(.) is the expectation operator; the higher order of wavenumbers are given by $K_n = K_0 b^n$.

After some mathematical manipulations, we have

$$\mathfrak{C}_z(\rho, \phi) = \pi\sigma^2 C^2 \sum_{n=0}^{N_f - 1} b^{2(s-2)n} \sum_{=-\infty}^{\infty} J_m(K_n\rho) S(m) e^{jm(\phi + \pi/2 - \bar{\beta}_n} \quad (2.11)$$

where

$$S(m) = \frac{1}{2\pi} \int_{-\pi}^{\pi} P_n(\psi_n) e^{-jm\psi_n} d\psi_n; \quad (2.12)$$

$J_m(\cdot)$ is Bessel function of order m. By noting $J_m(K_n\rho) = (-1)^m J_{-m}(K_n\rho)$, we may rewrite $\mathfrak{C}$ as

$$\mathfrak{C}_z(\rho, \phi) = \pi\sigma^2 C^2 \left\{ \sum_{n=0}^{N_f - 1} \frac{b^{2(s-2)n}}{2} J_0(K_n\rho) + \sum_{n=0}^{N_f - 1} b^{2(s-2)n} \sum_{m=1}^{\infty} (-1)^m J_{2m}(K_n\rho) S(2m) \cos 2m(\phi - \bar{\beta}_n \right\} \quad (2.13)$$

By considering the finite antenna beam-width, a Gaussian beam of the form $e^{-1/2(\rho/R_n)^2}$ is introduced to account for finite footprint, where $R_n = \varepsilon\Lambda_n$; $\varepsilon > 0$ and $\Lambda_n = \frac{2\pi}{K_n}$. Therefore,

$$\mathfrak{C}_z(\rho, \phi) = \pi\sigma^2 C^2 \left\{ \sum_{n=0}^{N_f - 1} \frac{b^{2(s-2)n}}{2} e^{-\rho^2/2R_n^2} J_0(K_n\rho) + \sum_{n=0}^{N_f - 1} b^{2(s-2)n} e^{-\rho^2/2R_n^2} \sum_{m=1}^{\infty} (-1)^m J_{2m}(K_n\rho) S(2m) \cos 2m(\phi - \bar{\beta}_n \right\} \quad (2.14)$$

The corresponding roughness spectra is

$$W(K, \Phi) = \frac{1}{2\pi}\int_0^\infty \int_0^{2\pi} \rho \mathfrak{C}_z(\rho, \phi) e^{-jK\rho\cos(\phi - \Phi)} d\phi d\rho$$
$$= \frac{\sigma^2 C^2}{2}\left\{ \sum_{n=0}^{N_f - 1} \frac{b^{2(s-2)n}}{2} R_n^2 e^{-\frac{R_n^2(K_n^2 + K^2)}{2}} I_0(R_n^2 K_n K) \right.$$
$$+ \sum_{n=0}^{N_f - 1} 4\pi b^{2(s-2)n} R_n^2 e^{-\frac{R_n^2(K_n^2+K^2)}{2}} \sum_{m=1}^{\infty} S(2m)\cos$$
$$\left. [2m(\Phi - \bar{\beta}_n)] I_{2m}(R_n^2 K_n K) \right\} \tag{2.15}$$

where $I_n(\cdot)$ is modified Bessel function of order n [15]

For sea, on possible choice of the random phase is $\bar{\beta}_n$ = mean wind direction, and $\psi_n = \beta_n - \bar{\beta}_n$ with probability density function

$$P(\psi_n) = g_n \left| \cos\frac{\psi_n}{2} \right|^{2e_n} \text{rect}\left[\frac{\psi_n}{2\pi}\right] \tag{2.16}$$

with

$$g_n = \frac{2^{2e_n - 1}\Gamma^2(1 + e_n)}{\pi\Gamma(1 + 2e_n)};\ e_n = \left(\frac{n_T + 1}{n + 1}\right)^{1/2};\ n_T = Int\left\{\frac{\ln(K_T / K_0)}{\ln(b)}\right\};\ K_T = \sqrt{\frac{g\rho_w}{\tau_s}} = 3.63 \text{ rad/cm}\ g$$

is gravitational acceleration = 981 cm/s^2, ρ_w is water density, τ_s is surface tension.

We note that in characterization of surface roughness, Mattia et al. [16] argued that the surface can be nonstationary and subsequently applied fractal concept to describe the surface. Instead of using correlation length characterizing the horizontal roughness scale, a fractal dimension was used to measure the surface roughness.

2.2.2 σ–ℓ Approach

Though fractal properties were investigated in the context of describing a random surface, we adopted a more classical, and more commonly used, roughness parameters, namely, the correlation length, which measures the horizontal roughness scale, and the RMS height, which describes the vertical roughness scale. The ratio of RMS height to correlation length is related to surface RMS slope.

Assuming that a randomly rough surface $z(x, y)$, a real stationary process, with zero mean and standard deviation σ, then

$$\left\langle z(x, y) z(x + \tau_x, y + \tau_y) \right\rangle = \sigma^2 \rho(\tau_x, \tau_y) \tag{2.17}$$

where $\sigma^2 = \langle z^2 \rangle - \langle z \rangle^2$ and z follows a Gaussian distribution

$$p(z) = \frac{1}{\sqrt{2\pi}\,\sigma} e^{-z^2/2\sigma^2} \tag{2.18}$$

For isotropic surface, we have $\tau_x = \tau_y = \tau$. In what follows, we first discuss the isotropic surface; the anisotropic surface can be treated similarly, but with more complex in directional dependence.

For the surface scattering problem, the joint probability density function between the two points on the surface is of interest and is

$$p(z, z') = \frac{\exp\{-(z^2 - \rho z z' + z'^2)/(2\sigma^2(1 - \rho^2)\}}{2\pi\sigma^2\sqrt{1 - \rho^2}} \tag{2.19}$$

The correlation function $\rho(\tau_x, \tau_y) \leftrightarrow \rho(\tau, \phi)$ appearing in Equations 2.17 and 2.19 is defined by

$$\rho(\tau) = \lim_{L\to\infty} \frac{1}{L} \int_0^{L-\tau} z(r)z(r+\tau)dr \tag{2.20}$$

where $r = \sqrt{x^2 + y^2}$; L is the surface length.

The surface spectrum and correlation function are related by Wiener–Khintchine Theorem, which states that the spectrum of a wide-sense stationary random process of zero mean and its autocorrelation function form a Fourier transform pair:

$$W(K, \varphi) = \int_0^{2\pi} \int_0^{\infty} \rho(\tau, \phi) e^{jK\tau\cos(\varphi - \phi)} \tau d\tau d\phi \tag{2.21}$$

For the isotropic surface, the autocorrelation function $\rho(\tau, \phi)$ is independent of direction ϕ and depending only on lag distance τ. The relation reduces to

$$W(K) = \int_0^{\infty} \rho(r) J_0(Kr) r dr \tag{2.22}$$

where J_0 is zeroth order Bessel function.

In reality, L is finite leading to a cut-off spectrum given by

$$W_m(K) = \begin{cases} 0, & 0 \le K < 2\pi/L \\ K_n/K^n, & 2\pi/L < K < \infty \end{cases} \tag{2.23}$$

where K_n is defined similarly to that of Equation 2.10. The corresponding measured height variance and correlation length are, respectively, calculated by the following equations:

$$\begin{aligned} \sigma_m^2 &= \frac{K_n L^{n-1}}{n-1} \\ \ell_m &= \frac{(n-1)^2 L}{2(2n-1)} \end{aligned} \tag{2.24}$$

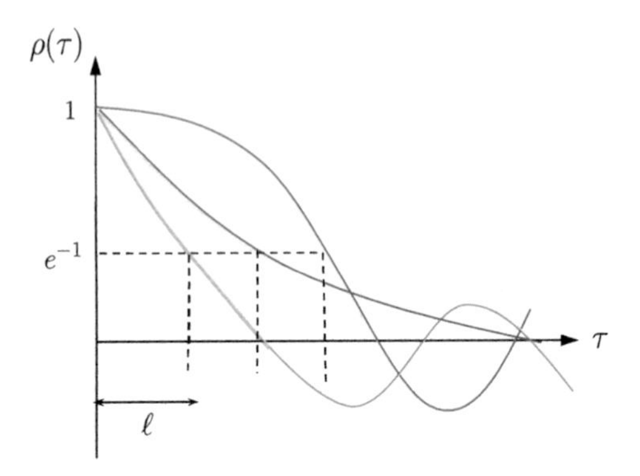

FIGURE 2.3 Correlation function and correlation length for an isotropic surface.

For simplicity but without loss of generality, we only discuss isotropic surface, i.e., note that for isotropic surface, $\rho(\tau, \phi) = \rho(\tau)$, in general, the correlation length ℓ is directionally dependent on anisotropic surfaces such as the sea. When the correlation function drops from unity to e^{-1}, the corresponding value of τ is defined as the correlation length ℓ, that is $\rho(\ell) = e^{-1}$. Figure 2.3 illustrates the correlation function as a function of lag distance τ. The standard deviation and correlation length are called intrinsic surface parameters under the $\sigma - \ell$ model.

The RMS slopes along x and y directions of the surface are given by

$$\begin{aligned} \sigma_{sx}^2 &= \int_{-\infty}^{\infty}\int_{-\infty}^{\infty} dk_x dk_y k_x^2 W(k_x, k_y) \\ \sigma_{sy}^2 &= \int_{-\infty}^{\infty}\int_{-\infty}^{\infty} dk_x dk_y k_y^2 W(k_x, k_y) \end{aligned} \tag{2.25}$$

For the isotropic surface, we have equal slope variances:

$$\sigma_s^2 = \sigma_{sx}^2 = \sigma_{sy}^2 = \pi \int_{-\infty}^{\infty} dk k^3 W(k) \tag{2.26}$$

2.3 CORRELATION FUNCTIONS AND ROUGHNESS SPECTRA

This subsection gives commonly used surface correlation functions and their surface spectra. For natural surfaces, because of formation mechanisms it may present a distinct correlation function that best describes its correlation properties. The following pairs are commonly used functions: Gaussian, Lorentzian, or better known as Exponential, x-power, power law, and Exponential-like.

Gaussian function:

$$\rho(\tau) = e^{-\tau^2/\ell^2} \tag{2.27a}$$

$$W(K) = \frac{\ell}{\pi} e^{-K^2\ell^2/2} \tag{2.27b}$$

Lorentzian (Exponential) function

$$\rho(\tau) = e^{-|\tau|/\ell}; \tag{2.28a}$$

$$W(k) = \frac{1}{\pi\ell(1 + k^2\ell^2)} \tag{2.28b}$$

x- Power

$$\rho(\tau) = \frac{1}{[1 + (\tau/\ell)^2]^{\mu}} \tag{2.29a}$$

$$W(k) = \frac{k^{1-\mu}\ell^{1-3\mu}}{2^{\mu-1}\Gamma(\mu)} K_{1-\mu}(k\ell) \tag{2.29b}$$

where $K_{1-\mu}$ modified Bessel function of order $1 - \mu$; Γ is Gamma function [15].

Generalized power law

$$\rho(\tau) = \frac{\sigma^2}{2^{p-2}\Gamma(p - 1)} \left[\frac{2b_p\tau}{a_p\ell}\right]^{p-1} K_{p-1}\left[\frac{2b_p\tau}{a_p\ell}\right] \tag{2.30a}$$

$$W(k) = \frac{\sigma^2\ell^2}{4\pi}(p - 1)\frac{a_p^2}{b_p^2}\left[1 + \frac{a_p^2}{b_p^2}\frac{k^2\ell^2}{4}\right]^{-p} \tag{2.30b}$$

where $a_p = \frac{\Gamma(p - 0.5)}{\Gamma(p)}$ and b_p is determined by the relation:

$$\left[\frac{2b_p}{a_p}\right]^{p-1} K_{p-1}\left[\frac{2b_p}{a_p}\right] = 2^{p-2}\Gamma(p - 1)e^{-1} \tag{2.30c}$$

where K_{p-1} is modified Bessel function of order $p - 1$; p is real number; $p \geq 1$.

It is easy to show that the preceding spectrum becomes Gaussian spectrum when power index of Equation 2.30b goes to infinite and reduces to the spectrum of exponential correlation function when is 1.5.

Exponential-like function

$$\rho(\tau) = \exp[-\tau(1 - e^{-\tau/x})/L] \tag{2.31a}$$

$$W^{(n)}(K) = \sum_{m=0}^{\infty}\left(\frac{L_e}{L}\right)^m \frac{L_e^2}{m!}\Gamma(m+2)_2F_1\left(\frac{2+m}{2}, \frac{3+m}{2}, 1, -KL_e\right) \tag{2.31b}$$

where $L_e = (xL)/(x + mL)$, $F_1(a, b, c; z)$ is the hypergeometric function [16]; and $L \geq x$.

To achieve a clear separation between a Gaussian shape around the origin and an exponential elsewhere, it is necessary to choose $L \gg x$. When x is small, the correlation length will be approximately equal to L. The transition region from exponential to Gaussian is proportional to x value, which controls the actual shape of the correlation function.

It has been reported in the past that for natural soil surfaces, high-frequency components of roughness scale exist and control the backscattering properties [9]. We consider an isotropic correlation function with an RMS slope that corresponds to surfaces with a spectrum containing an excessive amount of high-frequency spectral components. Thus, it can be used reliably in the low to moderate frequency region or equivalently in the small incident angle region in backscattering. This is because in the low-frequency region scattering is dependent on the shape of the surface spectrum and not on the RMS slope of the surface. As is well known, in the high-frequency limit scattering is proportional to the surface slope distribution. Thus, some error may be induced, if these correlation functions are used in the high-frequency calculations for surfaces that possess no RMS slope. However, it is possible for a correlation function to have the Gaussian property over a small region around the origin and possess an RMS slope but behaves like an exponential over large lag distance. We can see that the range over which the correlation function appears Gaussian-like is controlled by the ratio L/x. Hence, a change in L will change the width of the transition region. Over large lag distances, the function coincides with the exponential function. This means in backscattering computation, changing x is not expected to strongly influence the scattering behavior near the normal incidence. With the roughness spectra given in Equations 2.17–2.31b, it is not difficult to obtain the surface RMS slope, except the exponential one. For generalized power law, we can show that,

$$\sigma_s = \sqrt{\frac{2}{p-2}\frac{b_p}{a_p}}\left(\frac{\sigma}{\ell}\right), p > 2. \tag{2.32}$$

When $p \to \infty$, we have

$$\sigma_s = \sqrt{2}\left(\frac{\sigma}{\ell}\right), \tag{2.33}$$

which is the RMS slope for Gaussian spectrum.

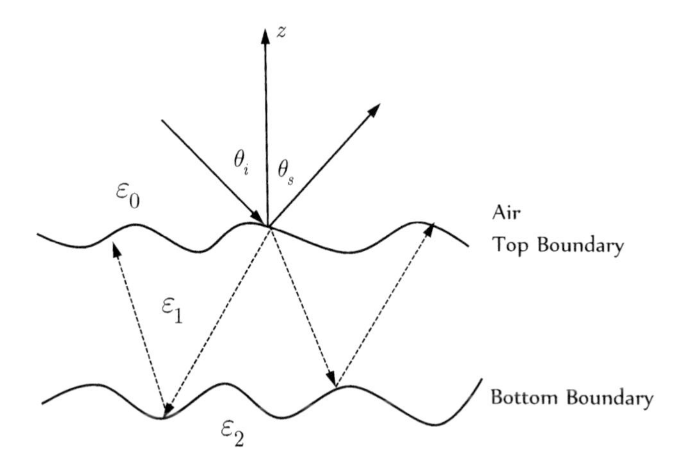

FIGURE 2.4 A double-layered rough surface with two sets of roughness scales.

2.4 MULTISCALE ROUGH SURFACE

Multiscale roughness occurs at many natural surfaces from sea to ground surfaces. A good example is the Sastrugi surface, which is random with many ridges induced by wind [1]. And the layered structure in Greenland is possible due to snow density fluctuations as snow accumulates. Thus, each layer has a distinct permittivity depending on the snow density and ice volume fraction. In particular, the coated top layer may contain more ice components due to wind erosion and melt. Then it is possible for the layer with larger permittivity to lie above a layer with smaller permittivity creating the internal total reflection. Since the layered structure is due to snow accumulation yearly, we consider the interfaces are randomly Sastrugi, and the ridges' heights are randomly distributed. The horizontal shift between top and bottom Sastrugi interfaces is also asymmetrically random. A profile of a double-layered random surface with height profiles z_1 and z_2 is shown in Figure 2.4, where z_1 and z_2 represent the top and the bottom surfaces, respectively. They can be either modeled by Equation 2.1 or other random surface types. The region zero above the top random Sastrugi surface is air, regions 1 and 2, separated by the bottom random interface, are denser and coarser snow, respectively.

One way is a view from the spectral analysis by noting the fact that the multiscale property makes the peak of surface spectra shifted away from its mean. Equivalently, its spatial correlation function exhibits a number of zero-crossings. For this reason, the modulation concept is appropriate to model the multiscale correlation function. The modulation can also be rationed from the fact the measured correlation function usually exhibits zero-crossing [17], which equally states the fact the maxima of surface spectra is shifted toward higher wavenumber. For radar observation, multiscale surfaces may be treated as a narrow-band signal.

When a randomly rough surface was riding on a large-scale surface, say a periodic surface, the resulting correlation function of such multiscale surface may be modeled as [17]

$$\rho_m(\tau) = \rho(\tau)\cos(k_m\tau) \tag{2.34}$$

where k_m is a modulating wavenumber determining the modulation depth; ρ is correlation function of the baseband surface, or the single scale surface, as given in Equations 2.17–2.31. The roughness spectrum can be evaluated by Equation 2.21.

Another possible, and more straightforward, way to describe the multiscale rough surface is a superposition of several single scale surfaces [18]:

$$z_m(x, y) = z_0(x, y) + \sum_{i=1}^{m} a_i z_i(x, y) \tag{2.35}$$

In the above expression, each scale of the surface may have different forms of the correlation function. A sum of Gaussian correlation functions with different correlation lengths and amplitudes was reported in [20] to simulate the backscattering from an exponential-like surface. In surface finish, a superposition of a Gaussian and an exponential function was proposed in [10].

Back to Equation 2.34, in the modulation process, the baseband correlation length L remains the same, but the correlation length of the modulated surface, and hence the multiscale surface, is changed. The actual length will be denoted as effective correlation length ℓ_e which is defined by $\rho(\ell_e) = e^{-1}$ and is dependent on the modulating wavenumber k_m, whose value cannot be arbitrary without limits for a given multiscale surface. Its upper bound is constrained by the physical property of roughness spectra, which should never be negative-valued.

Figure 2.5 shows the modulated correlation functions with $L = 1$ and various k_m. It can be seen that as k_m increases, the fluctuation of the correlation function is more drastic. It crosses the zero point faster and goes to more negative-valued for all the correlation functions considered. Therefore, the effective correlation length L decreases as k_m increases [17]. This states that if one measures the baseband correlation length L and defines it an effective correlation length to be observed by radar, it will produce false scattering return. In other words, the retrieved correlation length from such a backscattering return is really not the one normally measured in the field. Because the correlation length will be modified according to the magnitude of modulating wavenumber, hence, ignoring modulation effects definitely will be misleading in terms of parameter retrieval.

Sea surface is an anisotropic, non-Gaussian height, and multiscale rough surface due to the effect of local wind and the waves propagating from other parts of the sea. The modulation of different scales complicates the radar scattering from the sea surface. For multiscale rough sea surface, it is instructive to know that only a part of roughness scales (effective roughness) are responsible for backscattering at a given view angle and exploring frequency [19]. Different scales of roughness riding on the sea surface can also be calculated by different parts of sea spectral components, from which the effective roughness can be derived. Roughness scales responsible for radar scattering are usually characterized quantitatively by the correlation function and the surface RMS height. For illustration, let's take L-band (28 rad/m) as an example. The correlation functions and roughness parameters (RMS height

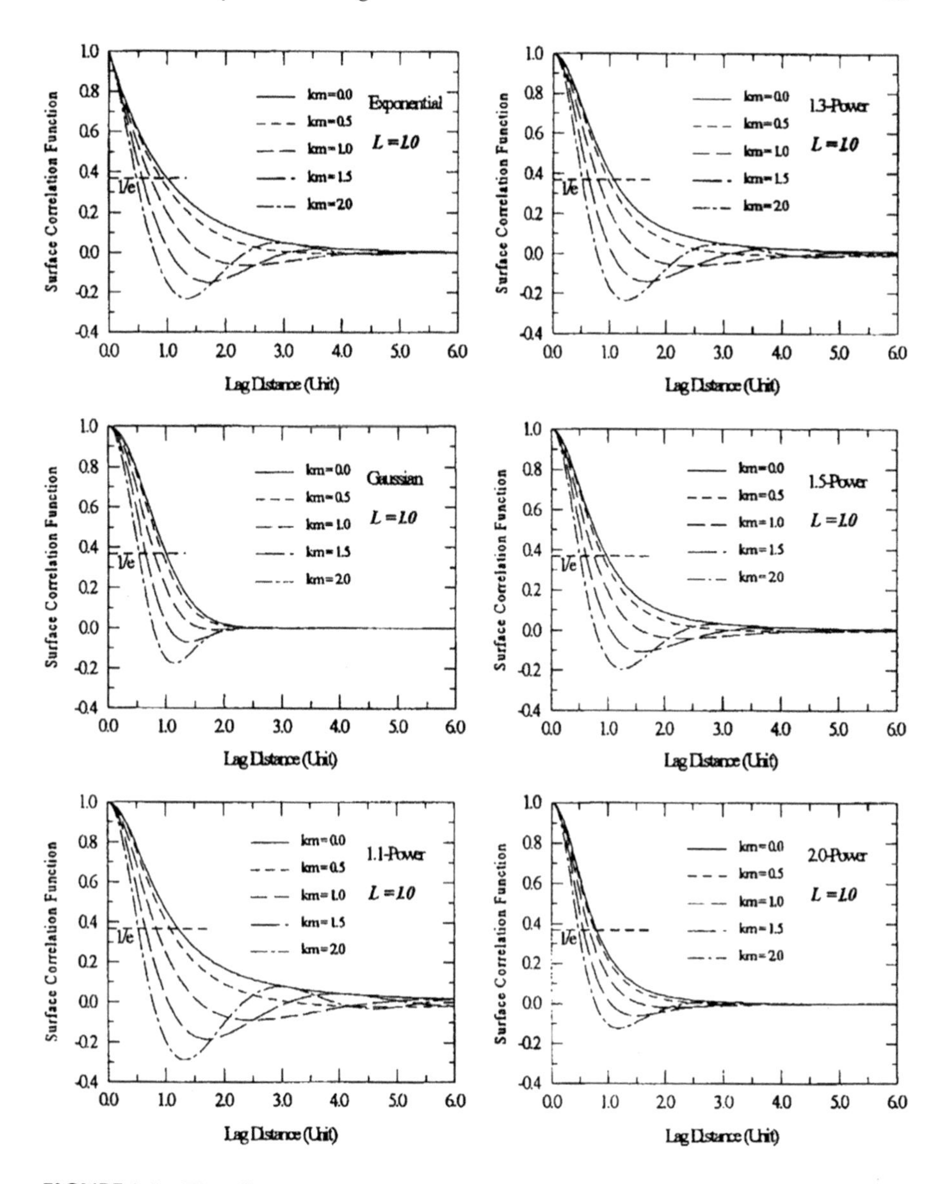

FIGURE 2.5 The effective correlation length changed by a modulating wavenumber (unit); different correlation functions are shown.

and correlation length) are computed by different spectral components: $k_e \in [0.1 * k, \infty)$, $[0.05 * k, \infty)$, and $(0, \infty)$ based on the Apel spectrum. The corresponding roughness parameters along the upwind- and crosswind directions at the wind speeds of 10 and 15 m/s are shown in Figure 2.6.

It can be seen from Figure 2.6 that the normalized correlation functions generated by high-wavenumber spectral components (see Figure 2.6(a)–(b)) are more oscillating with a decreasing amplitude, and the RMS height and correlation lengths

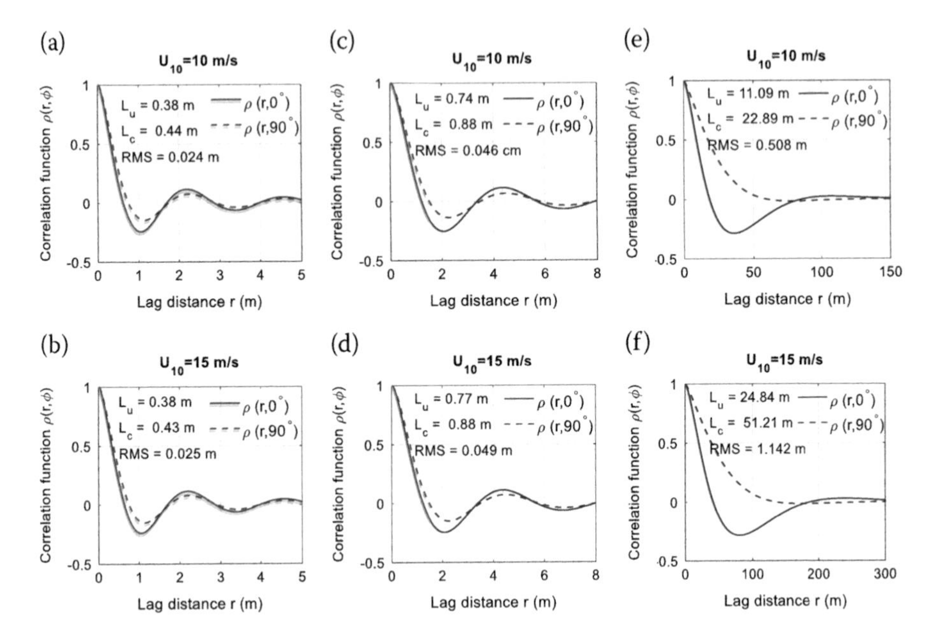

FIGURE 2.6 The normalized correlation functions and roughness parameters (RMS height and correlation length L_u, L_c in upwind and crosswind directions) generated by different wave spectral components k_e based on Apel spectrum model at L-band (1.26 GHz) for wind speeds of 10 m/s and 15 m/s, respectively. (a)–(b) $k_e \in [0.1*k, \infty)$; (c)–(d) $k_e \in [0.05*k, \infty)$; (e)–(f) $k_e \in (0,\infty)$).

in both upwind- and crosswind-directions are also small, which is in accord with that of short wind wave fields derived from the along-wind surface wave profiles [21]. With the addition of low-wavenumber spectral components, the roughness increases. When all the spectral components in sea spectrum are included to account for the roughness (see Figure 2.6(e)–(f)), the corresponding scatterer on the sea surface is very large with a tens of meters of correlation length and tens of centimeters or even one meter in RMS height. Meanwhile, the oscillation of the autocorrelation function of large-scale scatterer also flattens out and the roughness becomes larger as the wind speed increases. While, for the case of Figure 2.6(a)–(b) and Figure 2.6(c)–(d), the scatterer becomes rougher with a smaller correlation length and larger RMS height with the increase of wind speed.

2.5 UNCERTAINTIES OF ROUGHNESS PARAMETERS

In a statistical sense, $\mathbf{y} = \{\sigma, \ell\}$ is a random variable due to spatially and temporally varying properties, such that

$$\mathbf{y} = \mathbf{y}_t + \mathbf{y}_n \tag{2.36}$$

where $\mathbf{y}_t$ is true, $\mathbf{y}_n$ is error term.

In practice, the "truth" is never attainable, nor even measurable; it is always vague. Statistically, $\mathbf{y}_t$ and $\mathbf{y}_n$ may be assumed to be, as they usually are, uncorrelated, such that $\mathbf{y}$ is an unbiased estimate of $\mathbf{y}_t$, i.e.,

$$E(\mathbf{y}) = E(\mathbf{y}_t),\ \forall\ \mathbf{y}_n \sim N(0, \sigma_{\mathbf{y}_n}^2) \tag{2.37}$$

where E denotes statistical mean, $\sigma_{\mathbf{y}_n}^2$ is variance of $\mathbf{y}_n$. It turns out that to estimate the correlation length, we need to know the correlation function ρ. For natural surfaces the RMS height and the correlation length typically depend on the measured length. The correlation function of a random process is itself a random process and hence subject to statistical fluctuation. Let's take a look at the definition of the correlation function and RMS height. Let $z(x, y)$ be the random surface height, the covariance function $C(\tau) = \sigma^2\rho(\tau)$ is

$$\begin{aligned} C(\tau_x, \tau_y) &= \left\langle z(x, y)z(x + \tau_x, y + \tau_y)\right\rangle \\ &= \lim_{\substack{L_x \to \infty \\ L_y \to \infty}} \frac{1}{L_x L_y} \iint z(x, y)z(x + \tau_x, y + \tau_y)dxdy \end{aligned} \tag{2.38}$$

where L_x, L_y are surface lengths along x and y directions, respectively. Ideally, L_x, L_y are infinite, at least electromagnetically speaking. Practically they are not. In the following discussion, we consider the isotropic surface. The estimate of Equation 2.38 is written as

$$\hat{C}(\tau) = \frac{1}{N} \sum_{i=1}^{N} z(x_i)z(x_i + \tau) \tag{2.39}$$

where N is total number of sampling.

It has been problematic to estimate the correlation functions from measured data if the mean level of each sample is not known [22]. The variance of Equation 2.30 can be readily obtained from [23].

$$var[\hat{C}(\tau)] = \sigma^2 + \frac{2}{N^2} \sum_{i=1}^{N-1} (N - i)C(i\Delta\tau) \tag{2.40}$$

where $\Delta\tau$ is the sampling spacing and it is assumed that $C(N\Delta\tau) \to 0$.

The variance of estimate RMS height σ is

$$var(\hat{\sigma}) = \frac{1}{L_s} \int_0^\infty \hat{C}^2(\xi)d\xi \tag{2.41}$$

Both variances are strongly dependent on the shape of the correlation function. The RMS height and correlation length increase with measurement trace length.

REFERENCES

1. Xu, P., Chen, K. S., Liu, Y, *et al.*, Full-wave simulation and analysis of bistatic scattering and polarimetric emissions from double-layered Sastrugi surfaces, *IEEE Transactions on Geoscience and Remote Sensing*, 55(1), 292–307, January 2017.
2. Fung, A. K., and Chen, K. S., *Microwave Scattering and Emission Models for Users*, Artech House, Norwood, MA, 2010.
3. Ulaby, F., and Long, D., *Microwave Radar and Radiometric Remote Sensing*, University of Michigan, Ann Arbor, MI, 2014.
4. Yang, Y., and Chen, K. S., Polarized backscattering from randomly anisotropic rough surface, *IEEE Transactions on Geoscience and Remote Sensing*, 57(9), 6608–6618, September 2019.
5. Mandelbrot, B. B., *The Fractal Geometry of Nature*, W. H. Freeman & C., New York, 1983.
6. Voss, R. F., *Random Fractal Forgeries,* in *Fundamental Algorithms for Computer Graphics*, R. A. Earnshaw ed., Springer Verlag, Berlin, 805–835, 1985.
7. Brown, S. R., and Scholz, C. H., Broad band study of the topography of natural rock surfaces, *Journal of Geophysical Research*, 90(B14), 12575–12582, December 1985.
8. Flandrin, P., On the spectrum of fractional brownian motions, *IEEE Transactions on Information Theory*, 35(1), 197–199, 1989.
9. Falconer, K. *Fractal Geometry*, John Wiley & Sons, Chichester, UK, 1990.
10. Church, E. L., Fractal surface finish, *Applied Optics*, 27(8), 1518–1526, 1988.
11. Franceschetti, G., Iodice, A., Maddaluno, S., and Riccio, D., A fractal-based theoretical framework for retrieval of surface parameters from electromagnetic backscattering data, *IEEE Transactions on Geoscience and Remote Sensing*, 38(2), 641–650, 2000.
12. Berizzi, F., and Dalle-Mese, E., Fractal analysis of the signal scattered from the sea surface. *IEEE Transactions on Antennas and Propagation*, 47(2), 324–338, 1999.
13. Franceschetti, G., Iodice, A., Maddaluno, S., and Riccio, D., Scattering from natural rough surfaces modelled by fractional brownian motion two-dimensional processes, *IEEE Transactions on Antennas and Propagaton*, 47(9), 1405–1415, 1999.
14. Berizzi, F., and Dalle-Mese, E., Scattering from a 2D sea fractal surface: Fractal analysis of the scattered signal. *IEEE Transactions on Antennas and Propagation*, 50(7), 912–925, 2002.
15. Mattia, F., Le Toan, T., and Davidson, M., An analytical, numerical, and experimental study of backscattering from multiscale soil surfaces. *Radio Science*, 36(1), 119–135, 2001.
16. Abramowitz, M., and Stegun, I. A., *Handbook of Mathematical Functions: With Formulas, Graphs, and Mathematical Tables*, Dover Publications, New York, 1965.
17. Chen, K. S., Wu, T. D., and Fung, A. K., A study of backscattering from multiscale rough surface. *Journal of Electromagnetic Waves and Applications*, 12(7), 961–979, 1998.
18. Fung, A. K., *Backscattering from Multiscale Rough Surfaces with Application to Wind Scatterometry*, Artech House, Norwood, MA, 2015.
19. Xie, D. F., Chen, K. S., and Zeng, J. Y., The frequency selective effect of radar backscattering from multiscale sea surface, *Remote Sensing*, 11(2), 1–18, 2019, doi:10.3390/rs11020160.
20. Fung, A. K., *Microwave Scattering and Emission Models and Their Applications*, Artech House, Norwood, MA, 1994.
21. Caulliez, G., and Guérin, C. A., Higher-order statistical analysis of short wind wave fields. *Journal of Geophysical Research: Oceans*, 117, C06002-1–C06002-14, 2012.
22. Freniére, E. R., O'Neill, E. L., and Walther, A., Problem in the determination of correlation functions. II, *Journal of the Optical Society of America*, 69(4), 634–635, 1979.
23. Davenport, W. B., and Root, W. L., *An Introduction to the Theory of Random Signals and Noise*, Wiley-IEEE Press, NJ, 1987.

3 Basics of Electromagnetic Wave

3.1 MAXWELL'S EQUATIONS

The macroscopic wave behavior in media, for which we are interested in radar sensing, must obey Maxwell's equations [1–4]. As there exist many excellent books, we do not attempt to repeat the treatments of Maxwell's equations here. Instead, we give a minimum level of material for self-contained basics of electromagnetic wave in the context of this book. Maxwell's equations are given by two forms relevant to electromagnetic waves: integral forms and differential forms. The integral forms describe the behavior of electric and magnetic fields over surfaces or around paths, whereas the differential forms apply to specific locations. Because of the direct link to wave equations, we only give the differential forms here:

$$\nabla \times \mathbf{E}(\vec{r}, t) + \frac{\partial}{\partial t}\mathbf{B}(\vec{r}, t) = 0 \tag{3.1a}$$

$$\nabla \times \mathbf{H}(\vec{r}, t) - \frac{\partial}{\partial t}\mathbf{D}(\vec{r}, t) = \mathbf{J}(\vec{r}, t) \tag{3.1b}$$

$$\nabla \cdot \mathbf{B}(\vec{r}, t) = 0 \tag{3.1c}$$

$$\nabla \cdot \mathbf{D}(\vec{r}, t) = \rho(\vec{r}, t) \tag{3.1d}$$

In the preceding equations, the quantities and their unit are **E** (electric field): Volts per meter (V/m); **B** (magnetic flux density): Teslas (T), or Webers per square meter (Wb/m^2); **H** (magnetic field): Amperes per meter (A/m), and **D** (electric flux density): Coulombs per square meter (C/m^2); ρ (electric charge density): Coulombs per cubic meter (C/m^3). Maxwell's equations are first-order differential equations in space $\vec{r}$ and time t. In short notation, $\vec{r}$ and t are usually neglected.

The following continuation equation is the conservation law of electric charge and current density:

$$\nabla \cdot \mathbf{J}(\vec{r}, t) + \frac{\partial}{\partial t}\rho(\vec{r}, t) = 0 \tag{3.2}$$

Because Maxwell's equations are linear, we can express them in either time-domain or frequency-domain, which is attained by Fourier series expansion with a sinusoidal basis. Hence, for a time-harmonic field, the field A is written as $A(\vec{r}, t) = \Re\mathfrak{e}\{A(\vec{r})e^{j\omega t}\}$, where ω is the angular frequency and $\Re\mathfrak{e}$ means real part. Knowing that $\frac{\partial}{\partial t} \to j\omega$, Equations 3.1a–3.1d can be equally expressed in spectral forms

$$\nabla \times \mathbf{E} = -j\omega\mathbf{B} \tag{3.3a}$$

$$\nabla \times \mathbf{H} = j\omega\mathbf{D} + \mathbf{J} \tag{3.3b}$$

$$\nabla \cdot \mathbf{D} = \rho \tag{3.3c}$$

$$\nabla \cdot \mathbf{B} = 0 \tag{3.3d}$$

$$\nabla \cdot \mathbf{J} + j\omega\rho = 0 \tag{3.3e}$$

The field equations describe the relationships between the source and mediating fields within that frame of reference. Maxwell's differential equations are valid for any system in uniform relative motion concerning the laboratory frame of reference in which we normally do our measurements. It is known that Maxwell's equations are well-posed because of [1–4]:

the model has at least one solution (existence);
the model has at most one solution (uniqueness);
the solution is continuously dependent on the data supplied;
small changes in the data supplied produce equally small changes in the solution.

If sources are given, $\mathbf{J}$ and ρ are known. Equations 3.3a, 3.3b, and 3.3e constitute six independent scalar equations; $\mathbf{E}$, $\mathbf{H}$, $\mathbf{B}$, $\mathbf{D}$ are 12 scalar unknowns. It is observed that in Maxwell's equations, we have 12 scalar unknowns—3 for each 4 vectors, but only 6 independent scalar equations, requiring 6 more equations—that is constitute relations.

3.2 CONSTITUTE RELATIONS

The wave behavior in media, either artificial or natural, is described by the constitute relations, which generally involve a set of constitutive parameters and a set of constitutive operators, where the constitutive parameters give constants of proportionality between the fields or components in a dyadic relationship, and the constitutive operators include linear, nonlinear, and integro-differential operators. Understanding the wave behavior in and through the media is essential in many disciplines. For radio propagation, rains, vegetative components, and ionospheric layers in one way or another behave differently under the radio waves that go through them. And by these changes, if any, it is the objective of the remote sensing of such random media.

The connections between $\mathbf{E}$ and $\mathbf{D}$; $\mathbf{H}$ and $\mathbf{B}$ in Maxwell's equations are through medium's permittivity ε, permeability μ, and conductivity σ. In general, ε, μ, σ are not constant but in matrix form. In remote sensing of geophysical media, the properties of ε, μ, σ are of profound importance. Examples include the dielectric constants of sea ice, snow, soil, vegetation canopy constituents, and so on. Those dielectric constants can be time dispersive, frequency dispersive, and may be also dependent on temperature, pressure, capacity, etc. A very compact and systematic form of the constitute relations and the associated physical properties are due in [3]. We give them here for easy reference.

$$\begin{bmatrix} c\mathbf{D} \\ \mathbf{H} \end{bmatrix} = \bar{\bar{\mathbf{C}}} \begin{bmatrix} \mathbf{E} \\ c\mathbf{B} \end{bmatrix} \tag{3.4}$$

where

$$\bar{\bar{\mathbf{C}}} = \begin{bmatrix} \bar{\bar{\mathbf{P}}} & \bar{\bar{\mathbf{L}}} \\ \bar{\bar{\mathbf{M}}} & \bar{\bar{\mathbf{Q}}} \end{bmatrix} \tag{3.5}$$

Depending on the form of the tensor, the media are classified to be inhomogeneous if $\bar{\bar{\mathbf{C}}}$ is a function of space coordinate; nonstationary if $\bar{\bar{\mathbf{C}}}$ is a function of time; time dispersive if $\bar{\bar{\mathbf{C}}}$ is a function of time derivatives; spatial dispersive if $\bar{\bar{\mathbf{C}}}$ is a function of spatial derivatives; and nonlinear if $\bar{\bar{\mathbf{C}}}$ is a function of the electromagnetic field.

Some of the medium usually encountered in remote sensing of the environment include:

Free-space

$$\bar{\bar{\mathbf{P}}} = \bar{\bar{\mathbf{Q}}} = \bar{\bar{\mathbf{I}}}/\eta_0,\ \bar{\bar{\mathbf{L}}} = \bar{\bar{\mathbf{M}}} = 0, \tag{3.6}$$

where $\eta_0 = \sqrt{\varepsilon_0/\mu_0}$ is intrinsic impedance of free space, $\varepsilon = \varepsilon_0 = 4\pi \times 10^{-7}$ H/m; $\mu = \mu_0 = (36\pi)^{-1} \times 10^{-12}$ F/m.

Linear isotropic medium

$$\bar{\bar{\mathbf{P}}} = \frac{\varepsilon_r}{\eta_0}\bar{\bar{\mathbf{I}}}, \quad \bar{\bar{\mathbf{Q}}} = \frac{1}{\eta_0\mu_r}\bar{\bar{\mathbf{I}}}; \quad \bar{\bar{\mathbf{L}}} = \bar{\bar{\mathbf{M}}} = 0, \tag{3.7}$$

where $\varepsilon = \varepsilon_r\varepsilon_0$, $\mu = \mu_r\mu_0$.

Linear anisotropic medium

$$\bar{\bar{\mathbf{P}}} = c\bar{\bar{\varepsilon}}, \quad \bar{\bar{\mathbf{Q}}} = \frac{1}{c\bar{\bar{\mu}}}; \quad \bar{\bar{\mathbf{L}}} = \bar{\bar{\mathbf{M}}} = 0, \tag{3.8}$$

where $\bar{\bar{\varepsilon}}$ is permittivity tensor, and $\bar{\bar{\mu}}$ permeability tensor; $c = 1/\sqrt{\varepsilon_0\mu_0}$.

Readers are referred to consult relevant books [3] for more treatments of natural and man-made media.

3.3 WAVE REFLECTION AND TRANSMISSION AT A PLANE BOUNDARY

3.3.1 Laws of Reflection and Refraction

When wave propagating through the medium encounters a discontinuous dielectric boundary, the phenomena of reflection and transmission occurs. Here the boundary, we refer to a plane boundary. If the boundary is rough, as we described in the previous chapter, the wave reflection and transmission still give rise by physical law, only at a local coordinate. That is, the laws of reflection and refraction apply everywhere on the surface as long as there is a tangent plane on which the reflection and transmission exist, and Snell's law applies. In remote sensing of rough surface, it is perceptive to understand the physical laws that govern the reflection and transmission from the boundary so as to perceive the information about the boundary properties—both geometric and dielectric, from which the geophysical parameters can be inferred. As will be shown in the next chapter, the reflection and transmission for a rough surface are dependent on the local incidence angle, which is determined by the incident direction and surface unit normal [5].

As illustrated in Figure 3.1, a uniform horizontal-polarized plane wave incident upon the boundary, the incident, reflected, and transmitted electric fields are of the forms:

$$\begin{aligned} \mathbf{E}^i &= \hat{y}E_0 e^{-j\vec{k}_i \cdot \vec{r}} \\ \mathbf{E}^r &= \hat{y}R_h E_0 e^{j\vec{k}_r \cdot \vec{r}} \\ \mathbf{E}^t &= \hat{y}T_h E_0 e^{-j\vec{k}_t \cdot \vec{r}} \end{aligned} \tag{3.9}$$

where E_0 is the amplitude of the incident electric field, R_h and T_h are reflection and transmission coefficients to be determined by imposing the boundary conditions on the tangential and normal fields across the boundary. The propagating vectors, or the wave vectors, appearing in the spatial phase are given by

$$\vec{k}_i = \hat{k}_i k_i = \hat{x}k_{ix} + \hat{z}k_{iz};\ \vec{k}_r = \hat{k}_r k_r = \hat{x}k_{rx} - \hat{z}k_{rz};\ \vec{k}_t = \hat{k}_t k_t = \hat{x}k_{tx} + \hat{z}k_{tz} \tag{3.10}$$

Designating the relative permittivity and relative permeability in the incident and transmitted media as ε_i, μ_i and ε_t, μ_t, respectively, the wavenumbers in the incident, reflected, and transmitted media are, by dispersion relations:

$$k_i = \sqrt{k_{ix}^2 + k_{iz}^2} = \omega\sqrt{\mu_i \varepsilon_i} \tag{3.11a}$$

$$k_r = \sqrt{k_{rx}^2 + k_{rz}^2} = \omega\sqrt{\mu_i \varepsilon_i} \tag{3.11b}$$

$$k_t = \sqrt{k_{tx}^2 + k_{tz}^2} = \omega\sqrt{\mu_t \varepsilon_t} \tag{3.11c}$$

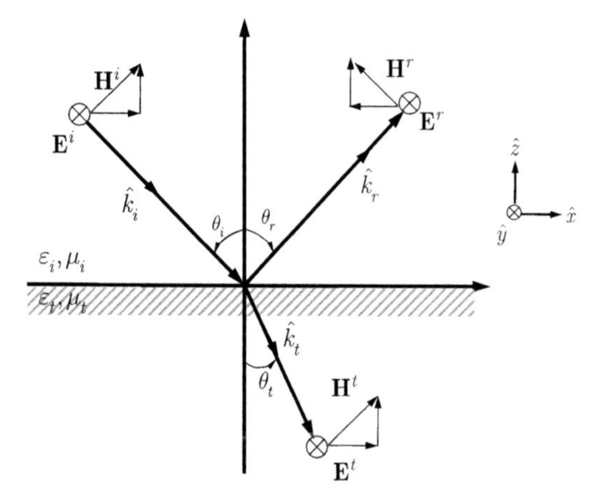

FIGURE 3.1 Wave reflection and transmission of a plane boundary that separates two media.

The components of the wave vectors can be written in more explicit expressions:

$$\begin{aligned} k_{ix} &= k_i \sin\theta_i,\ k_{iz} = k_i \cos\theta_i \\ k_{rx} &= k_i \sin\theta_r,\ k_{rz} = k_i \cos\theta_r \\ k_{tx} &= k_t \sin\theta_t,\ k_{tz} = k_t \cos\theta_t \end{aligned} \tag{3.12}$$

where the angles of reflection and transmission obey the laws of reflection and refraction. If the wave propagates in the free space, the plane of constant amplitude and plane of constant phase are coincident. From the continuity of the tangential magnetic fields

$$\begin{aligned} \hat{n} \times (\mathbf{E}_1 - \mathbf{E}_2) &= 0 \\ \hat{n} \times (\mathbf{H}_1 - \mathbf{H}_2) &= \mathbf{J}_s \\ \hat{n} \cdot (\mathbf{B}_1 - \mathbf{B}_2) &= 0 \\ \hat{n} \cdot (\mathbf{D}_1 - \mathbf{D}_2) &= \rho_s \end{aligned} \tag{3.13}$$

and upon the amplitude and phase matching, we come up with two equations with two unknowns

$$\begin{cases} 1 + R_h = T_h \\ 1 + R_v = \dfrac{\mu_t k_{tz}}{\mu_t k_{iz}} T_v \end{cases} \tag{3.14}$$

It follows that the reflection coefficient and transmission coefficient are given by

$$R_h = \frac{\mu_t k_i \cos\theta_i - \mu_i k_{tz}}{\mu_t k_i \cos\theta_i + \mu_i k_{tz}} = \frac{\eta_t \cos\theta_i - \eta_i \cos\theta_t}{\eta_t \cos\theta_i + \eta_i \cos\theta_t}, \tag{3.15a}$$

$$T_{\perp} = \frac{2\mu_2 k_{iz}}{\mu_2 k_{iz} + \mu_1 k_{tz}} = \frac{2\eta_2 \cos\theta_i}{\eta_2 \cos\theta_i + \eta_1 \cos\theta_t} \tag{3.15b}$$

By duality, the reflection coefficient for the vertical polarization takes the expression:

$$R_v = \frac{\varepsilon_i k_{tz} - \varepsilon_t k_{iz}}{\varepsilon_t k_{iz} + \varepsilon_i k_{tz}} = \frac{\eta_i \cos\theta_i - \eta_t \cos\theta_t}{\eta_t \cos\theta_i + \eta_t \cos\theta_t}, \tag{3.16a}$$

$$T_v = \frac{2\varepsilon_i k_{iz}}{\varepsilon_t k_{iz} + \varepsilon_i k_{tz}} \frac{\eta_t}{\eta_i} = \frac{2\eta_t \cos\theta_i}{\eta_i \cos\theta_i + \eta_t \cos\theta_t} \tag{3.16b}$$

In the preceding expressions, $k_{tz} = Re\{k_{tz}\} + jIm\{k_{tz}\}$, with its real part and imaginary part given by

$$Re\{k_{tz}\} = \frac{1}{\sqrt{2}}\left[Re\{k_t^2\} - k_i^2 \sin^2\theta_i + \sqrt{\left(Re\{k_t^2\} - k_i^2 \sin^2\theta_i\right)^2 + \left(Im\{k_t^2\}\right)^2}\right]^{1/2} \tag{3.17a}$$

$$Im\{k_{tz}\} = -\frac{1}{\sqrt{2}}\left[-\left(Re\{k_t^2\} - k_i^2 \sin^2\theta_i\right) + \sqrt{\left(Re\{k_t^2\} - k_i^2 \sin^2\theta_i\right)^2 + \left(Im\{k_t^2\}\right)^2}\right]^{1/2} \tag{3.17b}$$

and the characteristic impedances of the medium are $\eta_i = \sqrt{\mu_i/\varepsilon_i}$; $\eta_t = \sqrt{\mu_t/\varepsilon_t}$. If wave incident in free space, $\eta_0 = \sqrt{\mu_0/\varepsilon_0} = 376.7\Omega$.

The polarized reflectivity and transmissivity are used to indicate the power reflection and transfer. For horizontal polarization, the reflectivity and transmissivity are

$$\Gamma_h = |R_h|^2 \tag{3.18a}$$

$$\Im_h = \frac{Re\{\cos\theta_t/\eta_t\}}{Re\{\cos\theta_i/\eta_i\}} |T_h|^2 \tag{3.18b}$$

For vertical polarization, the reflectivity and transmittivity are

$$\Gamma_v = |R_v|^2 \tag{3.19a}$$

$$\Im_v = \frac{Re\{\eta_t \cos \theta_t\}}{Re\{\eta_i \cos \theta_1\}} |T_v|^2 \tag{3.19b}$$

By energy conservation at the boundary, we have $\Gamma + \Im = 1$.

Before moving on, we show the dependence of reflection coefficients on the dielectric constant assuming $\mu_r = 1$. The Brewster angle is "pseudo-Brewster angle." As the imaginary part of the dielectric changes, the pseudo-Brewster angle shifts, and the reflection coefficient is further away from zero to different extents. In Figure 3.2(a), the dielectric constant of $\varepsilon_r = 5 - j0.5,\ 10 - j2,\ 15 - j3$ corresponds to soil moisture contents of $m_v = 10\%,\ 20\%,\ 28\%$, respectively [6]. For the purpose of illustration, we keep the real part of the dielectric constant at 15, and change only the imaginary part is plotted in Figure 3.2(b). The angular dependence for both H and V polarizations are quite similar, except the Brewster effects on the V polarization, in which the dip becomes shallower for the larger imaginary part of constant. That is, for a more lossy medium, the reflection is stronger around the Brewster angle for a uniform plane wave incidence. In such a case, the total reflection does not occur at a lossy medium.

3.3.2 Reflection and Transmission in a Layered Medium

In the preceding treatment, we assumed the transmitted medium being homogeneous. However, natural surfaces are generally inhomogeneous with permittivity being spatially non-uniform. An inhomogeneous rough surface, whose properties vary continuously along one axis of a rectangular coordinate system (e.g., z-axis) but do not change in planes perpendicular to this axis, is considered in this study. The Earth's atmosphere, sea, soil, snow-covered, and frozen ground surfaces are approximately inhomogeneous media. Their properties (e.g., velocities of electromagnetic waves, density, etc.) vary considerably along the vertical direction, whereas the variations along the horizontal direction are much less noticeable. When the moisture content of the soil surface dries up or wets down, the dielectric permittivity

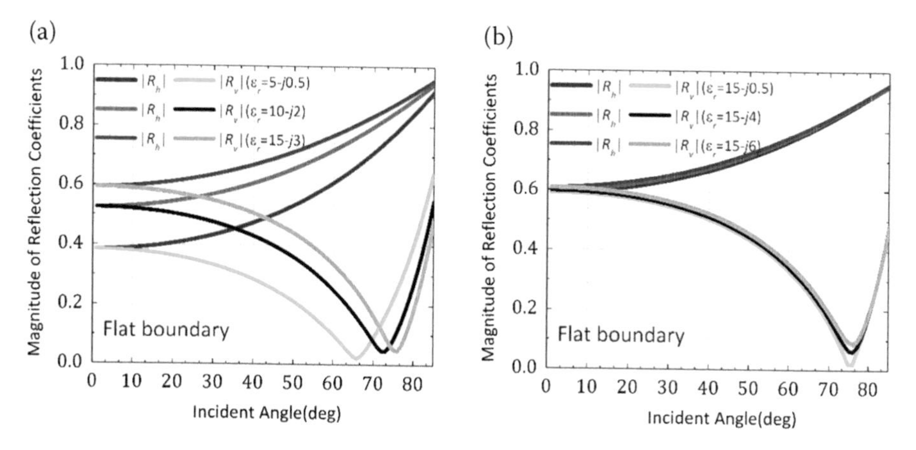

FIGURE 3.2 Magnitude of reflection coefficients for different dielectric constants.

becomes vertically non-uniform as illustrated in Figure 3.3, where the medium is assumed horizontally uniform and may be treated as a stack of thin stratified layers.

We discrete the lower medium into N stratified layers. For illustration, we take $N = 3$, as shown in Figure 3.4. The depth of the third layer extends to infinity. That is, the background is a half-space and no reflection from the bottom of the layer is accounted for. In three layers, we have two boundaries at $z = -d_1$ and $z = -d_2$. The z-component wavenumbers in each layer are k_{1z}, k_{2z}, k_{3z}, which together with the layer depth account for phase delays in each layer.

Denoting the reflection coefficient from layer ℓ to layer $\ell + 1$ as $R_{\ell,\ell+1}$, and the transmission coefficient as $T_{\ell,\ell+1}$, then the reflection and transmission process can be sketched in Figure 3.5. The multiple reflections and transmission between two adjacent layers occur continuously until decaying to zero. It follows that the total effective reflection coefficient is given by [1]

$$\tilde{R}_{12} = R_{12} + T_{12}R_{23}T_{21}e^{j2k_{2z}(d_2 - d_1)} + T_{12}R_{23}^2R_{21}T_{21}e^{j4k_{2z}(d_2 - d_1)} + \cdots . \quad (3.20)$$

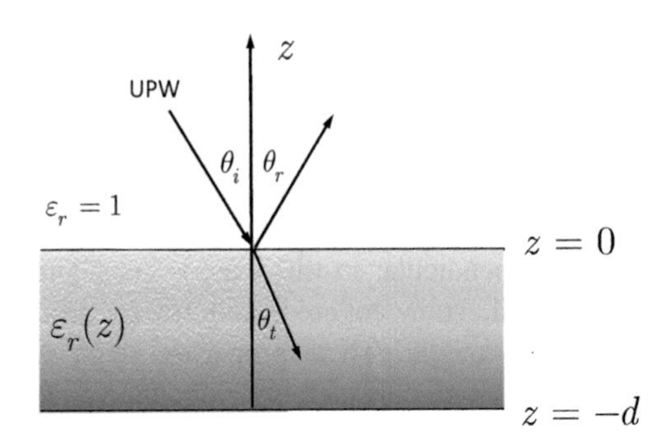

FIGURE 3.3 A uniform plane wave (UPW) incident upon an inhomogeneous boundary, where d is the medium depth.

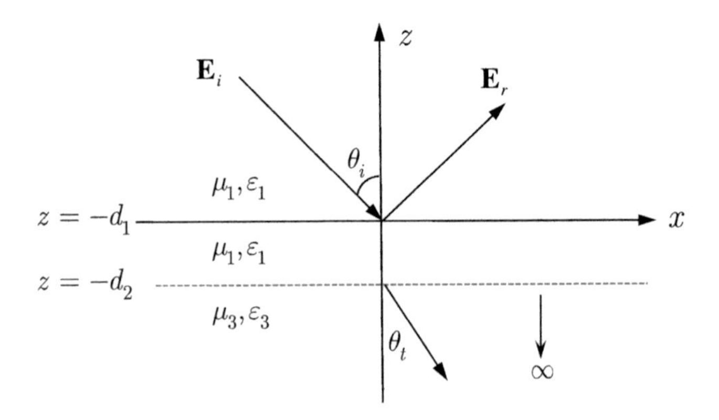

FIGURE 3.4 Reflection from a three-layer medium with a half-space background layer.

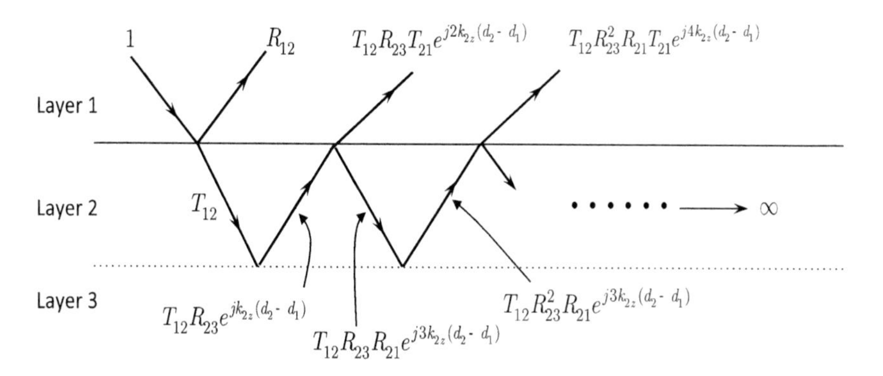

FIGURE 3.5 Reflection and transmission process of a multilayered medium.

which can be put into a compact form:

$$\tilde{R}_{12} = R_{12} + \frac{T_{12}R_{23}T_{21}e^{j2k_{2z}(d_2 - d_1)}}{1 - R_{21}R_{23}e^{j2k_{2z}(d_2 - d_1)}} \tag{3.21}$$

Following the preceding process and procedure, and noting the following relations:

$$\begin{aligned} T_{\ell,\ell+1} &= 1 + R_{\ell,\ell+1} \\ R_{\ell,\ell+1} &= -R_{\ell+1,\ell} \end{aligned} \tag{3.22}$$

it is ready to extend the number of layers to N to generalize Equation 3.21 to

$$\tilde{R}_{\ell,\ell+1} = \frac{R_{\ell,\ell+1} + \tilde{R}_{\ell+1,\ell+2}e^{j2k_{(\ell+1)z}(d_{\ell+1} - d_\ell)}}{1 + R_{\ell,\ell+1}\tilde{R}_{\ell+1,\ell+2}e^{j2k_{(\ell+1)z}(d_{\ell+1} - d_\ell)}} \tag{3.23}$$

As an example, the dielectric of the soil surface in dry up or wet down conditions may be described by a transitional layer [7], where the permittivity of the soil surfaces as a function of depth is mathematically given by [7–8]:

$$\varepsilon_r(z) = 1 + [\varepsilon_r(-\infty) - 1]\frac{\exp(-akz)}{1 + \exp(-akz)} \tag{3.24}$$

where k is the wavenumber; $\varepsilon_r(-\infty) = \varepsilon_\infty$ is the background dielectric constant; a is transition rate that controls the change rate of the dielectric constant in z direction; it is dependent on the available soil moisture depletion, among others. The real part of the dielectric constant always starts with $\varepsilon_r = 1$, which is free space (air), when eventually reaching ε_∞. The dielectric profile is roughly symmetrical about $z = 0$, beyond or below which the permittivity varies exponentially. Notice the continuous transition from the top of the surface ($z = 0$)

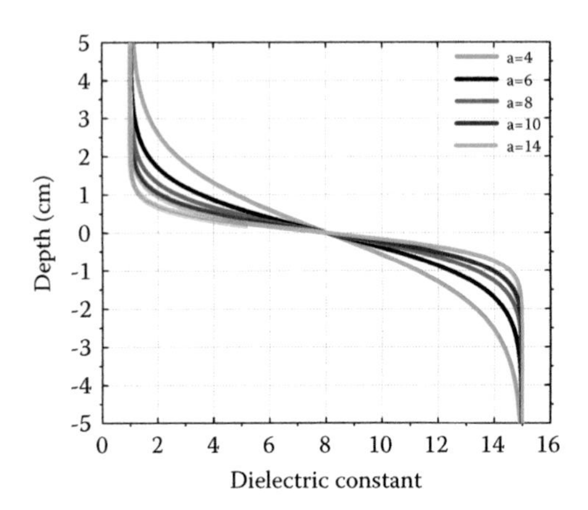

FIGURE 3.6 Dielectric profile of a soil surface as a function of depth z.

to the free space. Statistically, this transition physically makes sense in that the randomly rough surface for $z = 0$ represents the mean surface so that the dielectric permittivity is not discontinuous at $z = 0$. The background dielectric constant is determined by the field capacity, to a certain extent. As a numerical illustration, the dielectric profiles as a function of depth z with the transition rates from 4 to 14 are plotted in Figure 3.6.

The inhomogeneous dielectric profile in Figure 3.6 may be modeled by thin layers of piecewise constant regions, so that the subsurface can be equivalently treated as an N-layer medium. The dielectric profile may recover to a continuous profile when the thickness of each layer is infinitesimal. By applying the boundary condition at each layer, the complex reflection coefficients of rough surface on the top of the medium can be solved recursively by accounting for the subsurface reflection and refraction [7–8]. It turns out that for the dielectric profile given in Equation 3.24, the complex reflection coefficients for horizontal and vertical polarizations are given by:

$$R_h = \frac{\Gamma\left(jS\cos\theta_i\right)\Gamma\left[\left(-\frac{jS}{2}\right)\left(\cos\theta_i + \sqrt{\cos^2\theta_i + (\varepsilon_\infty - 1)}\right)\right]}{\Gamma\left(-jS\cos\theta_i\right)\Gamma\left[\left(\frac{jS}{2}\right)\left(\cos\theta_i - \sqrt{\cos^2\theta_i + (\varepsilon_\infty - 1)}\right)\right]} \times \frac{\Gamma\left[1 - \left(\frac{jS}{2}\right)\left(\cos\theta_i + \sqrt{\cos^2\theta_i + (\varepsilon_\infty - 1)}\right)\right]}{\Gamma\left[1 + \left(\frac{jS}{2}\right)\left(\cos\theta_i - \sqrt{\cos^2\theta_i + (\varepsilon_\infty - 1)}\right)\right]} \tag{3.25a}$$

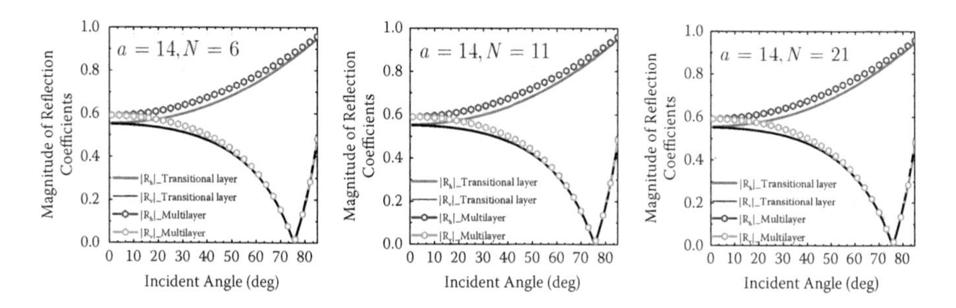

FIGURE 3.7 Reflection coefficients as a function of incident angle computed by recursive formulae (multilayer) and analytic formulae (transitional layer): (a) $a = 14, N = 6$, (b) $a = 14, N = 11$, (c) $a = 14, N = 21$.

$$R_v = \frac{\Gamma\left(jS\cos\theta_i\right)\Gamma\left[\left(-\frac{jS}{2}\right)\left(\cos\theta_i + \frac{1}{\varepsilon_\infty}\sqrt{\cos^2\theta_i + (\varepsilon_\infty - 1)}\right)\right]}{\Gamma\left(-jS\cos\theta_i\right)\Gamma\left[\left(\frac{jS}{2}\right)\left(\cos\theta_i - \frac{1}{\varepsilon_\infty}\sqrt{\cos^2\theta_i + (\varepsilon_\infty - 1)}\right)\right]} \times \frac{\Gamma\left[1 - \left(\frac{jS}{2}\right)\left((\varepsilon_\infty)^{\frac{1}{3}}\cos\theta_i + \sqrt{\cos^2\theta_i + (\varepsilon_\infty - 1)}\right)\right]}{\Gamma\left[1 + \left(\frac{jS}{2}\right)\left(\cos\theta_i - \sqrt{\cos^2\theta_i + (\varepsilon_\infty - 1)}\right)\right]} \quad (3.25b)$$

where $S = 2/a$ is relative thickness of layer, Γ is the Gamma function.

It is instructive to compare the reflection coefficients computed by Equations 3.25a and 3.25b with the recursive formulae for a multilayered medium [3]. We do so by digitizing the transitional layer as shown in Figure 3.6 into N discrete horizontally uniform layers. Figure 3.7 plots the H and V-polarized reflection coefficients of numerically computed and analytic formulae, at a transition rate of 14 for the different number of layers, $N = 6$, 11, and 21. It is found that the two sets of reflection coefficients, modeled by transitional layer and multilayer, are essentially the same, confirming the applicability of Equations 3.25a and 3.25b.

Another kind of useful dielectric profiles with the following exponential functional form, by controlling β value:

$$\begin{aligned} \varepsilon_r(z) &= \varepsilon_{rs} + \varepsilon_{rb}\frac{e^{\beta z} - 1}{e^{-\beta d} - 1}, \quad -d \le z \le 0 \\ \varepsilon_r(z) &= \varepsilon_r(-d), \; z \le -d \end{aligned} \quad (3.26)$$

where z is depth variable, and d is the total layer depth, all in meters; ε_{rs}, ε_{rb} are top and bottom dielectric constants, respectively. Here the dielectric constant may be referring to either the very near surface or subsurface water table, representing the stages of drying out of the surface, or the alternating stages of drying up and wetting down [8].

For numerical illustration, Figure 3.8 represents seven profiles by varying β values from –0.4 to 0.4, with $\varepsilon_{r0} = 2 - j0.2$, $\varepsilon_{rb} = 18 - j2.0$, $d = 20\ cm$. Note that profiles with same β values but opposite signs resemble centro-symmetric. When β approches to zero, the dielectric dependence on depth tends to be linear.

Figure 3.9 displays reflection coefficients for selected profiles in Figure 3.8. Recalled that in determining the reflection coefficients from boundary conditions of field continuation, it is done by matching the field amplitude and phase at the boundary. Figure 3.9 shows interesting results that two similar dielectric profiles can give rise to very distinct reflection coefficients, while different profiles may or may not produce different reflection coefficients having the same background dielectric constants. But if the background dielectric constant are different, similar dielectric profiles can result in two distinct reflection coefficients. For continuously varying dielectric in-depth, a sequence of phase matching is performed through the coherent propagation matrix. Hence, physically, it is possible to come up with very similar reflections coefficients with different dielectric profiles, as we numerically illustrate it in Figure 3.10.

Physically, the dielectric profile described by Equation 3.26 may be simply replaced by the moisture profile. We will look into this subject in later chapters.

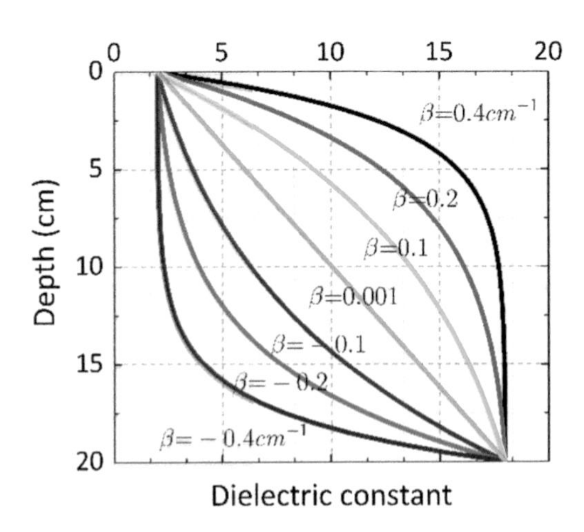

FIGURE 3.8 An exponential form of dielectric constant as a function of depth with different values of $\beta = -0.4, -0.2, -0.1, 0.001, 0.1, 0.2, 0.4$ which control the change rate of the dielectric variations.

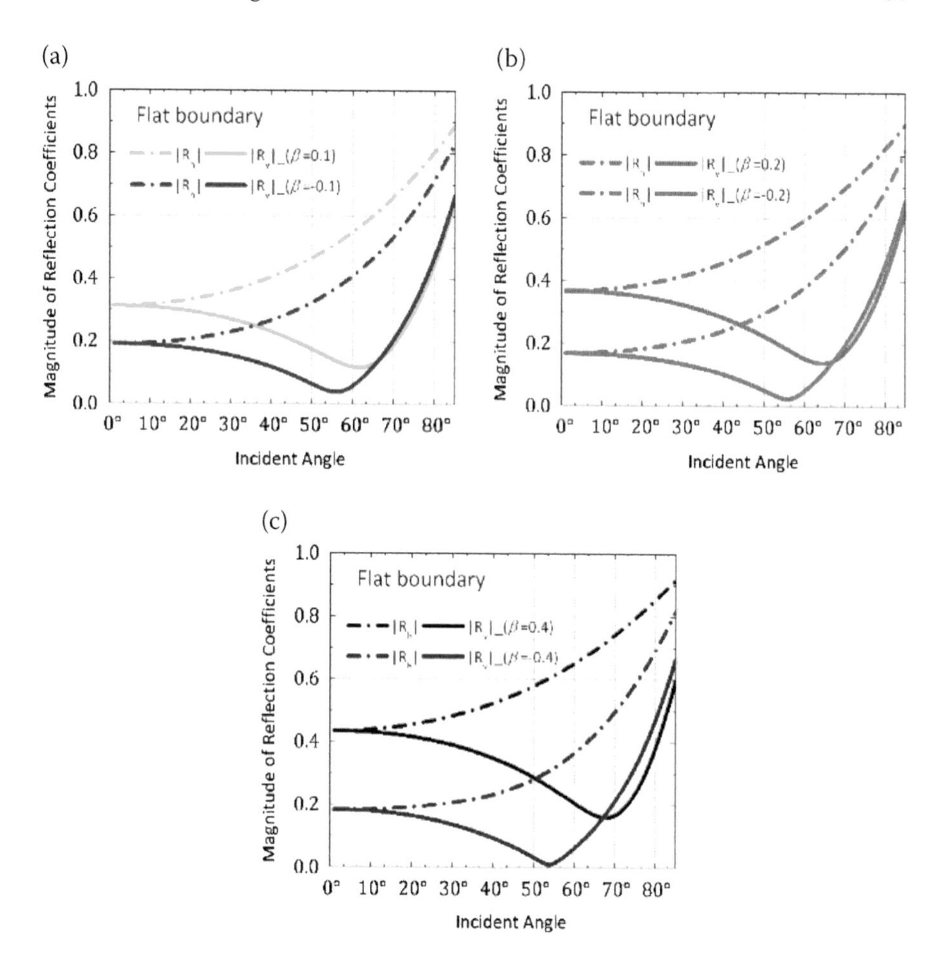

FIGURE 3.9 Magnitude of reflection coefficients at a flat boundary corresponding to the dielectric profiles shown in Figure 3.8: (a) $\beta = +0.1, -0.1$; (b) $\beta = +0.2, -0.2$; (c) $\beta = +0.4, -0.4$.

3.4 RADAR EQUATIONS FOR OBJECTS AND EXTENSIVE TARGETS

3.4.1 Radar Cross-Section

Target's radar cross-section, σ_{qp}, is a measure of target's power reflective capability and is a measurable quantity relating the incident power and reflected power on the target. It is commonly defined in the far-field conditions for a uniform plane wave incident upon the target.

$$\sigma_{qp} = 4\pi \lim_{R \to \infty} R^2 \frac{\Re e\{|\mathbf{E}_q^s|^2/\eta_s^*\}}{\Re e\{|\mathbf{E}_p^i|^2/\eta_i^*\}} \tag{3.27}$$

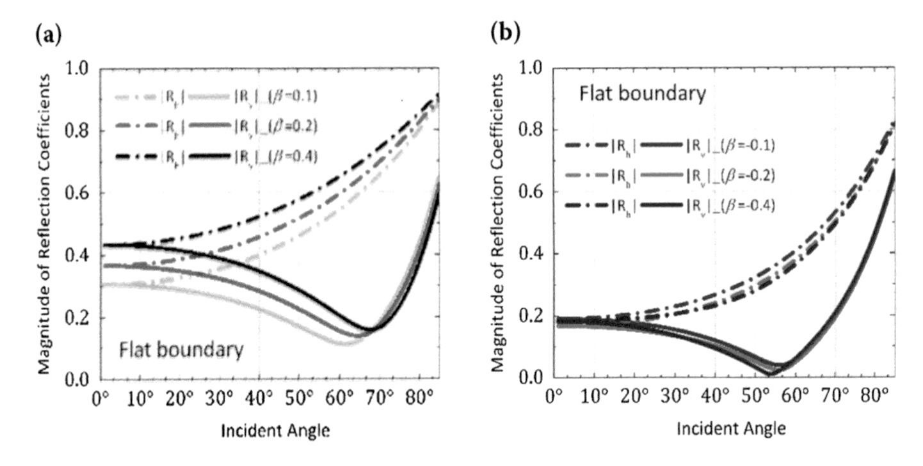

FIGURE 3.10 Magnitude of reflection coefficients at a flat boundary corresponding to the dielectric profiles shown in Figure 3.8: (a) $\beta = +0.1, +0.2, +0.4$; (b) $\beta = -0.1, -0.2, -0.4$.

where $\mathbf{E}_p^i$ is p-polarized incident field and $\mathbf{E}_q^s$ is q-polarized scattered field; η_s, η_i are intrinsic impedances of the media where incidence and scattering occur, respectively; $\Re e$ denotes real part, and $*$ the complex conjugate. For incidence in free space, $\eta_i = \eta_0$.

The bistatic radar equation from a point target takes the following equation:

$$P_r = \frac{P_t G_t(\theta_a, \phi_a) G_r(\theta_a, \phi_a) \lambda^2}{(4\pi)^3 R_t^2 R_r^2} \sigma_{qp} \tag{3.28}$$

where P_t is transmitting power and P_r the receiving power; $G_t(\theta_a, \phi_a)$ and $G_r(\theta_a, \phi_a)$ are transmitting and receiving antenna gains, respectively; λ is the incident wavelength, R_t, R_r are the ranges from the transmitter to the target, and from target to the receivers, respectively. It is understood that to consider the polarized RCS, transmitting and receiving antenna with proper polarizations are required.

To determine the RCS, there is a requirement of minimum range R [9]. Depending on the antenna physical size and wavelength, the range from near to far covers the (1) reactive near-field region which is evanescent modes and includes non-propagating field; (2) radiating near-field region, or the Fresnel zone, includes only propagating field components that are transverse radiating fields; (3) far-field region, or Fraunhofer zone, includes transverse radiating field components where pattern depending only on the angle of observation.

Roughly, the reactive near-field region is within 1 wavelength of range. Referring the Figure 3.11, a commonly accepted criteria of the far-field region are that (1) the phase variation across the target extent $\delta \le \pi/8$; (2) the transverse amplitude variation not to exceed e across the target when the target's maximum dimension is perpendicular to the direction of propagation, that is, $[E(R + L/2) - E(R - L/2)]/E(R) = \varepsilon \le 0.25dB$.

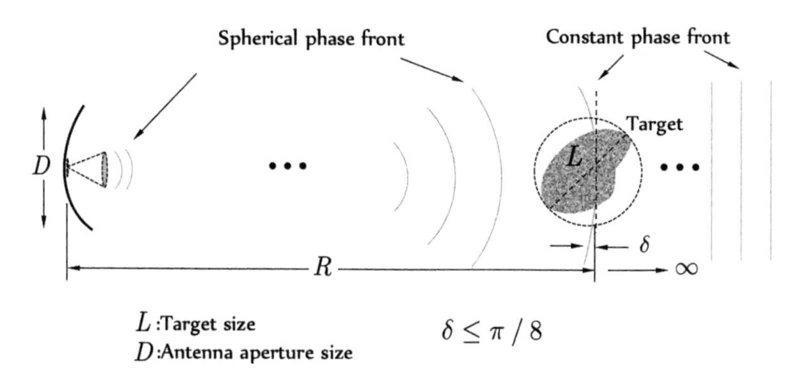

FIGURE 3.11 Far-field criteria for RCS measurement; δ is the phase variation across the target size L.

The RCS of simple and complex targets are obtained by exact solution or numerical solution of Maxwell's equations [10]. However, the RCS of much complex shape targets demands a heavy-loaded computation, or measurements and calibration [11]. We will not seek these topics further for they are beyond the scope of this book. For illustration, Figure 3.12 displays an X-band synthetic aperture radar image with a resolution of 1 meter over a coastal ocean off which there was a fishing boat. The enlarge image inside the rectangular shows a very complex pattern of RCS that came with strong interaction with the sea clutter. The fluctuation of RCS values makes it a random variable. Several common types of RCS and their probability density functions are summarized below for easy reference.

Nonfluctuating, Marcum, Swerling 0, or Swerling 5 Model:

$$p(\sigma) = \delta_D(\sigma - \bar{\sigma}) \tag{3.29}$$

where δ_D is Dirac delta function, and where $\bar{\sigma}$ is the mean RCS. It describes a constant backscattered power. The calibration sphere or corner reflector with no radar or target motion are typical examples.

Rayleigh/Chi-square of Degree 2 Model:

$$p(\sigma) = \frac{1}{\bar{\sigma}}\exp\left[\frac{-\sigma}{\bar{\sigma}}\right] \tag{3.30}$$

It models a collective of many randomly distributed and none dominant targets and is common as Swerling case 1 and 2 models. Note the Rayleigh model has equal mean and standard deviation of RCS, i.e., $\mathrm{var}(\sigma) = \bar{\sigma}^2$.

Rician Model:

$$p(\sigma) = \frac{1}{\bar{\sigma}}(1 + a^2)\exp\left[-a^2 - \frac{\sigma}{\bar{\sigma}}(1 + a^2)\right]I_0\left[2a\sqrt{(1 + a^2)(\sigma/\bar{\sigma})}\right] \tag{3.31}$$

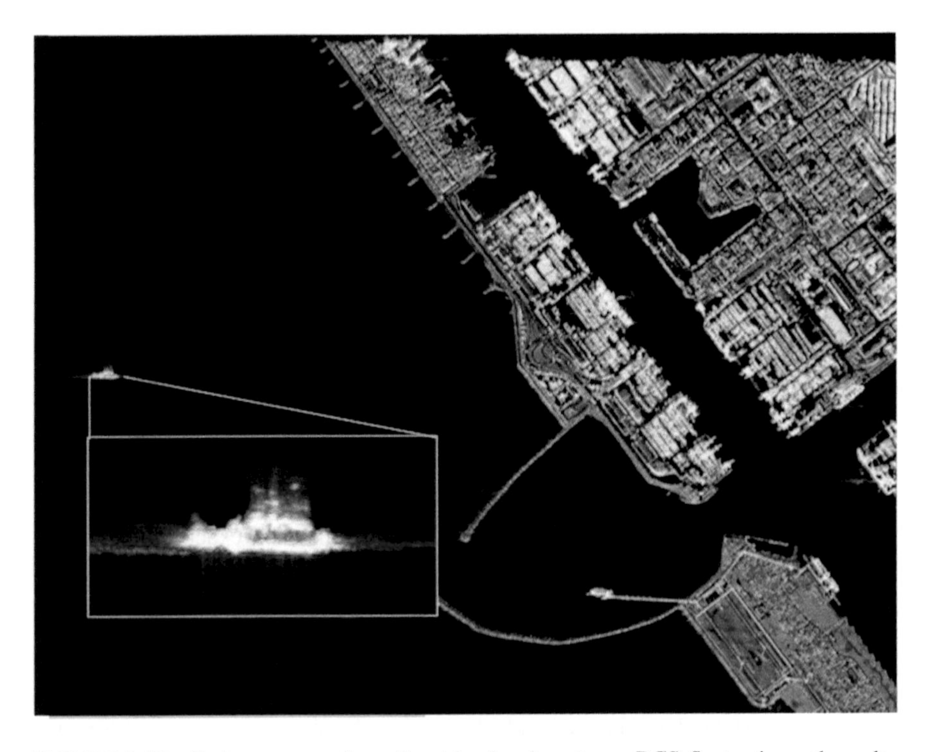

FIGURE 3.12 Radar cross-section of a ship showing strong RCS fluctuations; the radar image with resolution of 1 meter was acquired over costal ocean in 2003.

The Rician model describes a backscattered power from a collective of many random scatterers plus a dominant scatter; the ratio of dominant RCS to sum of small RCS is a^2. It can be shown that the variance of Rician RCS is

$$\mathrm{var}(\sigma) = \frac{(1 + 2a^2)}{(1 + a^2)^2}\bar{\sigma}^2 \tag{3.32}$$

Weibull model:

$$p(\sigma) = \kappa\beta\sigma^{\kappa - 1} \exp\left[-\beta\sigma^{\kappa}\right], \tag{3.33}$$

where the mean and variance of σ are:

$$\bar{\sigma} = \Gamma(1 + 1/\kappa)\beta^{-1/\kappa}, \tag{3.34a}$$

$$\mathrm{var}(\sigma) = \beta^{-2/\kappa}[\Gamma(1 + 2/\kappa) - \Gamma^2(1 + 1/\kappa)], \tag{3.34b}$$

and $\Gamma(\cdot)$ is Gamma function.

The Weibull model is used in an empirical fit to many measured target and clutter distributions.

Log-normal Model:

$$p(\sigma) = \frac{1}{\sqrt{2\pi}\, s\sigma}\exp\left[-\ln^2(\sigma/\sigma_m)/2s^2\right] \quad (3.35)$$

where the mean and variance of σ are given by:

$$\bar{\sigma} = \sigma_m \exp(s^2/2) \quad (3.36a)$$

$$\text{var}(\sigma) = \sigma_m^2 \exp(s^2)[\exp(s^2) - 1] \quad (3.36b)$$

where σ_m is median value of σ, and s is shape parameter.

Like the Weibull model, the log-normal model is in empirical fit to many measured target and clutter distributions and comes as the longest "tail" of all models just described. More about the RCS statistics models are referred to [12].

3.4.2 Scattering Coefficient

By definition, the polarized bistatic scattering coefficient for an extended or distributed target is the bistatic scattering cross-section per unit area, written as

$$\sigma_{qp}^0 = \frac{\left\langle \sigma_{qp} \right\rangle}{A} \quad (3.37)$$

where A the antenna illuminated area project onto the ground, and $\left\langle \sigma_{qp} \right\rangle$ is the ensemble average of collective RCSs within A.

When the target being incident upon is a distributed target, the radar equation becomes (see Figure 3.13)

$$\bar{P}_{r(p)} = P_{t(q)}\frac{\lambda^2}{(4\pi)^3}\int_A \frac{G_t(\theta_s, \phi_s)G_r(\theta_s, \phi_s)}{R_t^2 R_r^2}\sigma_{pq}^o dA \quad (3.38)$$

where q, p denote the transmitting and receiving polarizations; λ is the wavelength; $G_t(\theta_s, \phi_s)$ and $G_r(\theta_s, \phi_s)$ are antenna patterns of the transmitter and receiver in the direction (θ_s, ϕ_s); R_t and R_r are the ranges from the differential area dA to the transmitter and receiver, respectively. Because the scattering coefficient σ_{pq}^o is an ensemble average, the received power $\bar{P}_{r(p)}$ is the mean power. Unlike in monostatic, the illuminating area A is the projected overlay of the antenna patterns of transmitter and receiver, the shaded region shown in Figure 3.13.

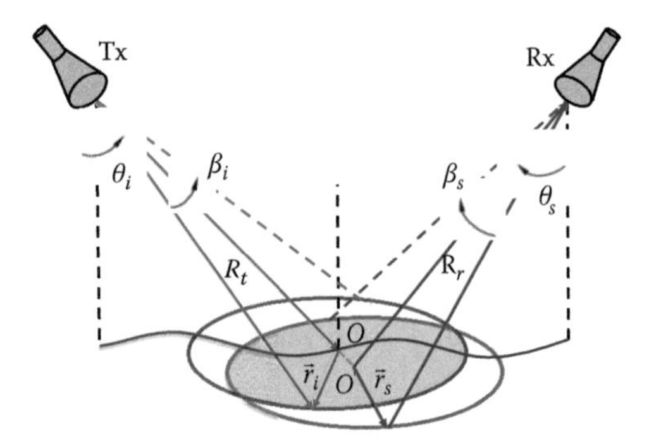

FIGURE 3.13 Geometry of bistatic scattering coefficient from a distributed target covered by the overlay of antenna patterns of transmitter and receiver, as shown in shaded area.

For beam-limited system, the transmitted power P_q^t within the beamwidth is a constant. Assuming the change in the $\sigma_{pq}^0(\theta_a, \phi_a)$ is negligible within A, the scattering coefficient can be evaluated by:

$$\sigma_{pq}^0(\theta_a, \phi_a) = \frac{\bar{P}_{r,p}}{P_{t,q}} \frac{(4\pi)^3 R_t^2 R_r^2}{\lambda^2} \iint_A \frac{1}{G_t(\theta_a, \phi_a) G_r(\theta_a, \phi_a)} dA \tag{3.39}$$

If we set the effective illumination area:

$$A_e = \iint_A G_t(\theta_s, \phi_s) G_r(\theta_s, \phi_s) dA \tag{3.40}$$

Then the bistatic scattering coefficient can be expressed as:

$$\sigma_{pq}^0(\theta_s, \phi_s) = \frac{\bar{P}_p^r}{P_q^t} \frac{(4\pi)^3 R_t^2 R_r^2}{\lambda^2 A_e} \tag{3.41}$$

In the external calibration, the calibration point target of known radar cross-section can be expressed as follow:

$$\sigma_{pq}^c(\theta_s, \phi_s) \frac{P_p^{rc}}{P_q^t} \frac{(4\pi)^3 R_{tc}^2 R_{rc}^2}{\lambda^2} \tag{3.42}$$

From Equations 3.41 and 3.42, the bistatic scattering coefficient can be calculated:

$$\sigma_{pq}^0(\theta_s, \phi_s) = \frac{P_p^r}{P_p^{rc}} \frac{R_t^2 R_r^2}{R_{tc}^2 R_{rc}^2} \frac{\sigma_c(\theta_s, \phi_s)}{A_e} \tag{3.43}$$

In the preceding equation, all the terms, except for A_e, on the right-hand side can be obtained by measurement. Based on the fact that the highest sidelobe level of the antenna pattern is much lower than the peak value of the main lobe measured by –3 dB beamwidth, in experimental measurements of the scattering coefficient, only the contributions from the bistatic footprint by the projected –3 dB beamwidth patterns A_e are considered as shown in the shaded region of Figure 3.13. This leads to an estimate of the scattering coefficient

$$\hat{\sigma}^o(\theta_s, \phi_s) = \frac{P_p^r \sigma_{pq}^c(\theta_s, \phi_s)}{P_p^{rc} A_e} \tag{3.44}$$

To fully account for the polarization dependence, we can express the scattering coefficient by noting that the scattered field and incident field are related by a scattering matrix **S** of the target in both amplitude and phase, as given by [13]

$$\mathbf{E}^r = \frac{e^{jkR}}{R}\mathbf{S}\mathbf{E}^i = \frac{e^{jkR}}{R}\begin{bmatrix} S_{qq} & S_{qp} \\ S_{pq} & S_{pp} \end{bmatrix}\mathbf{E}^i \tag{3.45}$$

where R is the range from the source to the target; S_{qp} the matrix elements are the complex scattering amplitude with p-polarized incidence and q-polarized scattering. In theory, the target is characterized fully in **S**. Once the fully polarized scattered field is measured, the measured scattering matrix **S** is obtained. Then, the radar scattering coefficient can be estimated according to the radar equation for extended target:

$$\sigma_{qp}^0 = \frac{4\pi}{A}|S_{qp}|^2 \tag{3.46}$$

Using polarization synthesis, scattering coefficient for arbitrary combination of incident and scattering polarizations can be obtained if a complete scattering matrix **S** is made available. Two special cases are often considered: co-polarization in which the polarizations of the incidence and scattering are the same, and the cross-polarization in which the polarizations of the incidence and scattering are orthogonal. For more details, please refer to [13].

REFERENCES

1. Ulaby, F. T., Moore, R. K., and Fung, A. K., *Microwave Remote Sensing*, Addison-Wesley, Reading, MA, 1982.
2. Tsang, L., Kong, J. A., and Shin, R. T., *Theory of Microwave Remote Sensing*, John Wiley, New York, 1985.
3. Kong, J. A., *Electromagnetic Wave Theory*, Wiley, New York, 1986.
4. Stratton, J. A., *Electromagnetic Theory*, McGraw-Hill, New York, 1941, reprinted by IEEE Press, New Jersey, 2007.
5. Ishimaru, A., *Electromagnetic Wave Propagation, Radiation, and Scattering: From Fundamentals to Applications*, Englewood Cliffs, NJ: Prentice-Hall, 1991, 2nd Edition reprinted by Wiley-IEEE Press, New York, 2017.

6. Ulaby, F. T., and Long, D. G., *Microwave Radar and Radiometric Remote Sensing*, University of Michigan Press, Ann Arbor, MI, 2014.
7. Brekhovskikh, L. M., *Waves in Layered Media*, 2nd Edition, Translated by R. T. Beyer, Academic Press, New York, 1980.
8. Fung, A. K., Dawson, M. S., Chen, K. S., Hsu, A. Y., Engman, E. T., O'Neill, P. O., and Wang, J., A modified IEM model for scattering from soil surface with application to soil moisture sensing, *IEEE IGARSS*, 2(2), 1297–1299, 1996.
9. Blacksmith, P., Hiatt, R. E., and Mack, R. B., Introduction to radar cross-section measurements, *Proceedings of the IEEE*, 53(8), 901–920, August 1965.
10. Ruck, G. T., ed., *Radar Cross Section Handbook*, Volumes 1 & 2, Kluwer Academic/Plenum Publishers, New York, 1970.
11. Knott, E. F., *Radar Cross Section Measurement*, Van Nostrand Reinhold, New York, 1993.
12. Richards, M. A., *Fundamentals of Radar Signal Processing*, McGraw-Hill, New York, 2005.
13. Lee, J. S., and Pottier, E., *Polarimetric Radar Imaging – From Basics to Applications*, CRC Press, Boca Raton, FL, 2009.

4 Analytical Modeling of Rough Surface Scattering

4.1 HUYGENS–FRESNEL PRINCIPLE

To facilitate our discussion on numerical or analytic modeling of scattered fields from a rough surface by solving a pair of integral-differentiate equations that governing the surface fields which result in radiated fields, we shall briefly introduce the Huygens–Fresnel principle, which states that every point on a wavefront is itself the source of spherical wavelets, and the secondary wavelets emanating from different points mutually interfere [1–5]. The sum of these spherical wavelets, taking amplitudes and phases into consideration, forms the wavefront. This principle is widely applied to analyze the problems of wave propagation both in the far-field limit and in near-field diffraction and also reflection. Recall that a Green's function is the impulse response of a point source. From linear system theory, once the impulse response function is known, the system response (the field) due to arbitrary sources is obtained by a convolution integral involving the source distribution and the impulse response function (the Green's function).

Suppose there is a source of current density $\mathbf{J}$ supported by the volume V, as depicted in Figure 4.1. The volume V is bounded by surface S. Then, the electric field in an unbounded, isotropic, homogeneous medium obeys the following vector wave equation:

$$(\nabla^2 + k^2)\left(\mathbf{E} + \frac{1}{j\omega\varepsilon}\mathbf{J}\right) + \frac{1}{j\omega\varepsilon}\nabla \times \nabla \times \mathbf{J} = 0 \tag{4.1}$$

For a point source, $\mathbf{J} = \bar{\bar{\mathbf{I}}}\delta(\vec{r} - \vec{r}')$, the solution to Equation 4.1 is the dyadic Green's function:

$$\mathbf{E} = \bar{\bar{\mathbf{G}}} = \left[\bar{\bar{\mathbf{I}}} + \frac{1}{k^2}\nabla\nabla\right]g(\vec{r} - \vec{r}') \tag{4.2}$$

with the Green's function $g(\vec{r} - \vec{r}') = \frac{e^{jk|\vec{r} - \vec{r}'|}}{4\pi|\vec{r} - \vec{r}'|}$.

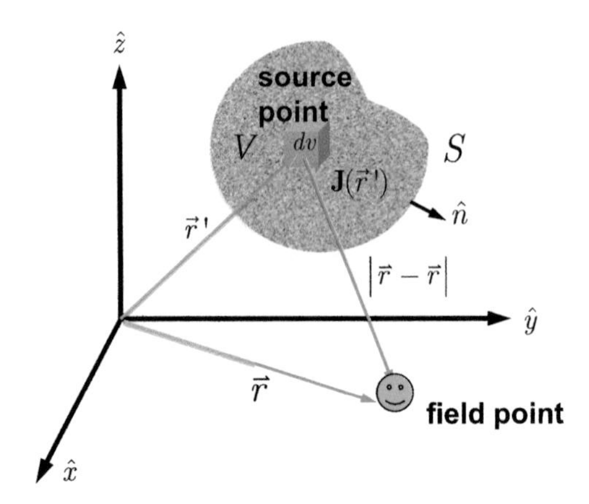

FIGURE 4.1 Radiated field from a source **J** inside the volume V.

In general, an arbitrary current density **J** can be treated as a collective sum of point sources. It follows that the solution of the electric field in the vector Equation 4.1 can be obtained by making use of Green's function and is given by [1–5]

$$\mathbf{E}(\vec{r}) = \frac{1}{j\omega\varepsilon}\iiint_V (\nabla \times \nabla \times \mathbf{J}(\vec{r}\,')g(\vec{r}, \vec{r}\,')dv' - \frac{1}{j\omega\varepsilon}\mathbf{J}(\vec{r}) \tag{4.3}$$

Using dyadic Green's function, Equation 4.3 can be equivalently written as

$$\mathbf{E}(\vec{r}) = -j\omega\varepsilon \iiint_V \bar{\bar{\mathbf{G}}}(\vec{r}, \vec{r}\,')\mathbf{J}(\vec{r}\,')dv' \tag{4.4}$$

For a source-free region V, namely, $\mathbf{J} = 0$, under the Huygens–Fresnel principle, the electric field **E** is expressed as:

$$\mathbf{E}(\vec{r}) = \oiint_S dS'\{j\omega\mu\bar{\bar{\mathbf{G}}}(\vec{r}, \vec{r}\,')\cdot[\hat{n} \times \mathbf{H}(\vec{r}\,')] + \nabla \times \bar{\bar{\mathbf{G}}}(\vec{r}, \vec{r}\,')\cdot[\hat{n} \times \mathbf{E}(\vec{r}\,')]\} \tag{4.5}$$

By duality, the magnetic field **H** is

$$\mathbf{H}(\vec{r}) = \oiint_S dS'\{-j\omega\varepsilon\bar{\bar{\mathbf{G}}}(\vec{r}, \vec{r}\,')\cdot[\hat{n} \times \mathbf{E}(\vec{r}\,')] + \nabla \times \bar{\bar{\mathbf{G}}}(\vec{r}, \vec{r}\,')\cdot[\hat{n} \times \mathbf{H}(\vec{r}\,')]\} \tag{4.6}$$

where the $\hat{n}$ is a unit vector normal to the surface. The electric and magnetic fields are only determined by the tangential fields, $\hat{n} \times \mathbf{E}$, $\hat{n} \times \mathbf{H}$, over the surface S enclosing V.

Equations 4.5 and 4.6 can be written more explicitly in vectoral formulation:

$$\mathbf{E}(\vec{r}) = \oiint_S ds' \{j\omega\mu g[\hat{n} \times \mathbf{H}(\vec{r}')] + \frac{j\omega\mu}{k^2}\nabla\nabla g\cdot[\hat{n} \times \mathbf{H}(\vec{r}')] + \nabla g \times [\hat{n} \times \mathbf{E}(\vec{r}')]\} \tag{4.7}$$

$$\mathbf{H}(\vec{r}) = \oiint_S ds'\{-j\omega\varepsilon g[\hat{n} \times \mathbf{E}(\vec{r}')] - \frac{j\omega\varepsilon}{k^2}\nabla\nabla g\cdot[\hat{n} \times \mathbf{E}(\vec{r}')] + \nabla g \times [\hat{n} \times \mathbf{H}(\vec{r}')]\} \tag{4.8}$$

The electric and magnetic fields, for closed surface S, according to the Stratton–Chu formula may be expressed as [1–5]:

$$\mathbf{E}(\vec{r}) = \oiint_S ds'\{j\omega\mu[\hat{n} \times \mathbf{H}(\vec{r}')]g(\vec{r}, \vec{r}') + [\hat{n}\cdot\mathbf{E}(r')]\nabla' g(\vec{r}, \vec{r}') + [\hat{n} \times \mathbf{E}(r')] \times \nabla' g(\vec{r}, \vec{r}')\} \tag{4.9}$$

$$\mathbf{H}(\vec{r}) = \oiint_S ds'\{-j\omega\varepsilon[\hat{n} \times \mathbf{E}(\vec{r}')]g(\vec{r}, \vec{r}') + [\hat{n}\cdot\mathbf{H}(r')]\nabla' g(\vec{r}, \vec{r}') + [\hat{n} \times \mathbf{H}(r')] \times \nabla' g(\vec{r}, \vec{r}')\} \tag{4.10}$$

Note that the normal fields $\varepsilon\hat{n}\cdot\mathbf{E}$, $\mu\hat{n}\cdot\mathbf{H}$ represent the surface charge densities. The normal fields in above equations can be identified by noting that:

$$\begin{aligned} \hat{n}'\cdot\mathbf{E} &= \frac{1}{j\omega\varepsilon}\nabla'_s\cdot(\hat{n}' \times \mathbf{H}) \\ \hat{n}'\cdot\mathbf{H} &= \frac{1}{-j\omega\mu}\nabla'_s\cdot(\hat{n}' \times \mathbf{E}) \end{aligned} \tag{4.11}$$

where ∇'_s is the surface divergence operator acting in source point coordinates with $\nabla_s = \hat{x}\frac{\partial}{\partial x} + \hat{y}\frac{\partial}{\partial y}$.

For the source-free region, compared to the three-dimensional volume integral equations of Equation 4.2 or Equation 4.3, Equations 4.9 and 4.10, the Fredholm integral equations of the first kind [2] are a two-dimensional problem with less complicated boundary conditions, which are to be solved for the scattering problem in the following section.

4.2 ELECTRIC FIELD INTEGRAL EQUATIONS (EFIE) AND MAGNETIC FIELD INTEGRAL EQUATIONS (MFIE)

When there is an incident field impinging upon the region V, we have a scattering problem. The total field is the sum of the incident field, which is known, and the scattered field, which is unknown and is to be determined. For the case of time-harmonic field, it is convenient to the integral equations governing the electric and magnetic fields in space-frequency form with $e^{-j\omega t}$ understood. For a source-free region as in our case, mathematically, we have [6]

$$\mathbf{E}(\vec{r}) = \Upsilon\mathbf{E}^i - \frac{\Upsilon}{4\pi}\oiint_S ds' \{j\omega\mu[\hat{n} \times \mathbf{H}(\vec{r}')]g(\vec{r}, \vec{r}') + [\hat{n}\cdot\mathbf{E}(r')]\nabla' g(\vec{r}, \vec{r}')$$
$$+ [\hat{n} \times \mathbf{E}(r')] \times \nabla' g(\vec{r}, \vec{r}')\} \quad (4.12)$$

$$\mathbf{H}(\vec{r}) = \Upsilon\mathbf{H}^i + \frac{\Upsilon}{4\pi}\oiint_S ds' \{-j\omega\varepsilon[\hat{n} \times \mathbf{E}(\vec{r}')]g(\vec{r}, \vec{r}') + [\hat{n}\cdot\mathbf{H}(r')]\nabla' g(\vec{r}, \vec{r}')$$
$$+ [\hat{n} \times \mathbf{H}(r')] \times \nabla' g(\vec{r}, \vec{r}')\} \quad (4.13)$$

where $\Upsilon = (1 - \Omega/4\pi)^{-1}$, $\Omega = \begin{cases} 0, & \vec{r} \notin S' \\ 2\pi, & \vec{r} \in S' \end{cases}$

It is of particular interest to find the fields outside the source region, that is, $\Upsilon = 1$. Then, from Equations 4.12 and 4.13, we have

$$\mathbf{E}(\vec{r}) = \mathbf{E}^i - \frac{1}{4\pi}\oiint_S ds' \{j\omega\mu[\hat{n}' \times \mathbf{H}(\vec{r}')]g(\vec{r}, \vec{r}') + [\hat{n}'\cdot\mathbf{E}(r')]\nabla' g(\vec{r}, \vec{r}')$$
$$+ [\hat{n}' \times \mathbf{E}(r')] \times \nabla' g(\vec{r}, \vec{r}')\} \quad (4.14)$$

$$\mathbf{H}(\vec{r}) = \mathbf{H}^i + \frac{1}{4\pi}\oiint_S ds' \{-j\omega\varepsilon[\hat{n}' \times \mathbf{E}(\vec{r}')]g(\vec{r}, \vec{r}') + [\hat{n}'\cdot\mathbf{H}(r')]\nabla' g(\vec{r}, \vec{r}')$$
$$+ [\hat{n}' \times \mathbf{H}(r')] \times \nabla' g(\vec{r}, \vec{r}')\} \quad (4.15)$$

It should be recognized that the total field is a sum of the incident field and the scattered field due to the presence of the surface S, which is treated as randomly rough in this book. The second term of the right-hand side of Equations 4.14 and 4.15 are the scattered electric field and scattered magnetic field, respectively.

We can take the operator on both sides of Equations 4.14 and 4.15 without affecting the field determination, to reach the following expressions:

$$\hat{n} \times \mathbf{E}(\vec{r}) = \hat{n} \times \mathbf{E}^i - \frac{1}{4\pi}\hat{n} \times \oiint_S ds' \{j\omega\mu[\hat{n}' \times \mathbf{H}(\vec{r}')]g(\vec{r}, \vec{r}')$$
$$+ [\hat{n}'\cdot\mathbf{E}(r')]\nabla' g(\vec{r}, \vec{r}') + [\hat{n}' \times \mathbf{E}(r')] \times \nabla' g(\vec{r}, \vec{r}')\} \quad (4.16)$$

$$\hat{n} \times \mathbf{H}(\vec{r}) = \hat{n} \times \mathbf{H}^i + \frac{1}{4\pi}\hat{n} \times \oiint_S ds' \{-j\omega\varepsilon[\hat{n}' \times \mathbf{E}(\vec{r}')]g(\vec{r}, \vec{r}')$$
$$+ [\hat{n}'\cdot\mathbf{H}(r')]\nabla' g(\vec{r}, \vec{r}') + [\hat{n}' \times \mathbf{H}(r')] \times \nabla' g(\vec{r}, \vec{r}')\} \quad (4.17)$$

Now, it is clearer to identify the above pair of integral equations as the Fredholm integral equations of the second kind.

Figure 4.2 illustrates the fields in two media separated by a closed surface S, where the unit normal vector $\hat{n}_2$ is pointing to medium 2, $\hat{n}_2 = -\hat{n}$. The medium is described by the permittivity (ε_1, ε_2) and permeability (μ_1, μ_2). If the surface is finite conducting, the incident wave in the upper medium (medium 1) is penetrable through the surface into the lower medium (medium 2), we have both scattered fields in the upper medium (medium 1) and transmitted field in the lower medium.

Following Equations 4.16 and 4.17, the total tangential fields in medium 1 and medium 2 can be given as, assuming that $\mu_1 = \mu_2 = \mu$:

$$\begin{aligned}\hat{n} \times \mathbf{E}_1(\vec{r}) = \hat{n} \times \mathbf{E}^i \ - \frac{1}{4\pi}\hat{n} \times \oiint_S ds' \{j\omega\mu[\hat{n}' \times \mathbf{H}_1(\vec{r}')]g(\vec{r}, \vec{r}') \\ + [\hat{n}' \cdot \mathbf{E}_1(r')]\nabla' g(\vec{r}, \vec{r}') + [\hat{n}' \times \mathbf{E}_1(r')] \times \nabla' g(\vec{r}, \vec{r}')\}\end{aligned} \tag{4.18}$$

$$\begin{aligned}\hat{n} \times \mathbf{H}_1(\vec{r}) = \hat{n} \times \mathbf{H}^i \ + \frac{1}{4\pi}\hat{n} \times \oiint_S ds' \{ -j\omega\varepsilon[\hat{n}' \times \mathbf{E}_1(\vec{r}')]g(\vec{r}, \vec{r}') \\ + [\hat{n}' \cdot \mathbf{H}_1(r')]\nabla' g(\vec{r}, \vec{r}') + [\hat{n}' \times \mathbf{H}_1(r')] \times \nabla' g(\vec{r}, \vec{r}')\}\end{aligned} \tag{4.19}$$

$$\begin{aligned}\hat{n}_2 \times \mathbf{E}_2(\vec{r}) = \frac{1}{4\pi}\hat{n}_2 \times \oiint_S ds' \{j\omega\mu[\hat{n}_2' \times \mathbf{H}_2(\vec{r}')]g_2(\vec{r}, \vec{r}') \\ + [\hat{n}_2' \cdot \mathbf{E}_2(r')]\nabla' g_2(\vec{r}, \vec{r}') + [\hat{n}_2' \times \mathbf{E}_2(r')] \times \nabla' g(\vec{r}, \vec{r}')\}\end{aligned} \tag{4.20}$$

$$\begin{aligned}\hat{n}_2 \times \mathbf{H}_2(\vec{r}) = \frac{1}{4\pi}\hat{n}_2 \times \oiint_S ds' \{-j\omega\varepsilon[\hat{n}_2' \times \mathbf{E}_1(\vec{r}')]g_2(\vec{r}, \vec{r}') \\ + [\hat{n}_2' \cdot \mathbf{H}_2(r')]\nabla' g_2(\vec{r}, \vec{r}') + [\hat{n}_2' \times \mathbf{H}_2(r')] \times \nabla' g(\vec{r}, \vec{r}')\}\end{aligned} \tag{4.21}$$

where g_1, g_2 are Green's functions for medium 1 and medium 2, respectively.

Equations 4.18–4.21 are of profound use in solving the two-dimensional scattering problems, together with imposing the boundary conditions that the tangential fields and normal fields be continuous across the boundary, that is,

$$\begin{aligned}\hat{n} \times (\mathbf{E}_1 - \mathbf{E}_2) &= 0, \\ \hat{n} \times (\mathbf{H}_1 - \mathbf{H}_2) &= 0.\end{aligned} \tag{4.22}$$

$$\hat{n} \cdot (\varepsilon_1 \mathbf{E}_1 - \varepsilon_2 \mathbf{E}_2) = 0 \tag{4.23}$$

It follows that by subtracting Equation 4.18 from Equation 4.20, and Equation 4.19 from Equation 4.21 yields

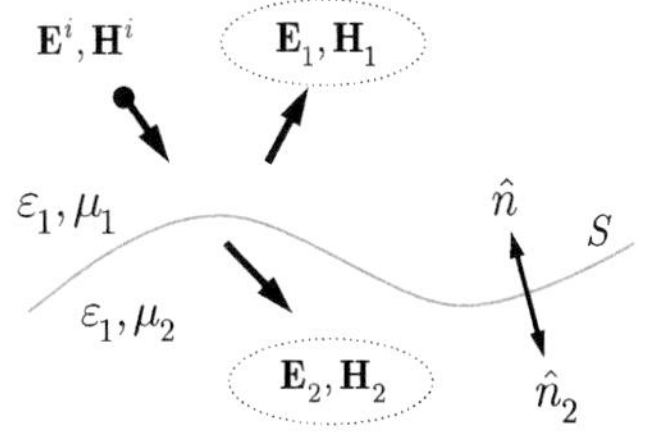

FIGURE 4.2 Fields in medium 1 and medium 2, separated by surface *S*.

$$\hat{n} \times \mathbf{E}^i = \frac{1}{4\pi}\hat{n} \times \oiint_S ds' \left\{ j\omega\mu[\hat{n}' \times \mathbf{H}][g_1 + g_2] - [\hat{n}' \times \mathbf{E}] \times \nabla'[g_1 + g_2] \right.$$
$$\left. + [\hat{n}' \cdot \mathbf{E}]\nabla'\left[g_1 + \frac{\varepsilon_1}{\varepsilon_2}g_2\right]\right\} \tag{4.24}$$

$$\hat{n} \times \mathbf{H}^i = -\frac{1}{4\pi}\hat{n} \times \oiint_S ds' \left\{ j\omega\varepsilon_1[\hat{n}' \times \mathbf{E}]\left[g_1 + \frac{\varepsilon_2}{\varepsilon_1}g_2\right] - [\hat{n}' \times \mathbf{H}] \times \nabla'[g_1 + g_2] \right.$$
$$\left. + [\hat{n}' \cdot \mathbf{H}(r')]\nabla'[g_1 + g_2]\right\} \tag{4.25}$$

Note that the above equations are valid for $\vec{r} \in S$.

Before proceeding, it is interesting to see if S is a perfectly conducting surface, on which the following boundary conditions are enforced:

$$\hat{n} \times \mathbf{E} = 0; \quad \hat{n} \cdot \mathbf{H} = 0 \tag{4.26}$$

We may work on the tangential fields with $\vec{r} \in S$, $\Omega = 2\pi$, $\Upsilon = 2$, to simplify Equations 4.24 and 4.25 to

$$0 = \hat{n} \times \mathbf{E}^i - \frac{1}{4\pi}\hat{n} \times \oiint_S ds' \{j\omega\mu[\hat{n}' \times \mathbf{H}(\vec{r}')]g(\vec{r}, \vec{r}') + [\hat{n}' \cdot \mathbf{E}(r')]\nabla' g(\vec{r}, \vec{r}')\} \tag{4.27}$$

$$\hat{n} \times \mathbf{H}(\vec{r}) = 2\hat{n} \times \mathbf{H}^i - \frac{1}{2\pi}\hat{n} \times \oiint_S ds'[\hat{n}' \times \mathbf{H}(r')] \times \nabla' g(\vec{r}, \vec{r}') \tag{4.28}$$

Solving one of Equations 4.27 and 4.28 would allow us to obtain the unknown tangential fields. Practically, for geometry of randomly rough surface, we prefer to solve the magnetic field from Equation 4.28 and obtain the electric field from the Maxwell's equations, without even involving Equation 4.27, which contains two terms inside the integral: $\hat{n}' \times \mathbf{H}$ and $\hat{n}' \cdot \mathbf{E} = (1/j\omega\varepsilon)\nabla'_s \cdot \hat{n}' \times \mathbf{H}$.

The importance of Equations 4.24, 4.25 or Equations 4.27, 4.28 is that once the surface tangential fields, the secondary sources, are solved, the scattered field in medium 1 or transmitted field in medium 2 can be readily obtained by the Huygens–Fresnel principle, as we already outlined in the previous section.

4.3 SOLUTIONS OF EFIE AND MFIE—NUMERICAL AND ANALYTICAL APPROACHES

So far, we have reviewed the integral equations governing the surface tangential fields. Numerical and analytical approaches are two typical solutions of the integral equations (EFIE & MFIE). The numerical evaluation of the forward-backward (FB) method is applied to scattering from a 1-D dielectric rough surface [7]. The Monte Carlo simulation based on the steepest descent fast multipole method (SDFMM) is used to compute the polarimetric scattering matrix for two-layered rough ground [8]. The numerical Maxwell model of three-dimensional (NMM3D) simulations was developed by Tsang et al. [9] to accelerate the Method of Moment (Mom) solutions. Though numerical methods, in general, may have the flexibility to deal with various types of rough surfaces, bulky computational resources are highly demanded. Here, we will seek the solutions of the integral equations with much emphasis on the analytical approach, as there are abundant publications in journal articles or books have devoted to numerically solving the integral equations.

We now express Equations 4.29 and 4.30 in more explicit forms by considering the polarization states of the fields. Referring to Figure 4.3, the plane wave incident electric and magnetic fields are written as

$$\mathbf{E}^i = \hat{p}E_0 \exp[-j(\vec{k_i}\cdot\vec{r})], \tag{4.29}$$

$$\mathbf{H}^i = \frac{1}{\eta}\hat{k}_i \times \mathbf{E}^i \tag{4.30}$$

where $j = \sqrt{-1}$; i denotes incident wave; $\hat{p}$ is the unit polarization vector; E_0 is the amplitude of the incident electric field, and η the intrinsic impedance of the upper medium, respectively. The position vector is $\vec{r} = x\hat{x} + y\hat{y} + z\hat{z}$ and wave vectors in incident and scattering directions are defined as follows, respectively

$$\vec{k_i} = k\hat{k}_i = \hat{x}k_{ix} + \hat{y}k_{iy} + \hat{z}k_{iz};\ k_{ix} = k\sin\theta_i\cos\phi_i,\ k_{iy} = k\sin\theta_i\sin\phi_i,\ k_{iz} = k\cos\theta_i \tag{4.31}$$

$$\vec{k_s} = k\hat{k}_s = \hat{x}k_{sx} + \hat{y}k_{sy} + \hat{z}k_{sz};\ k_{sx} = k\sin\theta_s\cos\phi_s,\ k_{sy} = k\sin\theta_s\sin\phi_s,\ k_{sz} = k\cos\theta_s \tag{4.32}$$

For linearly horizontal-polarized and vertical-polarized waves, the polarization vector $\hat{p}$, for incident and scattering waves, is defined as

$$\begin{aligned}
\hat{h}_i &= -\hat{x}\sin\phi_i + \hat{y}\cos\phi_i \\
\hat{v}_i &= \hat{h}_i \times \hat{k}_i = -(\hat{x}\cos\theta_i\cos\phi_i + \hat{y}\cos\theta_i\sin\phi_i + \hat{z}\sin\theta_i) \\
\hat{h}_s &= \hat{\phi} = -\hat{x}\sin\phi_s + \hat{y}\cos\phi_s \\
\hat{v}_s &= \hat{\theta} = \hat{h}_s \times \hat{k}_s = \hat{x}\cos\theta_s\cos\phi_s + \hat{y}\cos\theta_s\sin\phi_s - \hat{z}\sin\theta_s
\end{aligned} \tag{4.33}$$

Next, we give three typical analytical approach: Kirchhoff Approximation (KA), the Small Perturbation Method (SPM), and the Integral Equation Method (IEM).

4.3.1 Kirchhoff Approximation (KA)

The Kirchhoff method is under the assumption that reflection of every point on the rough surface is the tangent-plane approximation. That is, the rough surface can be considered as consisting of many inclined planes in the local region [4,5,10].

Under the tangent-plane approximation, the total field is the summation of the incident field and reflected field. The tangential surface fields can be depicted as

$$(\hat{n} \times \mathbf{E})_k = \hat{n} \times (\mathbf{E}^i + \mathbf{E}^r) \tag{4.34}$$

$$(\hat{n} \times \mathbf{H})_k = \hat{n} \times (\mathbf{H}^i + \mathbf{H}^r) \tag{4.35}$$

where the subscript k stands for the Kirchhoff approximation.

Since the incident field can be decomposed into local horizontal and vertical components, and the reflected and incident fields are related by the Fresnel reflection coefficient, the tangential Kirchhoff fields in Equations 4.34 and 4.35 can be expressed as [18,19]

$$(\hat{n} \times \mathbf{E}_h)_k = \hat{n} \times [(1 + R_h)\hat{p} - (R_h + R_v)(\hat{p} \cdot \hat{d})\hat{d}]E^i \tag{4.36}$$

$$(\hat{n} \times \mathbf{E}_v)_k = \hat{n} \times [(1 - R_v)\hat{p} + (R_v + R_h)(\hat{p} \cdot \hat{t})\hat{t}]E^i \tag{4.37}$$

$$\eta(\hat{n} \times \mathbf{H}_h)_k = \hat{n} \times \vec{k}_i \times [(1 - R_h)\hat{p} + (R_h + R_v)(\hat{p} \cdot \hat{d})\hat{d}]E^i \tag{4.38}$$

$$\eta(\hat{n} \times \mathbf{H}_v)_k = \hat{n} \times \vec{k}_i \times [(1 + R_v)\hat{p} - (R_h + R_v)(\hat{p} \cdot \hat{t})\hat{t}]E^i \tag{4.39}$$

In Equations 4.36–4.39, $\hat{p}$ is unit polarized vector of the incident wave, $\vec{k}_i$ is the wave vector of the incident wave, $\hat{n}$ is the unit normal vector of the rough surface; $\hat{t}, \hat{d}$ are unit vectors defined by the local reference coordinate system;

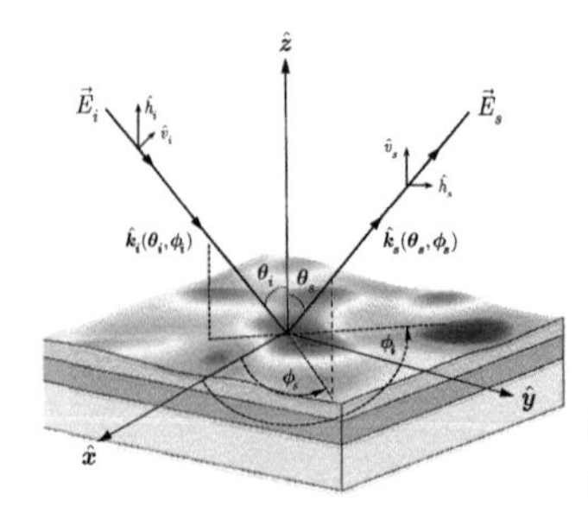

FIGURE 4.3 Geometry of wave scattering from a rough surface.

R_h, R_v are the Fresnel reflection coefficients for vertical and horizontal polarization, respectively.

Hence, the scattered field is

$$E_{qp}^{s} = K \int \{\hat{q} \cdot [\hat{k}_s \times (\hat{n} \times \mathbf{E}_p) + \eta(\hat{n} \times \mathbf{H}_p)]\} e^{j\left(\vec{k}_s - \vec{k}_i\right) \cdot \vec{r}} ds \tag{4.40}$$

4.3.2 Small Perturbation Method (SPM)

The SPM for the rough surface scattering was derived by Rice [11]. In terms of the electromagnetic wavelength, the SPM approximation assumes that the surface heights and slopes are smaller. The surface fields are expanded in a perturbation series of surface height. The zero-order solution of SPM is consistent with the scattered field for a plane interface. The first order represents the incoherent scattered filed as [4].

$$\sigma_{qp}^{o} = 8 \, |k^2 \sigma \cos \theta_i \cos \theta_s \alpha_{qp}|^2 \; W(k_x + k \sin \theta_i, k_y) \tag{4.41a}$$

$$\begin{aligned} k_x &= -k \sin \theta_s \cos \phi_s \\ k_y &= -k \sin \theta_s \sin \phi_s \end{aligned} \tag{4.41b}$$

$$W(k_x, k_y) = \frac{1}{2\pi} \int_{-\infty}^{\infty} \int_{-\infty}^{\infty} \rho(u, v) e^{-jk_x u - jk_y v} du dv \tag{4.41c}$$

In Equation 4.41, σ is the surface RMS height, α_{qp} is the polarization amplitude, which is the function of incident angle θ, scattering angle θ_s, scattering azimuth angle ϕ_s, complex relative dielectric constant ε_r, and relative magnetic permeability μ_r.

$$\alpha_{hh} = \frac{\left\{\left[\sqrt{\mu_r \varepsilon_r - \sin^2 \theta_s}\sqrt{\mu_r \varepsilon_r - \sin^2 \theta} \cos \phi_s - \mu_r \sin \theta \sin \theta_s\right] (\mu_r - 1) - \mu_r^2 (\varepsilon_r - 1) \cos \phi_s\right\}}{\left(\mu_r \cos \theta_s + \sqrt{\mu_r \varepsilon_r - \sin^2 \theta_s}\right)\left[\mu_r \cos \theta + \sqrt{\mu_r \varepsilon_r - \sin^2 \theta}\right]} \tag{4.42a}$$

$$\alpha_{hv} = \frac{\left[\mu_r (\varepsilon_r - 1)\sqrt{\mu_r \varepsilon_r - \sin^2 \theta} - \varepsilon_r (\mu_r - 1)\sqrt{\mu_r \varepsilon_r - \sin^2 \theta_s}\right] \sin \phi_s}{\left(\mu_r \cos \theta_s + \sqrt{\mu_r \varepsilon_r - \sin^2 \theta_s}\right)\left[\varepsilon_r \cos \theta + \sqrt{\mu_r \varepsilon_r - \sin^2 \theta}\right]} \tag{4.42b}$$

$$\alpha_{vh} = \frac{\left[\varepsilon_r(\mu_r - 1)\sqrt{\mu_r\varepsilon_r - \sin^2\theta} - \mu_r(\varepsilon_r - 1)\sqrt{\mu_r\varepsilon_r - \sin^2\theta_s}\right]\sin\phi_s}{\left(\varepsilon_r\cos\theta_s + \sqrt{\mu_r\varepsilon_r - \sin^2\theta_s}\right)\left[\mu_r\cos\theta + \sqrt{\mu_r\varepsilon_r - \sin^2\theta}\right]} \tag{4.42c}$$

$$\alpha_{vv} = \frac{\left\{\left[\sqrt{\mu_r\varepsilon_r - \sin^2\theta_s}\sqrt{\mu_r\varepsilon_r - \sin^2\theta}\cos\phi_s - \varepsilon_r\sin\theta\sin\theta_s\right] (\varepsilon_r - 1) - \varepsilon_r^2(\mu_r - 1)\cos\phi_s\right\}}{\left(\varepsilon_r\cos\theta_s + \sqrt{\mu_r\varepsilon_r - \sin^2\theta_s}\right)\left[\varepsilon_r\cos\theta + \sqrt{\mu_r\varepsilon_r - \sin^2\theta}\right]} \tag{4.42d}$$

The first-order perturbative solution of SPM only accounts for the single scattering. The multiple scattering is considered in the high order solution (e.g., second- and third-order solution), which report in [13]. Fourth- and higher-order small-perturbation solution for scattering from dielectric rough surfaces are reported in [14].

4.3.3 Small Slope Approximation (SSA)

The SSA in wave scattering by rough surface was proposed by Voronovich [15,16]. This method solves the scattering problem of a rough surface by combining the perturbation theory with the tangent-plane approximation.

The total field is the summation of incident filed and scattered field

$$\mathbf{E} = \mathbf{E}_i + \mathbf{E}_s \tag{4.43}$$

The incident field can be represented as

$$\mathbf{E}_i = \hat{p}k_z^{-1/2}\exp(j\hat{k}_\rho\cdot\vec{r}_\rho + jk_z z) \tag{4.44}$$

The scattered field can be obtained as

$$\mathbf{E}_s = \int k_{sz}^{-1/2}\exp(j\hat{k}_{s\rho}\cdot\vec{r}_\rho - j\hat{k}_{sz}z)S(\hat{k}_{s\rho}, \hat{k}_\rho)d\hat{k}_{s\rho} \tag{4.45}$$

where the scattering amplitude is

$$S(\hat{k}_{s\rho}, \hat{k}_\rho) = \frac{1}{(2\pi)^2}\int z(\vec{r})\ \Phi[\hat{k}_{s\rho}, \hat{k}_\rho, \vec{r}, z(\vec{r})]\exp[-j(\hat{k}_{s\rho} - \hat{k}_\rho)\cdot\vec{r} - j(k_z + k_{sz})z(\vec{r})]d\vec{r} \tag{4.46}$$

with

$$\Phi = \delta(\vec{\xi})\Phi^{(0)} + \int \delta(\vec{\xi} - \vec{\xi_1})\Phi^{(1)}(\vec{\xi_1})z(\vec{\xi_1})d\vec{\xi_1} + \int \delta(\vec{\xi} - \vec{\xi_1} - \vec{\xi_2})\Phi^{(2)}(\vec{\xi_1}, \vec{\xi_2})z(\vec{\xi_1})z(\vec{\xi_2})d\vec{\xi_1}\, d\vec{\xi_2} \tag{4.47}$$

Here, if we keep first-order Taylor series expansion of Φ, the first scattering amplitude is

$$S^{(1)}\left(\hat{k}_{sp}, \hat{k}_{\rho}\right) = \frac{2\sqrt{k_z k_{sz}}}{k_z + k_{sz}} \frac{B_1\left(\hat{k}_{sp}, \hat{k}_{\rho}\right)}{(2\pi)^2} \int \exp\left[-j\left(\hat{k}_{sp} - \hat{k}_{\rho}\right)\cdot\vec{r} - j(k_z + k_{sz})z(\vec{r})\right] d\vec{r} \tag{4.48}$$

The first-order scattering coefficient of SSA takes the form

$$\sigma_{qp}(\hat{k}_{sp}, \hat{k}_{\rho}) = \langle S^{(1)}(\hat{k}_{sp}, \hat{k}_{\rho})S^{(1)*}(\hat{k}_{sp}, \hat{k}_{\rho})\rangle - |\langle S^{(1)}(\hat{k}_{sp}, \hat{k}_{\rho})\rangle|^2 \tag{4.49}$$

$$\sigma_{qp}^{(1)}(\hat{k}_{sp}, \hat{k}_{\rho}) = \frac{1}{\pi}\left| \frac{2k_z k_{sz} B_1\left(\hat{k}_{sp}, \hat{k}_{\rho}\right)}{k_z + k_{sz}} \right|^2 \int \exp[-j(\hat{k}_{sp} - \hat{k}_{\rho})\cdot\vec{r} - (k_z + k_{sz})^2\rho(0)] \{\exp[-(k_z + k_{sz})^2\rho(\vec{r})] - 1\}\, d\vec{r} \tag{4.50}$$

The related coefficients of Equations 4.44–4.50 can be found in [15,16]. The higher-order SSA is given in [17].

4.3.4 Integral Equation Model (IEM)

The original Integral Equation Model (IEM) for a dielectric rough surface proposed by Fung et al. [18,19] has been satisfactorily used in the area of remote sensing of terrain and sea. The IEM can be considered as a connection between the KA and SPM. The IEM model includes single and multiple scattering terms, though, for small to moderate surface slope, single scattering is generally sufficiently accurate in predicting co-polarized returns. In depolarized backscattering, single scattering disappears and multiple scattering is responsible for the total return.

4.4 ADVANCED INTEGRAL EQUATION MODELS

The Advanced Integral Equation Model (AIEM) [20,21], is an extended version of IEM. For an easier explanation of the scattering mechanism, new expressions for multiple scattering up to second order are given. Comparisons with numerical simulation by NNM3D [9] and soil field measurements [22] are made to validate the AIEM multiple scattering in backscattering.

4.4.1 Single and Multiple Scattering

In IEM/AIEM model, the far-zone scattered field is expressed as a sum of the Kirchhoff and the complementary scattered fields

$$E_{qp}^{s} = E_{qp}^{k} + E_{qp}^{c} \tag{4.51}$$

where the Kirchhoff field is given by

$$E_{qp}^{k} = CE_0 \int f_{qp}\, e^{j\left(\vec{k}_s - \vec{k}_i\right)\vec{r}}\, dxdy \tag{4.52}$$

where the incident and scattering wave vectors $\vec{k}_i = \{k_x, k_y, k_z\}$ and $\vec{k}_s = \{k_{sx}, k_{sy}, k_{sz}\}$ are defined (4.31) and (4.32); $C = -\frac{jk}{4\pi R}\exp[-jkR]$, E_0 is the incident filed amplitude, and f_{qp} is the Kirchhoff coefficient with q, p denoting the polarization ($q, p = h, v$).

The complementary scattered field, propagating through upper and lower media with both upward and downward directions, is written as

$$E_{qp}^{c} = \frac{CE_0}{8\pi^2} \int \left\{ \mathcal{F}_{qp}^{\pm} e^{ju(x - x')+jv(y - y') - jq_1|z - z'|} e^{j\vec{k}_s\vec{r} - \vec{k}_i\vec{r}'} \right.$$
$$\left. + \mathcal{F}_{qp}^{\pm} e^{ju(x - x')+jv(y - y') - jq_2|z - z'|} e^{j\vec{k}_s\vec{r} - \vec{k}_i\vec{r}'} \right\} dxdydx'dy'dudv, \tag{4.53}$$

where $\vec{r}$, $\vec{r}'$ are respectively the source and field point coordinates of the surface; z' represents the surface height at (x', y') on the surface z with $z' \in z$ (see Figure 4.4).

In Equations 4.52 and 4.53, the Kirchhoff coefficient f_{qp} and complementary field coefficient $\mathcal{F}_{qp}$, $\mathcal{F}_{qp}$ are functions of the Fresnel reflection coefficients, σ^2 is the variance of the surface height, k is incident wavenumber, θ_i, ϕ_i are the incident angle and incident azimuth angle, θ_s, ϕ_s are the scattering angle and scattering azimuth angle; $\mathbf{W}^{(n)}(k_x, k_y)$ is the surface roughness spectrum of the surface related to the nth power of the surface correlation function by the two-dimensional Fourier transform, assuming that the surface height follows a Gaussian distribution.

It is seen at this point that in the IEM/AIEM model development, a once-iterated solution, using the Kirchhoff current as initial guess, to the integral equations governing surface tangential fields is obtained. From the estimated surface fields, then the scattered field can be calculated and is written as a sum of the Kirchhoff field and complementary field as given by Equation 4.51, with f_{qp} denoting the Kirchhoff field coefficient, which is related to incident field only, and $\mathcal{F}_{qp}$, $\mathcal{G}_{qp}$ denoting the complementary field coefficients, where $\mathcal{F}_{qp}$ represents the propagation in upper medium and $\mathcal{G}_{qp}$ is the propagation in lower medium. Figure 4.4 illustrates a physical interpretation of the field coefficients f_{qp}, $\mathcal{F}_{qp}$, $\mathcal{G}_{qp}$, where the signs are designated as

$$\mathcal{F}_{qp},\ \mathcal{G}_{qp} = \begin{cases} \mathcal{F}^{+}_{qp},\ \mathcal{G}^{+}_{qp};\ z > z' \\ \mathcal{F}^{-}_{qp},\ \mathcal{G}^{-}_{qp};\ z < z' \end{cases} \tag{4.54}$$

The Kirchhoff field only accounts for single scattering, while the complementary field includes multiple scattering. The higher-order beyond the second-order terms are only important for extremely steep surface and at near grazing angle. In what follows, we consider the multiple scattering up to the second order.

After lengthy but straightforward manipulations, the field coefficients $f_{qp}\mathcal{F}_{qp}$, $\mathcal{G}_{qp}$ can be obtained. Consequently, the mean scattered power in far-zone is

$$P_s = \langle E^s_{qp} - \langle E^s_{qp}\rangle\rangle\langle E^{s*}_{qp} - \langle E^{s*}_{qp}\rangle\rangle \tag{4.55}$$

where the asterisk denotes the complex conjugate. Now by substituting Equations 4.52 and 4.53 into Equations 4.51 and 4.55, we realize that the scattered power is a sum of Kirchhoff term, cross term, and the complementary term. From the previous argument regarding the physical interpretation of Figure 4.4, it is known that the multiple scattering is given rise by the cross and complementary terms only. It follows that corresponding to the scattered power, the scattering coefficient can be written as a sum of the single scattering $\sigma_{qp}(\mathrm{s})$ and the multiple scattering $\sigma_{qp}(\mathrm{m})$:

$$\sigma^o_{qp} = \sigma_{qp}(\mathrm{s}) + \sigma_{qp}(\mathrm{m}) \tag{4.56}$$

The single scattering term remain the same as in [18,19], and reads as

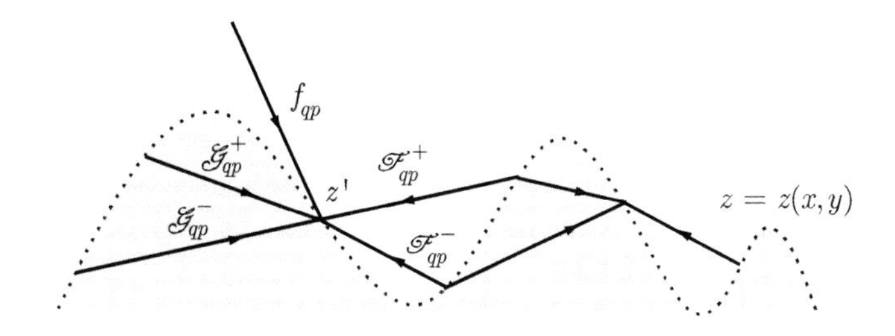

FIGURE 4.4 Physical interpretation of the field coefficients f_{qp}, $\mathcal{F}_{qp}$, $\mathcal{G}_{qp}$ when computing the scattered field due to the surface tangential field at surface point $z'(z' \in z)$.

$$\sigma_{qp}(\mathrm{s}) = \sigma_{qp}^{k} + \sigma_{qp}^{kc} + \sigma_{qp}^{c} \tag{4.57}$$

where σ_{qp}^{k} is the Kirchhoff term, σ_{qp}^{kc} is the cross term, and σ_{qp}^{c} is the complementary term. For easy of reference, a more explicit form of Equation 4.57 is given by [21]

$$\sigma_{qp}(\mathrm{s}) = \frac{k^2}{2} e^{-\sigma^2\left(k_{sz}^2 + k_z^2\right)} \sum_{n=1}^{\infty} \frac{\sigma^{2n}}{n!} \left|\mathbf{I}_{qp}^{n}\right|^2 \mathbf{W}^{(n)}\left(k_{sx} - k_x, k_{sy} - k_y\right) \tag{4.58}$$

$$\begin{aligned}
\mathbf{I}_{qp}^{n} = {} & (k_{sz} + k_z)^n f_{qp} e^{-\sigma^2 k_z k_{sz}} \\
& + \frac{1}{4}\Bigg\{ \mathcal{F}_{qp}^{+}\left(-k_x, -k_y\right)(k_{sz} - k_z)^n e^{-\sigma^2\left(k_z^2 - k_z k_{sz} + k_z k_z\right)} \\
& + \mathcal{G}_{qp}^{+}\left(-k_x, -k_y\right)(k_{sz} - k_{tz})^n e^{-\sigma^2\left(k_{tz}^2 - k_{tz} k_{sz} + k_{tz} k_z\right)} \\
& + \mathcal{F}_{qp}^{-}\left(-k_x, -k_y\right)(k_{sz} + k_z)^n e^{-\sigma^2\left(k_z^2 + k_z k_{sz} - k_z k_z\right)} \\
& + \mathcal{G}_{qp}^{-}\left(-k_x, -k_y\right)(k_{sz} + k_{tz})^n e^{-\sigma^2\left(k_{tz}^2 + k_{tz} k_{sz} - k_{tz} k_z\right)} \\
& + \mathcal{F}_{qp}^{+}\left(-k_{sx}, -k_{sy}\right)(k_z + k_{sz})^n e^{-\sigma^2\left(k_{sz}^2 - k_{sz} k_{sz} + k_{sz} k_z\right)} \\
& + \mathcal{G}_{qp}^{+}\left(-k_{sx}, -k_y\right)(k_z + k_{tsz})^n e^{-\sigma^2\left(k_{tsz}^2 - k_{tsz} k_{sz} + k_{tsz} k_z\right)} \\
& + \mathcal{F}_{qp}^{-}\left(-k_{sx}, -k_{sy}\right)(k_z - k_{sz})^n e^{-\sigma^2\left(k_{sz}^2 + k_{sz} k_{sz} - k_{sz} k_z\right)} \\
& + \mathcal{G}_{qp}^{-}\left(-k_{sx}, -k_{sy}\right)(k_z - k_{tsz})^n e^{-\sigma^2\left(k_{tsz}^2 + k_{tsz} k_{sz} - k_{tsz} k_z\right)} \Bigg\}
\end{aligned} \tag{4.59}$$

In single scattering, as given in Equation 4.58, the complementary field coefficients are specified by the incident and scattering wave vector; they are a function of spectral wavenumbers under the radiation conditions.

Following the procedure in [23], the derivations of these coefficients for upward and downward propagators in the upper medium and lower medium are straightforward but much more intricate. Keeping up to double bounce terms and after quite tedious steps, the multiple scattering coefficient can be derived and compactly expressed as a sum of three parts:

$$\sigma_{qp}(\mathrm{m}) = \sum_{l=1}^{3} \sigma_{qp}^{kc_l}(\mathrm{m}) + \sum_{i=1}^{8} \sigma_{qp}^{c_i}(\mathrm{m}) + \sum_{j=9}^{14} \sigma_{qp}^{c_j}(\mathrm{m}) \tag{4.60}$$

where the first part of the right-hand side of Equation 4.60 is the cross term

$$\begin{aligned}\sigma_{qp}^{kc_l}(\mathrm{m}) = \frac{k^2}{8\pi} Re\,[f_{qp}^{\;*} \iint & \{\mathcal{F}_{qp}^{+}(u, v) g^{kc_l}(u, v, q_1) \\ & + \mathcal{F}_{qp}^{-}(u, v) g^{kc_l}(u, v, -q_1)\; \mathcal{G}_{qp}^{+}(u, v) g^{kc_l}(u, v, q_2) \\ & + \mathcal{G}_{qp}^{-}(u, v) g^{kc_l}(u, v, -q_2)\}\, du dv]\end{aligned} \tag{4.61}$$

The factors g^{kc_l}, $l = 1, 2, 3$, inside the double integral, are explicitly given in Appendix 4A. The factors appearing in Equation 4.61 entail double summations over the surface roughness spectra at different spectral components, which are naturally resonant with incident wavenumber and scattering wavenumber, and hence responsible for the multiple scattering. To restrict to radiation modes, it requires that $k_z - q > 0$, $k_z - q' > 0$, $k_{sz} - q > 0$, $k_{sz} - q' > 0$.

f_{qp} are given as

$$\begin{aligned} f_{hh} = & [(1 + R_h)\hat{v}_s \cdot (\hat{n} \times \hat{h}) + (1 - R_h)\hat{h}_s \cdot (\hat{n} \times \hat{v})] D_1 \\ & - (R_v + R_h)(\hat{h} \cdot \hat{d})[\hat{h}_s \cdot (\hat{n} \times \hat{t}) + \hat{v}_s \cdot (\hat{n} \times \hat{d})] D_1 \end{aligned} \tag{4.62a}$$

$$\begin{aligned} f_{hv} = & [(1 - R_v)\hat{v}_s \cdot (\hat{n} \times \hat{v}) - (1 + R_v)\hat{h}_s \cdot (\hat{n} \times \hat{h})] D_1 \\ & + (R_v + R_h)(\hat{v} \cdot \hat{t})[\hat{v}_s \cdot (\hat{n} \times \hat{t}) - \hat{h}_s \cdot (\hat{n} \times \hat{d})] D_1 \end{aligned} \tag{4.62b}$$

$$\begin{aligned} f_{vh} = & [(1 - R_h)\hat{v}_s \cdot (\hat{n} \times \hat{v}) - (1 + R_h)\hat{h}_s \cdot (\hat{n} \times \hat{h})] D_1 \\ & - (R_v + R_h)(\hat{h} \cdot \hat{d})[\widehat{v_s} \cdot (\hat{n} \times \hat{t}) - \hat{h}_s \cdot (\hat{n} \times \hat{d})] D_1 \end{aligned} \tag{4.62c}$$

$$\begin{aligned} f_{vv} = & -[(1 - R_v)\hat{h}_s \cdot (\hat{n} \times \hat{v}) + (1 + R_v)\hat{v}_s \cdot (\hat{n} \times \hat{h})] D_1 \\ & - (R_v + R_h)(\hat{v} \cdot \hat{t})[\hat{h}_s \cdot (\hat{n} \times \hat{t}) + \hat{v}_s \cdot (\hat{n} \times \hat{t})] D_1 \end{aligned} \tag{4.62d}$$

The second and third parts in Equation 4.60 are given rise from the complementary terms:

$$\sigma_{qp}^{c_i}(\mathrm{m}) = \frac{k^2}{64\pi} \iint \Bigg\{ \mathcal{F}_{qp}^{+}(u,v)\mathcal{F}_{qp}^{+*}(u',v')g^{c_i}\left(u,v,q_1,q'_1\right)$$

$$+ \mathcal{F}_{qp}^{+}(u,v)\mathcal{F}_{qp}^{-*}(u',v')g^{ci}\left(u,v,q_1,-q'_1\right)\mathcal{F}_{qp}^{-}(u,v)\mathcal{F}_{qp}^{+*}(u',v')$$

$$g^{c_i}\left(u,v,-q_1,q'_1\right) + \mathcal{F}_{qp}^{-}(u,v)\mathcal{F}_{qp}^{-*}(u',v')g^{c_i}$$

$$\left(u,v,-q_1,-q'_1\right)\mathcal{F}_{qp}^{+}(u,v)\mathcal{G}_{qp}^{+*}(u',v')g^{c_i}\left(u,v,q_1,q'_2\right)$$

$$+ \mathcal{F}_{qp}^{+}(u,v)\mathcal{G}_{qp}^{-*}(u',v')g^{c_i}\left(u,v,q_1,-q'_2\right)\mathcal{F}_{qp}^{-}(u,v)\mathcal{G}_{qp}^{+*}(u',v')$$

$$g^{c_i}\left(u,v,-q_1,q'_2\right) + \mathcal{F}_{qp}^{-}(u,v)\mathcal{G}_{qp}^{-*}(u',v')g^{c_i}$$

$$\left(u,v,-q_1,-q'_2\right)\mathcal{G}_{qp}^{+}(u,v)\mathcal{F}_{qp}^{+*}(u',v')g^{c_i}\left(u,v,q_2,q'_1\right)$$

$$+ \mathcal{G}_{qp}^{+}(u,v)\mathcal{F}_{qp}^{-*}(u',v')g^{ci}\left(u,v,q_2,-q'_1\right)\mathcal{G}_{qp}^{-}(u,v)\mathcal{F}_{qp}^{+*}(u',v')$$

$$g^{c_i}\left(u,v,-q_2,q'_1\right) + \mathcal{G}_{qp}^{-}(u,v)\mathcal{F}_{qp}^{-*}(u',v')g^{c_i}$$

$$\left(u,v,-q_2,-q'_1\right)\mathcal{G}_{qp}^{+}(u,v)\mathcal{G}_{qp}^{+*}(u',v')g^{c_i}\left(u,v,q_2,q'_2\right)$$

$$+ \mathcal{G}_{qp}^{+}(u,v)\mathcal{G}_{qp}^{-*}(u',v')g^{c_i}\left(u,v,q_2,-q'_2\right)\mathcal{G}_{qp}^{-}(u,v)\mathcal{G}_{qp}^{+*}(u',v')$$

$$g^{c_i}\left(u,v,-q_2,q'_2\right) + \mathcal{G}_{qp}^{-}(u,v)\mathcal{G}_{qp}^{-*}(u',v')g^{c_i}\left(u,v,-q_2,-q'_2\right)\Bigg\}$$

$$du dv \tag{4.63a}$$

$$\sigma_{qp}^{c_j}(\mathrm{m}) = \frac{k^2}{64\pi} \iint \Bigg\{ \mathcal{F}_{qp}^{+}(u, v)\mathcal{F}_{qp}^{+*}(u', v')g^{c_j}\left(u', v', q_1, q'_1\right)$$

$$+ \mathcal{F}_{qp}^{+}(u, v)\mathcal{F}_{qp}^{-*}(u', v')g^{c_j}\left(u', v', q_1, -q'_1\right)\mathcal{F}_{qp}^{-}(u, v)\mathcal{F}_{qp}^{+*}(u', v')$$

$$g^{c_j}\left(u', v', -q_1, q'_1\right) + \mathcal{F}_{qp}^{-}(u, v)\mathcal{F}_{qp}^{-*}(u', v')g^{c_j}$$

$$\left(u', v', -q_1, -q'_1\right)\mathcal{F}_{qp}^{+}(u, v)\mathcal{G}_{qp}^{+*}(u', v')g^{c_j}\left(u', v', q_1, q'_2\right)$$

$$+ \mathcal{F}_{qp}^{+}(u, v)\mathcal{G}_{qp}^{-*}(u', v')g^{c_j}\left(u', v', q_1, -q'_2\right)\mathcal{F}_{qp}^{-}(u, v)\mathcal{G}_{qp}^{+*}(u', v')$$

$$g^{c_j}\left(u', v', -q_1, q'_2\right) + \mathcal{F}_{qp}^{-}(u, v)\mathcal{G}_{qp}^{-*}(u', v')g^{c_j}$$

$$\left(u', v', -q_1, -q'_2\right)\mathcal{G}_{qp}^{+}(u, v)\mathcal{F}_{qp}^{+*}(u', v')g^{c_j}\left(u', v', q_2, q'_1\right)$$

$$+ \mathcal{G}_{qp}^{+}(u, v)\mathcal{F}_{qp}^{-*}(u', v')g^{c_j}\left(u', v', q_2, -q'_1\right)\mathcal{G}_{qp}^{-}(u, v)\mathcal{F}_{qp}^{+*}(u', v')$$

$$g^{c_j}\left(u', v', -q_2, q'_1\right) + \mathcal{G}_{qp}^{-}(u, v)\mathcal{F}_{qp}^{-*}(u', v')g^{c_j}$$

$$\left(u', v', -q_2, -q'_1\right)\mathcal{G}_{qp}^{+}(u, v)\mathcal{G}_{qp}^{+*}(u', v')g^{c_j}\left(u', v', q_2, q'_2\right)$$

$$+ \mathcal{G}_{qp}^{+}(u, v)\mathcal{G}_{qp}^{-*}(u', v')g^{c_j}\left(u', v', q_2, -q'_2\right)\mathcal{G}_{qp}^{-}(u, v)\mathcal{G}_{qp}^{+*}(u', v')$$

$$g^{c_j}\left(u', v', -q_2, q'_2\right) + \mathcal{G}_{qp}^{-}(u, v)\mathcal{G}_{qp}^{-*}(u', v')g^{c_j}\left(u', v', -q_2, -q'_2\right)\Bigg\}$$

$$du'dv' \tag{4.63b}$$

Note that factors g^{c_i}, g^{c_j} in Equations 4.63a and 4.63b are grouped in such a way of convenient notation so that their arguments are given by

$$\begin{cases} u' = u,\ v' = v,\ q' = q;\ i = 1 \\ u' = -k_x - k_{sx} - u,\ v' = -k_y - k_{sy} - v;\ i = 2 \\ u' = -k_{sx},\ v' = -k_{sy},\ q' = k_{sz};\ i = 3, 4, 5 \\ u' = -k_x,\quad v' = -k_y,\quad q' = k_z;\ i = 6, 7, 8 \\ u = -k_{sx},\quad v = -k_{sy},\quad q = k_{sz};\ j = 9, 10, 11 \\ u = -k_x,\quad v = -k_y,\quad q = k_z;\ j = 12, 13, 14 \\ q_1 = \sqrt{k^2 - u^2 - v^2},\ q'_1 = \sqrt{k^2 - u'^2 - v'^2} \\ q_2 = \sqrt{\varepsilon_r k^2 - u^2 - v^2},\ q'_2 = \sqrt{\varepsilon_r k^2 - u'^2 - v'^2} \end{cases} \tag{4.64}$$

The factors g^{c_i}, g^{c_j} appearing in Equation 4.63 are explicitly derived and given in Appendix 4A.

Unlike in single scattering, the complementary field coefficients are not directly specified by the incident and scattering wave vector; they are a function of spectral wavenumbers under the radiation conditions. The associated factors g^{c_i}, g^{c_j} are also a function of the spatial wavenumber, which are not uniquely determined by the incident and scattering wave. Physically, the surface fields contributing to the multiple scattering come from secondary sources in all possible directions over the surface, again, under radiation conditions.

Combining Equations 4.61 and 4.63 into Equation 4.60 for multiple scattering, numerical computations of double integrals over double summations are necessary. The complementary field coefficients $\mathcal{F}_{qp}$, $\mathcal{G}_{qp}$ are given in Appendix 4B and their associated factors in Appendix 4C.

The scattering matrix S_c of circularly polarized waves is related to the scattering matrix S_{FSA} for linearly polarized waves in the forward scattering alignment (FSA) convention by the following transformation

$$S_c = \begin{bmatrix} S_{RR} & S_{RL} \\ S_{LR} & S_{LL} \end{bmatrix} = \frac{1}{2}\begin{bmatrix} 1 & -j \\ 1 & j \end{bmatrix} S_{FSA} \begin{bmatrix} 1 & 1 \\ j & -j \end{bmatrix} \tag{4.65}$$

where the subscripts RL means that the polarization state incident field is left-hand circularly polarized (L) wave, and scattered field is right-hand circularly polarized (R) wave. The backscatter alignment (BSA) convention is related to FSA via

$$S_{FSA} = \begin{bmatrix} 1 & 0 \\ 0 & -1 \end{bmatrix} S_{BSA} \tag{4.66}$$

In the backscatter alignment (BSA) convenient, the scattering matrix is

$$S_{BSA} = \begin{bmatrix} S_{vv} & S_{vh} \\ S_{hv} & S_{hh} \end{bmatrix} \tag{4.67}$$

After some matrix manipulations, the elements of matrix S_c in the FSA convention are expressed as

$$S_{RR} = (S_{vv} + jS_{hv} + jS_{vh} - S_{hh})/2 \tag{4.68a}$$

$$S_{LR} = (S_{vv} - jS_{hv} + jS_{vh} + S_{hh})/2 \tag{4.68b}$$

$$S_{RL} = (S_{vv} + jS_{hv} - jS_{vh} + S_{hh})/2 \tag{4.68c}$$

$$S_{LL} = (S_{vv} - jS_{hv} - jS_{vh} - S_{hh})/2 \tag{4.68d}$$

The scattering coefficients can be obtained by

$$\sigma^o_{\alpha\beta} = \frac{4\pi}{A}\langle |S_{\alpha\beta}|^2 \rangle \tag{4.69}$$

where the $\langle\ \rangle$ denotes the ensemble average; subscripts α and β the polarization states; A the antenna illuminated area.

The normalized radar cross-section (NRCS) for circular polarization can be readily given in terms of linear polarizations as follows

$$\begin{aligned}\sigma^0_{RR} = \tfrac{1}{4}\{&\sigma^0_{vv} + \sigma^0_{hh} + \sigma^0_{vh} + \sigma^0_{hv} - 2[Re(\sigma^0_{vvhh} - \sigma^0_{vhhv}) \\ &- Im(\sigma^0_{vvvh} + \sigma^0_{vvhv} + \sigma^0_{vhhh} + \sigma^0_{hvhh})]\}\end{aligned} \tag{4.70a}$$

$$\begin{aligned}\sigma^0_{LR} = \tfrac{1}{4}\{&\sigma^0_{vv} + \sigma^0_{hh} + \sigma^0_{vh} + \sigma^0_{hv} + 2[Re(\sigma^0_{vvhh} - \sigma^0_{vhhv}) \\ &+ Im(\sigma^0_{vvvh} - \sigma^0_{vvhv} - \sigma^0_{vhhh} + \sigma^0_{hvhh})]\}\end{aligned} \tag{4.70b}$$

$$\begin{aligned}\sigma^0_{RL} = \tfrac{1}{4}\{&\sigma^0_{vv} + \sigma^0_{hh} + \sigma^0_{vh} + \sigma^0_{hv} + 2[Re(\sigma^0_{vvhh} - \sigma^0_{vhhv}) \\ &- Im(\sigma^0_{vvvh} - \sigma^0_{vvhv} - \sigma^0_{vhhh} + \sigma^0_{hvhh})]\}\end{aligned} \tag{4.70c}$$

$$\begin{aligned}\sigma^0_{LL} = \tfrac{1}{4}\{&\sigma^0_{vv} + \sigma^0_{hh} + \sigma^0_{vh} + \sigma^0_{hv} - 2[Re(\sigma^0_{vvhh} - \sigma^0_{vhhv}) \\ &+ Im(\sigma^0_{vvvh} + \sigma^0_{vvhv} + \sigma^0_{vhhh} + \sigma^0_{hvhh})]\}\end{aligned} \tag{4.70d}$$

In Equations 4.70a–4.70d, *Re* denotes the real part, *Im* the imaginary part.

4.4.2 Multiple Scattering Contributions

To examine the multiple scattering contribution, we plot the single scattering (AIEM(S)), multiple scattering (AIEM (M)), and total scattering (AIEM) in Figure 4.5(a_1, b_1, c_1), for exponentially correlated surfaces with a fixed $k\ell$ of 3.0, ε_r of $15.0 - j2.0$ and $k\sigma$ varying from 0.1 to 0.5. It can be seen from the plots that the multiple scattering has more impact on lager RMS slope; it contributes not much at small incident angles until $k\sigma$ exceeding 0.1. As $k\sigma$ increases to 0.5, the multiple scattering becomes equally important over all incident angles. Figure 4.5(a_2, b_2, c_2)

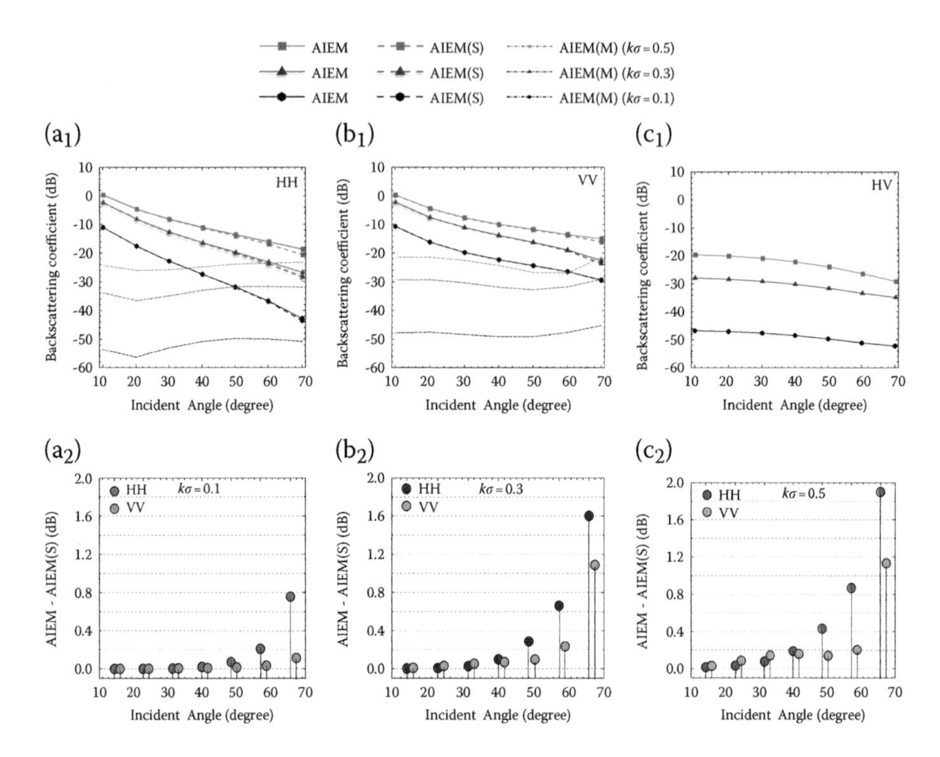

FIGURE 4.5 Angular behavior of the backscattering predicted by AIEM single scattering (AIEM(S)), multiple scattering (AIEM(M)), and total scattering (AIEM) coefficient for $k\ell = 3.0$, $\varepsilon_r = 15.0 - j2.0$ with rms height of $k\sigma = 0.1,\ 0.3,\ 0.5$: ($a_1$) HH ($b_1$) VV and ($c_1$) HV polarization and the deviation between single scattering and total scattering for both HH and VV polarizations: (a_2) $k\sigma = 0.1$, (b_2) $k\sigma = 0.3$, and (c_2) $k\sigma = 0.5$.

shows the deviation between single scattering and total scattering. It can be seen that the multiple scattering has more impact on HH than VV, and as$k\sigma$ increases, the contribution of multiple scattering is lager. Besides, the multiple scattering is enhanced as the incident angle increases. Especially, the deviation between the single scattering and total scattering is about 2.0 dB for HH polarization and 1.2 dB for VV polarization when $k\sigma$ is 0.5 and the incident angle is 70°.

It is worth examining the dependence of multiple scattering on the surface dielectric constant. In doing so, Figure 4.6(a_1, b_1, c_1) shows the backscattering coefficient from an exponentially correlated surface with a fixed $k\sigma$ of 0.5, $k\ell$ of 0.3 and ε_r changing. It can be observed that the contribution of multiple scattering tends to increase as permittivity increase for both co-polarizations and de-polarizations. Besides, we can see that the result of total scattering and single scattering are very close at the smaller incident angles for HH and VV polarization. They begin to deviate when at larger incidence. Figure 4.6(a_2, b_2, c_2) shows the difference between the single scattering and total scattering to demonstrate that the multiple scattering, again, has more impact on HH than VV, and the multiple scattering is enhanced as the incident angle increases.

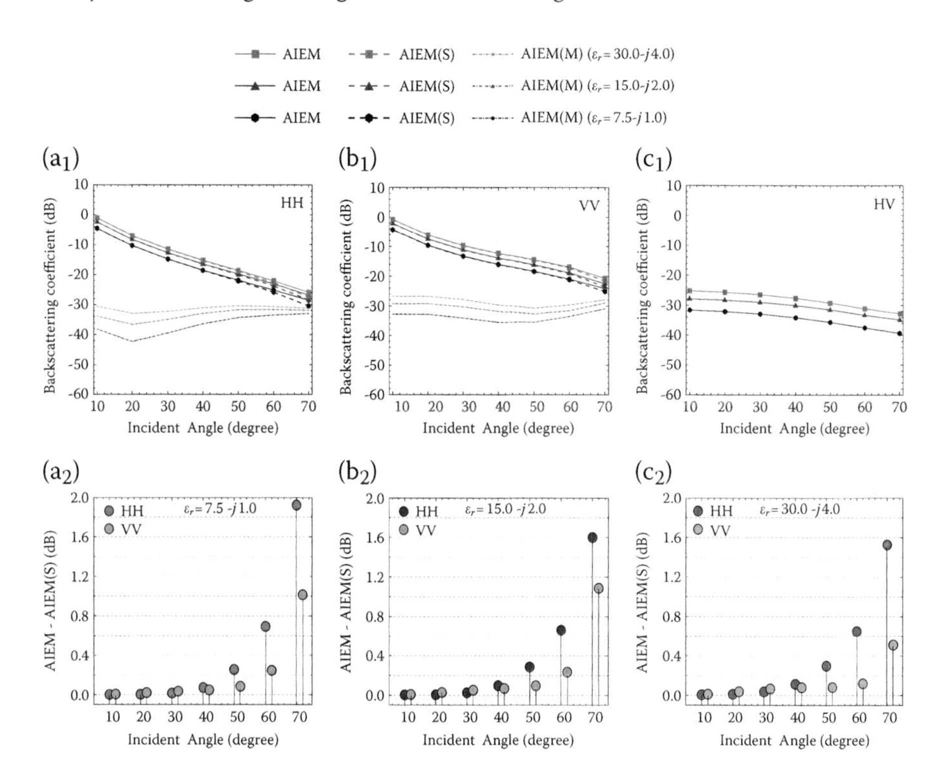

FIGURE 4.6 Angular patterns of the backscattering coefficient predicted by AIEM single scattering (AIEM(S)), multiple scattering (AIEM(M)), and total scattering (AIEM) at $k\ell = 3.0$, $k\sigma = 0.3$ with difference permittivity $\varepsilon_r = 7.5 - j1.0, 15.0 - j2.0, 30.0 - j4.0$ for (a_1) HH (b_1) VV and (c1) HV polarization and the deviation between single scattering and total scattering for both HH and VV polarization: (a_2) $\varepsilon_r = 7.5 - j1.0$, ($b_2$) $\varepsilon_r = 15.0 - j2.0$ (c_2) $\varepsilon_r = 30.0 - j4.0$.

In Figure 4.7, both like- and cross-polarized scattering coefficients are plotted versus ellipticity angle and orientation angle with two surface roughness and dielectric constants. The normalized correlation length and dielectric constant are set as $kl = 2.07$, $\varepsilon_r = 10 - j1.7$. The effect of surface roughness, by changing normalized RMS heights: 0.5 and 1.0 is illustrated. The incident angle is fixed at 30°. Another case showing a dielectric constant effect on scattering behaviors is presented in Figure 4.8, where the surface roughness parameters $kl = 2.07$, $k\sigma = 1.0$, with two dielectric constants of $5.0 - j0.6$ and $15 - j2.2$, keeping the incident angle to 30°. For smaller RMS height and dielectric constant, the maximum appears at VV polarization. The LL and RR polarizations nearly approach to zero. For cross-polarizations, the maximum locates at circular polarization, LR, and RL and the minimum appears at linear polarization. As the RMS height and dielectric constant increase, the scattering coefficients both for like- and cross-polarizations are enhanced, as they should be. Note that the maxima are now shifted to the somewhere between VV and HH polarizations. Moreover, the scattering coefficients for HV and VH polarizations are no longer zero. The minima are moved to around HV and VH polarizations. The shifts of the maxima and minima are due to the fact that the multiple scattering is enhanced with increasing RMS height (or slope for a fixed *kl*) and dielectric constant. The results indicate that multiple scattering is not negligible to investigate the fully polarimetric scattering characteristic. Therefore, in the following section, multiple scattering is included in all calculations.

4.4.3 Validation of AIEM Model

4.4.3.1 Comparison with Numerical Simulations

The comparison of the scattering coefficient between the AIEM and numerical results of NMM3D [9] are shown in Figure 4.9 for an exponentially correlated surface with $\varepsilon_r = 9.0 - j2.5,\ 15.0 - j3.5,\ 30.0 - j4.5$ at incident angle of 40°. Here four different slopes are given in Figure 4.9, which are ¼ (a_1, b_1, c_1), 1/7 (a_2, b_2, c_2), 1/10 (a_3, b_3, c_3), 1/15 (a_4, b_4, c_4). The normalized RMS height varies from 0.021 to 0.21. The correlation plots between AIEM and NMM3D for HH, VV and HV polarizations are shown in Figure 4.10. It can be observed that AIEM predictions are generally in good agreement with NMM3D simulations for both co- and cross-polarizations. The root mean square error (RMSE) and correlation

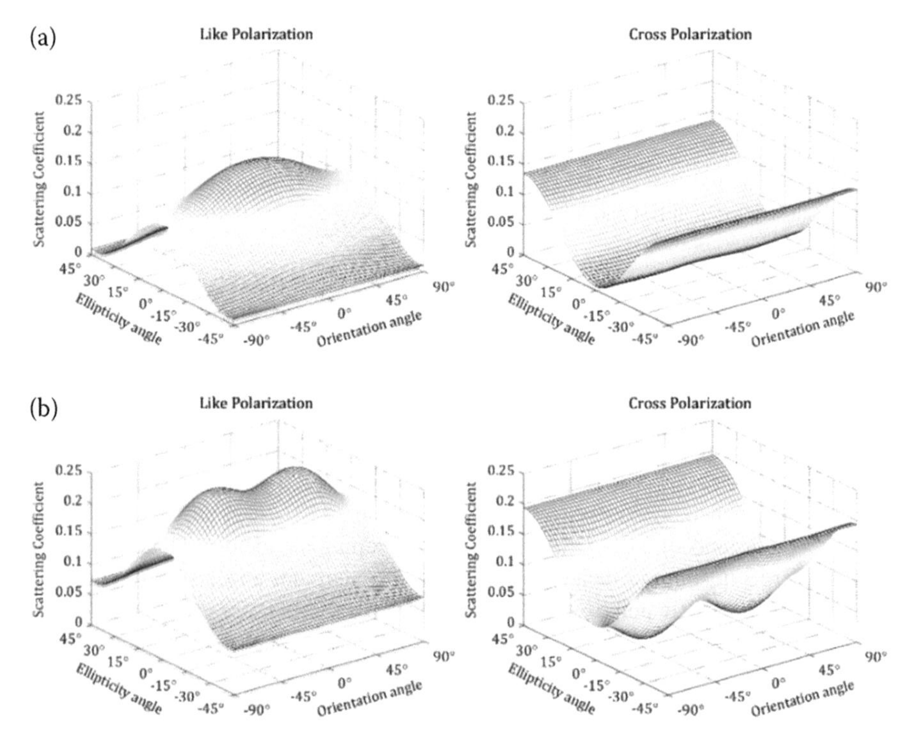

FIGURE 4.7 Scattering coefficients as function of polarization states with two surface roughness: $kl = 2.07,\ \varepsilon_r = 10 - j1.7,\ \theta_i = 30°$(a): $k\sigma = 0.5$; (b) $k\sigma = 1.0$.

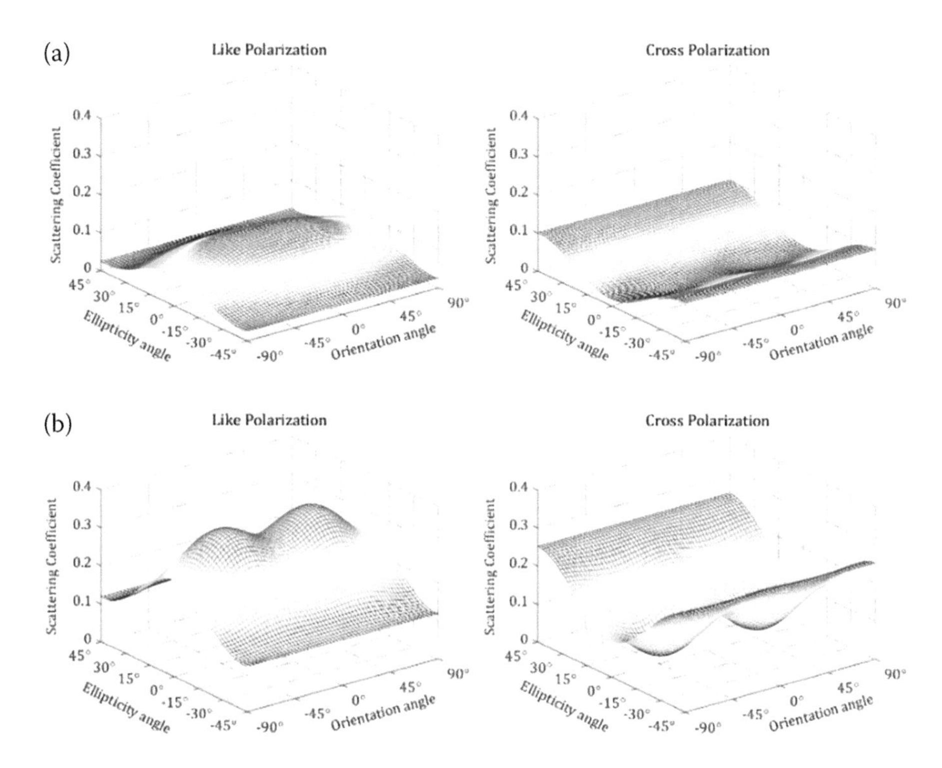

FIGURE 4.8 Scattering coefficients as function of polarization states with two dielectric constants: $kl = 2.07$, $k\sigma = 1.0$, $\theta_i = 30°$(a): $\varepsilon_r = 5.0 - j0.6$; (b): $\varepsilon_r = 15 - j2.2$.

coefficient (r) of AIEM compared with NMM3D are given in Table 4.1, where, as we can read, the correlation coefficients for both co- and cross-polarizations are all greater than 0.96 and RMSE is about 1.5 dB, persistently implying an almost perfect match between AIEM and NMM3D in predicting backscattering coefficients for co- and cross-polarizations.

To access the performance of AIEM model in predicting the bistatic scattering, comparisons between AIEM, simulations by MoM (Method of Moment), and SSA (small slope approximation with first-order approximation, SSA-1) [24] for Gaussian correlated surface are given in Figure 4.11. Two sets of surface roughness are: $kl = 6.28$, $k\sigma = 1.57$ and $kl = 12.56$, $k\sigma = 3.14$, with permittivity $25 + j3.0$, and 45° of incident angle. From these comparisons, we see that the AIEM model predictions match quite well with the MoM simulations. Notice that strong spikes appearing in specular direction is due to the coherent scattering, which was excluded in both AIEM and SSA.

4.4.3.2 Comparison with Measurement Data

We now compare the backscattering coefficients predicted by AIEM with measured POLARSCAT data [22] for three exponentially correlated surfaces, S1, S2, and S3, each surface having wet and dry conditions corresponding to different dielectric

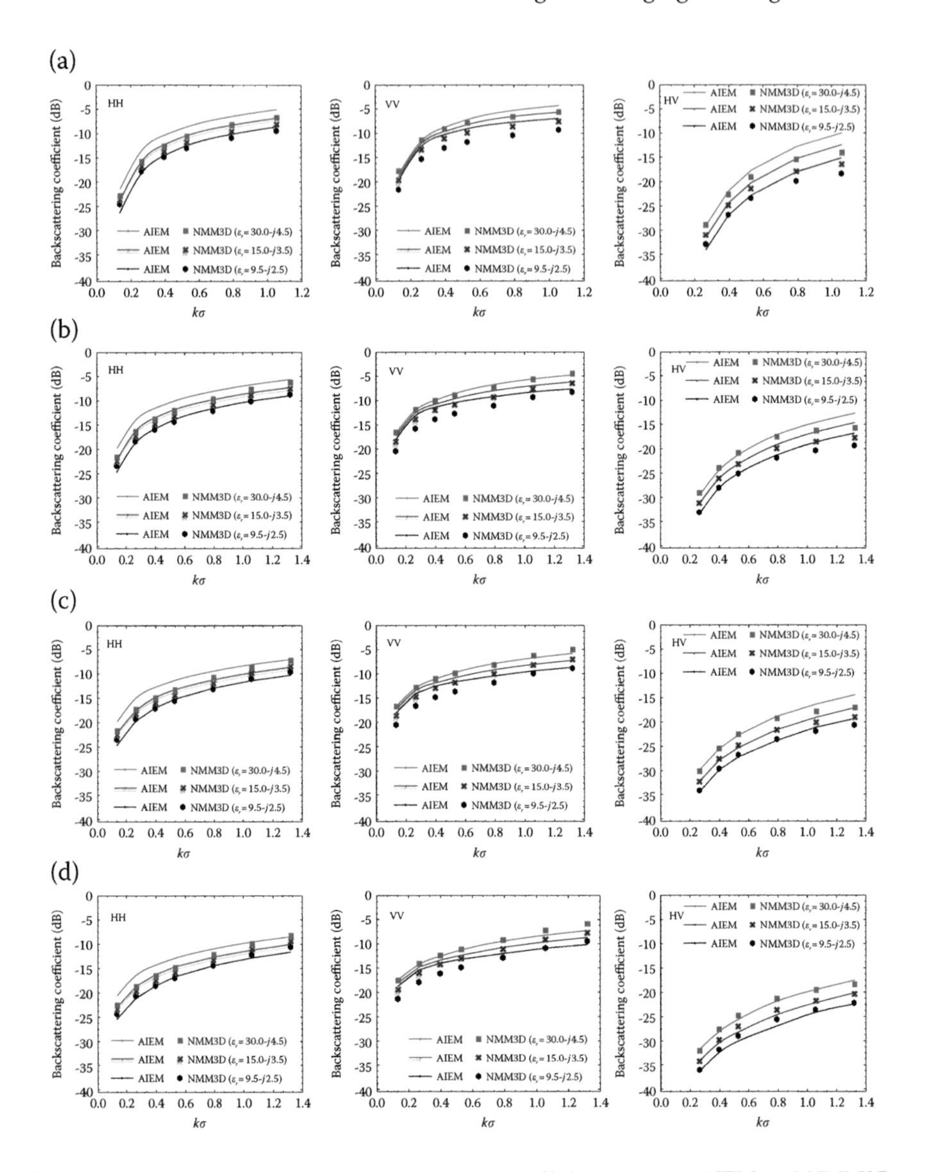

FIGURE 4.9 Comparison of the backscattering coefficient between AIEM and NMM3D for an exponentially correlated surface with $\varepsilon_r = 9.0 - j2.5, 15.0 - j3.5, 30.0 - j4.5$ at incident angle of 40° for co- and cross-polarizations. The corresponding rms slope are 1/4 (a), 1/7 (b), 1/10 (c), and 1/15 (d).

constants. POLARSCAT is a polarimetric scatterometer operating at L, C, X bands (with center frequencies of 1.25, 4.75, and 9.5 GHz, respectively) at incident angles of 10°~70° for HH, VV and HV/VH polarizations. More POLARSCAT characteristics are given in [22]. Among three surfaces, **S3** has the largest RMS height and slope.

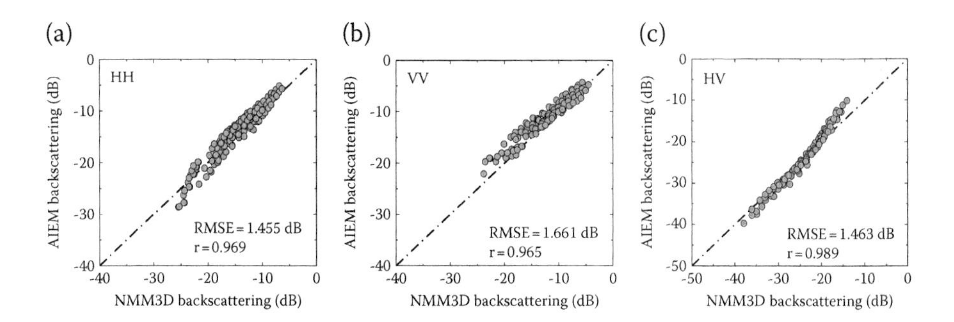

FIGURE 4.10 Correlation plot between AIEM and NNM3D for an exponentially correlated surface with incident angle of 40° for co- and cross-polarized backscattering.

TABLE 4.1
Comparison of the Predicted Backscattering between AIEM and NMM3D

	HH	VV	HV
RMSE	1.455dB	1.661dB	1.463dB
r	0.969	0.965	0.989

The correlation plots of backscattering coefficients between the AIEM and the measured data for HH, VV, and HV polarizations are shown in Figures 4.12 and 4.13. Figure 4.12 shows the data for three frequency bands, and Figure 4.13 shows the data by separating the wet and dry conditions. The co-polarized backscattering coefficients predicted by AIEM are in excellent agreement with the measured data for both HH and VV polarizations. The inclusion of multiple scattering helps improve the accuracy of HH polarization, though only slightly. Overall, the correlation plot points are very close to the 1:1 line, suggesting that the AIEM is in excellent agreement with measured data for both level and trend. The root mean square error (RMSE) and correlation coefficient(r) of AIEM compared with measured data are given for different frequency bands in Table 4.2 and for different wetness conditions in Table 4.3, respectively. The correlation coefficients for both co- and cross-polarizations are all larger than 0.89 and RMSEs are about 2.6 dB, confirming a good match between the AIEM and the measured data.

Now that the model predicted and measured Muller matrices are compared in this section. The Muller matrix from a rough surface with the ground truth of $kl = 2.62$, $k\sigma = 0.126$ and $m_v = 0.126$ measured by scatterometer reads [25]:

$$\mathbf{M}_m = \begin{bmatrix} 0.0012 & 0.00002 & 0 & 0 \\ 0.00002 & 0.0008 & 0 & 0 \\ 0 & 0 & 0.0010 & -0.0002 \\ 0 & 0 & 0.0002 & 0.0010 \end{bmatrix} \tag{4.71}$$

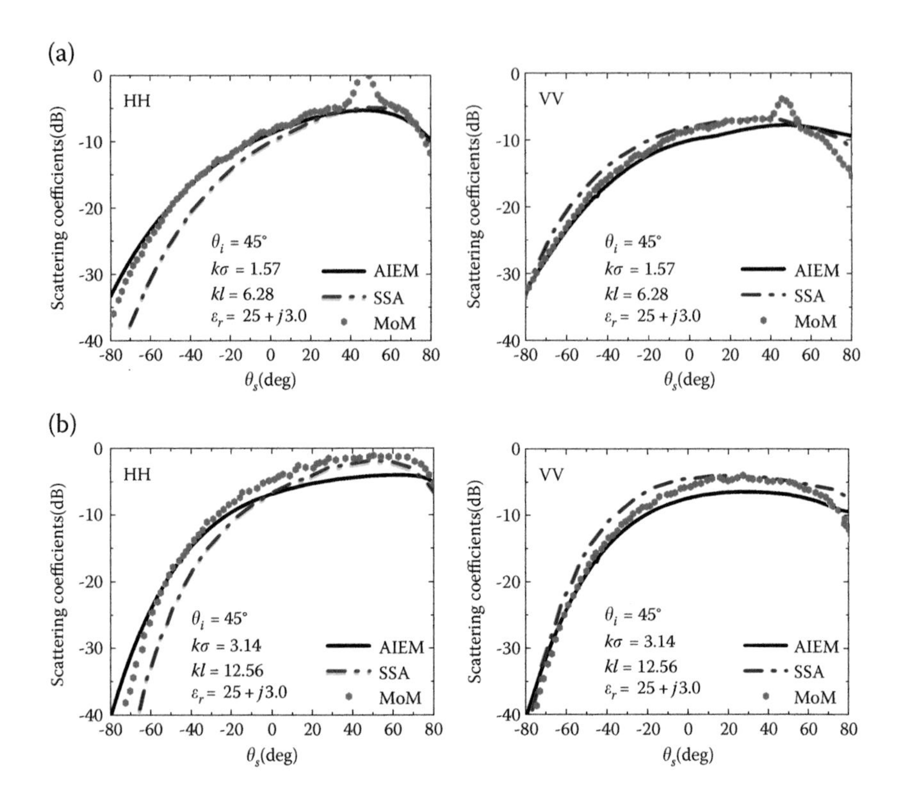

FIGURE 4.11 Comparison of bistatic scattering between AIEM model and numerical results of SSA and MoM for dielectric rough surface with Gaussian correlated function, and 45° of incident angle $\varepsilon_r = 25 + j3.0$: (a) $kl = 6.28$, $k\sigma = 1.57$, (b) $kl = 12.56$, $k\sigma = 3.14$.

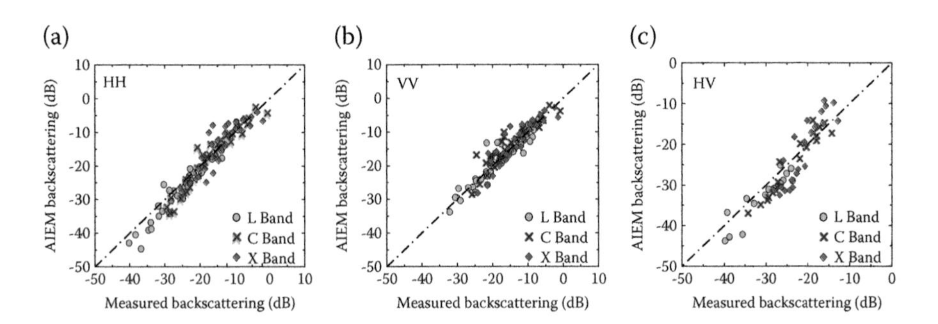

FIGURE 4.12 Correlation plots between AIEM and measured data for co- and depolarized backscattering coefficients at L/C/X bands for (a) HH, (b) VV, (c) HV.

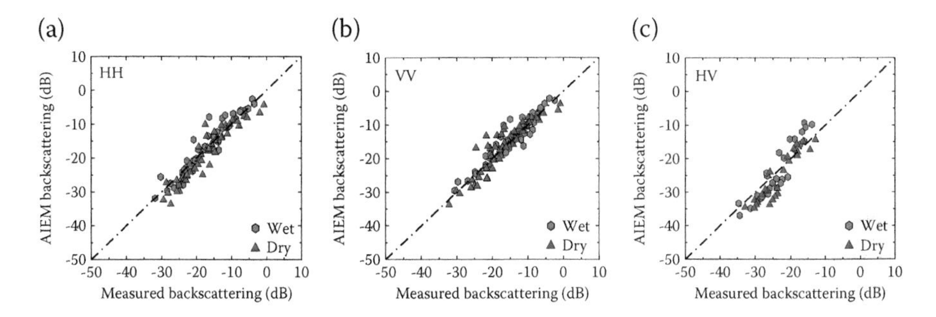

FIGURE 4.13 Correlation plots between AIEM and measured data for co- and depolarized backscattering coefficients at wet and dry soil for (a) HH, (b) VV, (c) HV.

TABLE 4.2
Correlation between AIEM Predictions and Measured Data for Different Frequencies

Polarizations	L-Band (f=1.25 GHz)		C-Band (f=4.75 GHz)		X-Band (f=9.5 GHz)	
	RMSE (dB)	*r*	*RMSE* (dB)	*r*	*RMSE* (dB)	*r*
HH	2.829	0.974	2.633	0.961	3.235	0.895
VV	2.551	0.935	2.596	0.921	2.608	0.904
HV	2.869	0.915	2.078	0.962	3.595	0.895

TABLE 4.3
Correlation between AIEM Predictions and Measured Data for Different Surface Wetness

Polarizations	Wet Soil		Dry Soil	
	RMSE(dB)	*r*	*RMSE*(dB)	*r*
HH	2.410	0.954	2.835	0.943
VV	2.192	0.947	2.933	0.918
HV	3.706	0.931	3.827	0.923

The model predicted Muller matrix gives

$$\mathbf{M}_{model} = \begin{bmatrix} 0.001252 & 0.000026 & 0 & 0 \\ 0.000026 & 0.000714 & 0 & 0 \\ 0 & 0 & 0.00095 & -0.00001 \\ 0 & 0 & 0.00001 & 0.00095 \end{bmatrix} \tag{4.72}$$

From the above, we see that the model predicted Muller matrix is in excellent agreement with the measured one.

4.4.3.3 Comparison between POLARSCAT, NMM3D, SSA2, AIEM, and SPM2

In this section, we compute the depolarized backscattering coefficients by AIEM at L-band 40° from exponentially correlated surface samples at three different roughness scales and various dielectric constants. An attempt is made to compare the measured data by POLARSCAT [22] and predictions by NMM3D [9], SSA2 [16], SPM2 [13], and AIEM. Numeric values are given in Table 4.4. At first glance, the POLARSCAT measured data has oscillatory behavior over the increasing dielectric constant. Such a phenomenon is not uncommon in practical measurements. NMM3D predicts well the dielectric behavior, at higher dielectric, but gives a smaller dynamic range of the depolarized backscattering, compared to SSA2, SPM2, and AIEM. Overall, the AIEM predictions are close to NMM3D results compared to those of SSA2 and SPM2, and give reasonably well match to POLARSCAT data for all surface samples.

4.5 NUMERICAL EXAMPLES FOR SOIL AND OCEAN SURFACES

4.5.1 Scattering Behaviors from Soil Surface

4.5.1.1 Radar Backscattering Behaviors

4.5.1.1.1 Surface Roughness Effects

The polarized backscattering coefficient as a function of azimuthal angle for two anisotropic surfaces in HH, VV, and HV polarization is shown in Figure 4.14, respectively. For each case, incidence angles of 20°, 40°, and 60° are chosen. In these figures, the radar frequency is 5 GHz, and the dielectric constant sets to 15.0 − j1.0. Three surfaces, being weakly and strongly anisotropic, are constructed by fixing the correlation length in y direction at 3.5 cm, and changing the one in x direction from 0.5 to 2.5 cm.

The effect of correlation length on HH polarized and VV polarized backscattering coefficients is illustrated in Figure 4.14. From these observations, the effects of correlation length on HH and VV polarizations in both the angular trend and level are quite similar. It is evident that the scattering curves approximate to circle for weakly anisotropic surface. The curve tends to be elliptical, and a concave appears at up/down direction as the anisotropy increases. When the surface becomes strongly anisotropic, the scattering pattern in polar graphs is shifting from concentric circular to distorted *Butterfly shape*. As the view azimuthal angle increases, the scattering coefficient first decreases at 0° to 90° and then increases at 90° to 180°, and the scattering curve has almost perfect symmetry about 180° in the azimuthal variation over the full 360°. The maxima of the backscattering coefficient locate at up/down directions. The modulation depth is related to the degree of anisotropy. It suggests that the backscattering has a stronger dependence on small to moderate anisotropy at smaller incidence, meaning that the modulation depth varies

TABLE 4.4
Comparison of Depolarized Backscattering Coefficients at L-Band 40° for POLARSCAT, NMM3D, SSA2, AIEM, and SPM2

(a) RMS Height = 0.94 cm, Correlation Length = 6.90 cm

N	ε_r	POLARSCAT(dB)	NMM3D(dB)	SSA2(dB)	SPM2(dB)	AIEM(dB)
1	2.89–j0.15	–38.23	–38.03	–46.87	–49.65	–44.87
2	3.61–j0.24	–37.12	–37.37	–44.15	–47.04	–42.04
3	4.18–j0.31	–36.38	–36.9	–42.65	–45.57	–40.47
4	4.31–j0.32	–33.78	–36.79	–42.36	–45.26	–40.14
5	5.40–j0.44	–35.41	–36.00	–40.43	–43.32	–38.11
6	6.42–j0.55	–36.81	–35.35	–39.16	–42.02	–36.79

(b) RMS Height = 1.78 cm, Correlation Length = 8.30 cm

N	ε_r	POLARSCAT(dB)	NMM3D(dB)	SSA2(dB)	SPM2(dB)	AIEM(dB)
1	2.94–j0.16	–29.69	–28.99	–36.97	–38.62	–34.52
2	3.32–j0.20	–28.04	–28.66	–35.43	–37.12	–32.90
3	3.64–j0.24	–29.64	–28.39	–34.39	–36.11	–31.83
4	3.82–j0.27	–30.35	–28.25	–33.87	–35.64	–31.33
5	4.15–j0.30	–29.86	–27.98	–33.04	–34.79	–30.44
6	6.42–j0.55	–24.43	–24.11	–26.23	–27.75	–23.61
7	11.27–j1.00	–23.22	–23.40	–25.34	–26.80	–22.70
8	14.19–j1.26	–22.99	–23.27	–25.18	–26.63	–22.53

(c) RMS Height = 3.47 cm, Correlation Length = 11.00 cm

N	ε_r	POLARSCAT(dB)	NMM3D(dB)	SSA2(dB)	SPM2(dB)	AIEM(dB)
1	2.94–j0.16	–25.98	–21.04	–28.88	–27.9042	–25.5626
2	3.34–j0.21	–22.48	–20.95	–27.26	–26.3585	–23.9141
3	3.64–j0.24	–24.72	–20.86	–26.29	–25.3835	–22.8917
4	4.15–j0.30	–23.41	–20.7	–24.94	–24.0599	–21.5187
5	4.63–j0.36	–23.36	–20.52	–23.93	–23.0654	–20.5056
6	11.27–j1.00	–20.48	–17.55	–18.16	–17.0389	–14.6118
7	14.19–j1.26	–18.83	–16.65	–17.13	–15.9226	–13.4462

inversely with the incident angle. Such modulation property agrees with observing in the field measurements [26]. The reason behind this modulation depth change is perhaps due to scattering reduction in smaller incident angle, resulting from the wave interactions with the surface spectrum via Bragg resonance.

Now, inspecting the effects of RMS height is in order. The backscattering as a function of incident angle, with viewing azimuthal angles of 0°, 45° and 90°, is shown in Figure 4.15. The co-polarized backscattering, both HH and VV, shows a steep angular trend at azimuthal angle of 0°, and 90°, but relatively flat at an azimuthal angle of 45°. At a smaller incident angle, the azimuthal dependence becomes weaker. It can be seen that a drop-off of scattering appears for 0° and 45° azimuthal angle at smaller incidence. That is when the incident angle is less than 20°, the scattering increases with the increase of azimuthal angle. However, the scattering decreases as the azimuthal

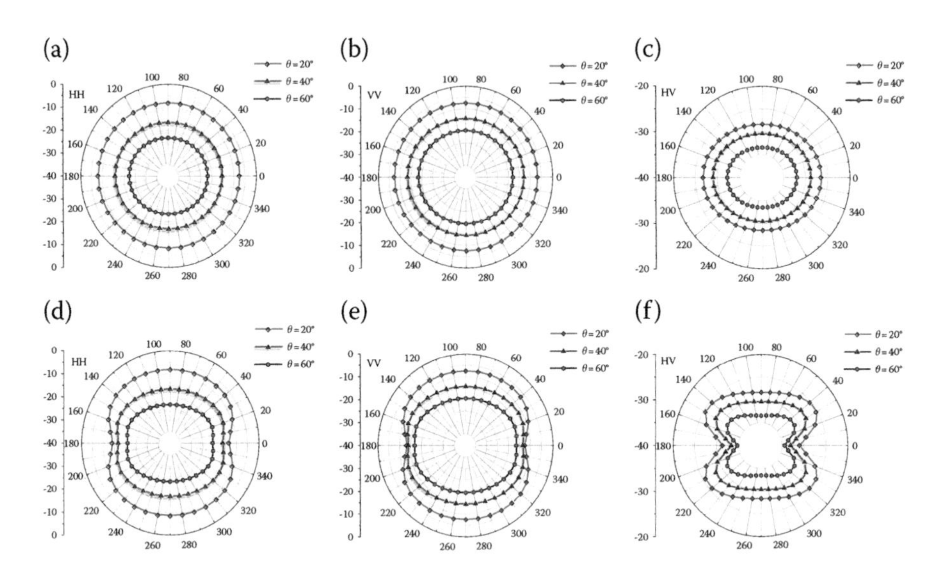

FIGURE 4.14 Effects of correlation length for anisotropic surface with $\varepsilon_r = 15.0 - j1.0$, $\sigma_z = 0.3$ cm, (a, b, c) weak anisotropy $L_x = 2.5$ cm, $L_y = 3.5$ cm, (d, e, f): strong anisotropy $L_x = 0.5$ cm, $L_y = 3.5$ cm. (a, d) HH polarization, (b, e) VV polarization, (c, f) HV polarization.

angle increases when the incident angle is greater than 20°. The azimuthal variation of cross-polarized backscattering tends to level off as compared to co-polarizations. A closer look reveals that the cross-polarized scattering at an azimuthal angle of 90° is in between those at azimuthal angles of 0° and 45°, regardless of the incident angle. For the rougher surface, as shown in Figure 4.15(d–f), the above observations remain the same azimuthal tendencies. The scattering coefficient monotonically decreases with increasing incident angle. Overall, the impact of RMS height between the co- and cross-polarizations are different. The backscattering coefficients of co- and cross-polarizations increase with the increasing RMS height, as expected. Besides, their angular trends are similar except the crossings move to the larger incident angle for co-polarization.

4.5.1.1.2 Dielectric Constant Effects

The co- and cross-polarized scattering behaviors as a function of viewing azimuthal angle with two distinct dielectric constants are plotted for incident angles of 20°, 40°, and 60°. As shown in Figure 4.16, the azimuthal variations are more oscillating in virtue of the stronger anisotropy. Figure 4.16(a–c) gives the results of small dielectric constant, and Figure 4.16(d–f) shows the results of large dielectric constant, for HH, VV, and HV polarizations. In comparison, as the dielectric constant increases, the scattering coefficient is enhanced, but the azimuthal dependence remains similar. It also should be noted that, for HV polarization, the difference of scattering magnitudes among incident angles of 20°, 40°, and 60° gradually decreases with increasing dielectric constant, as physically expected. To further examine the effects of dielectric

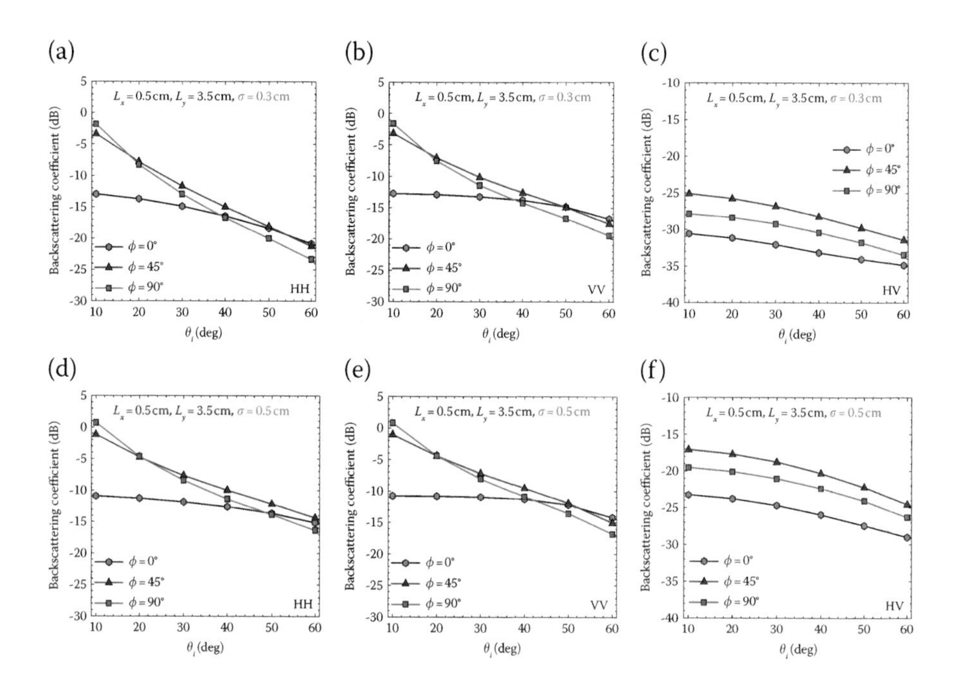

FIGURE 4.15 The co- and cross-polarized scattering behaviors as a function of incident angle from strongly anisotropic surface with different *rms* heights $L_x = 0.5$ cm,$L_y = 3.5$ cm, $\varepsilon_r = 15.0 - j1.0$: (a, b, c) $\sigma = 0.3$ cm; (d, e, f) $\sigma = 0.5$ cm. (a, d) HH polarization, (b, e) VV polarization, (c, f) HV polarization.

constant for anisotropic surface, the azimuthal patterns of HH/VV ratio are illustrated in Figure 4.17, where the solid symbols denote the result of small dielectric constant, and the hollow symbols the result of large dielectric constant. Overall the HH/VV ratio is less than 1 and decreases with the increasing incident angle. Also, the HH/VV ratio for small dielectric constant is greater than that for the large dielectric constant. That is, as the dielectric constant increases, the scattering increases both in HH and VV polarization, but the increased amplitude is more so in VV. These findings bring nothing new but confirm the physical facts. In comparison, the backscattering shows stronger azimuthal dependence on anisotropy at a lower dielectric constant. As the enhancement of anisotropy, the dips gradually occur at the up/down directions, seen in Figure 4.17(c).

4.5.1.2 Radar Bistatic Scattering Behaviors

In this section, fully polarimetric bistatic scattering both for homogeneous and inhomogeneous rough surface are analyzed. Here, the dielectric profiles of this inhomogeneous rough surface are given in Chapter 3 (see Figure 3.6). Note that all the last terms on the right-hand side of Equations 4.71a–4.71d are relatively small, such that the LL polarized scattering coefficient is very close to the RR polarized. Similarly, LR and RL polarizations are close to each other. Hence, only RR for like-polarized scattering and LR for cross-polarized scattering are discussed in this study. As shown in Figure 4.18, the scattering coefficients as a function of azimuthal angle

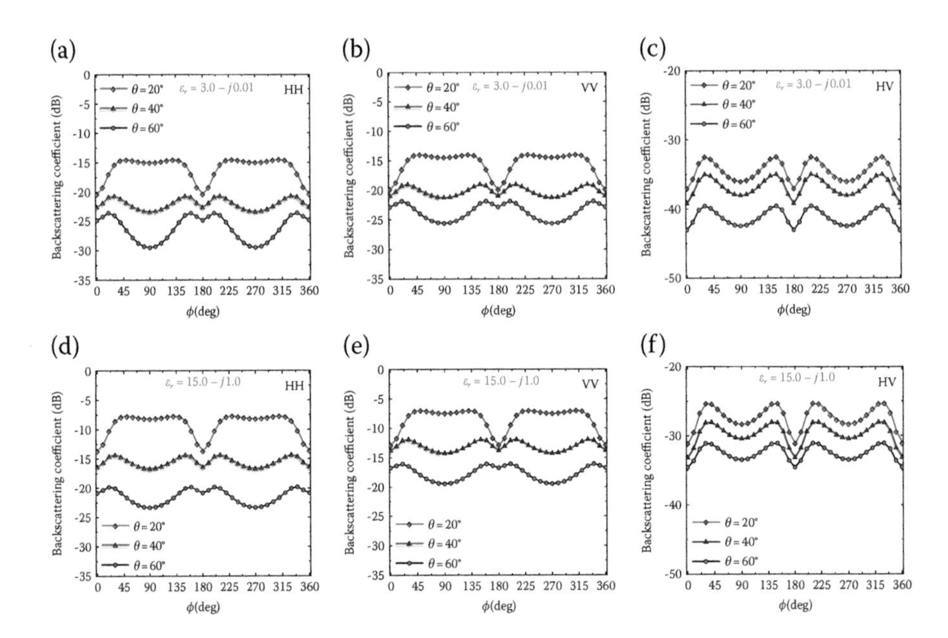

FIGURE 4.16 The co- and cross-polarized scattering behaviors as a function of view azimuthal angle from strongly anisotropic surface with different dielectric constants $L_x = 0.5$ cm, $L_y = 3.5$ cm, $\sigma = 0.3$ cm; (a, b, c): $\varepsilon_r = 3.0 - j0.01$; (d, e, f): $\varepsilon_r = 15.0 - j1.0$. (a, d) HH polarization, (b, e) VV polarization, (c, f) HV polarization.

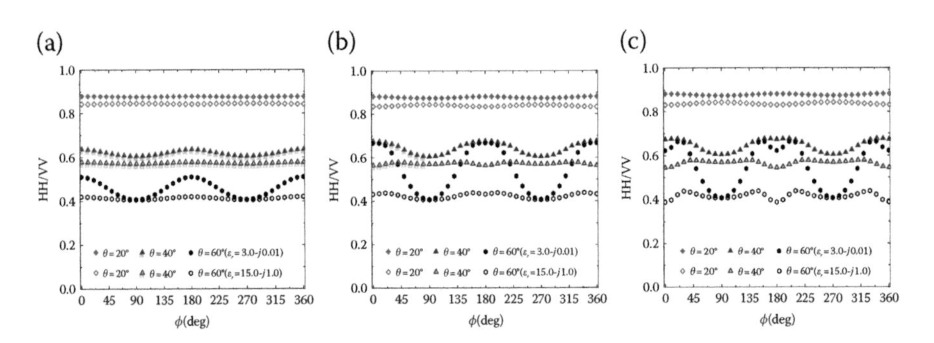

FIGURE 4.17 HH/VV ratio for small and large dielectric constants: $\varepsilon_r = 3.0 - j0.01$, $\varepsilon_r = 15.0 - j1.0$. (a) weakly anisotropic surface $L_x = 2.5$ cm, $L_y = 3.5$ cm, $\sigma = 0.3$ cm, (b) moderately anisotropic surface $L_x = 1.5$ cm, $L_y = 3.5$ cm, $\sigma = 0.3$ cm, (c) strongly anisotropic surface $L_x = 0.5$ cm, $L_y = 3.5$ cm, $\sigma = 0.3$ cm.

are presented with surface parameters $kl = 5$, $k\sigma = 0.5$, $\varepsilon_r = 9 - j0.5$. The incident angle is 40°, and the scattering angle is fixed at 20°. The simulation results for homogeneous surface are shown with the solid lines, and the results for inhomogeneous surfaces are designated by symbolled solid lines. The presence of surface inhomogeneity generally leads to several interesting features that do not appear in the homogeneous surface. In general, the scattering coefficients for both linear and circular polarizations are enhanced. The varying trend of scattering curves between linear and circular polarizations are quite different. For linear polarization, a sharp dip appears at both HH and VV polarizations for homogeneous surfaces. However, for inhomogeneous surface, the dynamic range of HH and VV polarized scattering coefficients over the azimuth angles are profoundly reduced. Furthermore, the smaller valley for VV polarization gradually moves to the larger azimuthal angle, and that for HH polarization shifts to smaller azimuthal angle [see Figure 4.18(a)]. Note that the scattering coefficient for VV polarization is greater than HH in the forward region for inhomogeneous surface, but HH polarization is relatively large for a homogenous case. In Figure 4.18(b), the HV and VH polarized scattering coefficients increase significantly, especially in the backward region. It is worth noting that the HV polarization is greater than VH, and the relationship between them is reversed for a homogenous surface. In Figure 4.18(c), for an inhomogeneous rough surface, the scattering coefficients both for LR and RR polarizations are enhanced, the increase of RR polarization is greater than the LR. Hence the difference between LR and RR polarizations decreases.

In Figure 4.19, the bistatic scattering coefficients as a function of scattering angle and azimuthal angle for homogeneous and inhomogeneous rough surfaces are presented. The scattering patterns are exhibited in Figure 4.19(a) for a homogeneous case and in Figure 4.19(b) for an inhomogeneous case, respectively. Figure 4.19(c) displays the difference, in dB, of homogeneous and inhomogeneous cases. From these scattering patterns, there exists a richer feature, in the sense of angular dependence, in the forward region than that in the backward region, especially a stronger scattering strength appears near the specular direction $\theta_i = 40°$. By contrast, the HH and VV polarized scattering coefficients show strong azimuth dependence. The stronger scattering coefficient appears at the incident plane ($\phi_s = 0°$) and weaker scattering coefficient locates at the cross-plane ($\phi_s = 90°$). In the backward region, the scattering coefficient is relatively larger for a smaller incident angle ($\theta_s < 40°$). Both HV and VH polarized scattering coefficients are stronger in the cross-plane, and the backscattering coefficient

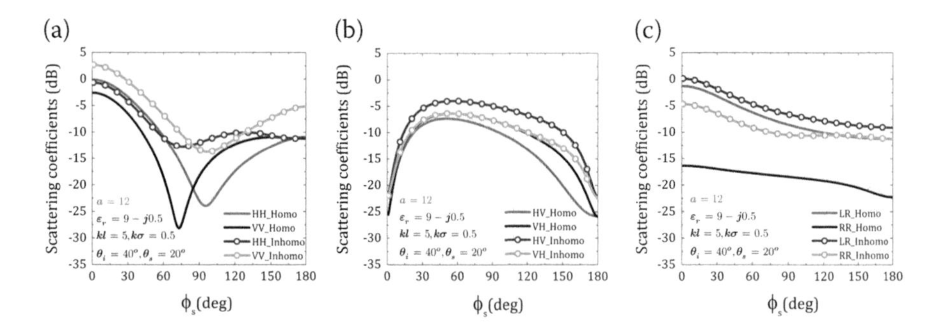

FIGURE 4.18 Full bistatic scattering difference between homogeneous and inhomogeneous rough surfaces with exponential correlated function for $kl = 5$, $k\sigma = 0.5$, $\varepsilon_r = 9 - j0.5$, $\theta_i = 40°$, $\theta_s = 20°$, $a = 12$. (a) HH, VV polarization, (b) HV, VH polarization. (c) LR, RR polarization.

($\theta_s = 40°$, $\phi_s = 180°$) is relatively weaker. For circular polarization, the scattering features both for LR and RR polarizations exhibit the weak azimuth dependence. The scattering coefficient is more uniform except for the specular direction. According to Figure 4.19(c), the difference of scattering coefficients between homogenous and inhomogeneous cases are illustrated. In virtue of inhomogeneity, the HH polarized scattering coefficient increases 5~12 dB in the backward region, especially in the cross plane. By comparison, the notable increase of VV polarized scattering coefficient is about 15 dB near the cross-plane of the forward region. The increase of scattering coefficients both for HV and VH polarizations appears in the backward region. However, the difference is that the major increase of HV polarization occurs at the smaller scattering angle (e.g., $\theta_s < 40°$), and yet the increase of VH polarization is mainly concentrated on the larger scattering angle region ($\theta_s > 40°$). For circular polarization, the scattering increasing in RR polarization is greater than that in LR, and the increase of RR polarized scattering coefficient mostly occurs at the smaller scattering angle (e.g., 0°~20°). The LR polarization increases less than 5 dB in this case.

4.5.1.2.1 Transition Rate Effects

For inhomogeneous rough surfaces, the bistatic scattering is also a function of transition rate, which controls the change rate of the dielectric constant with depth. To demonstrate this effect, we set the transition rates to 6 and 12 and plotted the bistatic scattering curves in Figure 4.20. The surface parameters are $kl = 5$, $k\sigma = 0.5$, $\varepsilon_r = 9 - j0.5$, with the incident angle of 40° and 20° of scattering angle. As a reference, the scattering curves for homogeneous rough surfaces are also given here. For a relatively small transition rate (e.g., $a = 6$), the HH polarized scattering coefficient is closer to VV, and HV becomes close to the VH. As the transition rate increases, the variation of the scattering curves for both linear and circular polarizations can be summarized as the following points. The scattering coefficients are enhanced both for linear and circular polarizations. For linear polarization, the difference of VV and HH polarizations becomes larger due to the fact VV increases faster than HH; the valley of VV moves to larger azimuth angle, and yet that of HH trends to smaller azimuth angle; the HV polarization increases faster than VH polarization does, such that the difference between HV and VH polarizations is gradually enhanced; the scattering coefficient increases more significantly in the backward region for both HV and VH polarizations. For circular polarization, the scattering coefficients are strengthened for both LR and RR polarizations, and the difference between LR and RR polarizations decreases.

Figure 4.21 presents the hemispherical plots of the full polarization bistatic scattering on the whole scattering plane, with the color scale indicating the scattering coefficients in dB. The effect of transition rate is investigated with surface parameters being selected as $kl = 5$, $k\sigma = 0.5$, $\varepsilon_r = 9 - j0.5$. Figure 4.21(a) demonstrates the results of a smaller transition rate (e.g., $a = 6$), and Figure 4.21(b) shows the results with the transition rate setting as $a = 12$. The incident angle is fixed at 40°. With the increasing transition increases, the scattering coefficients for both linear and circular polarizations are enhanced on the whole scattering plane. In the forward region, the scattering coefficient of VV polarization increases faster than that of HH.

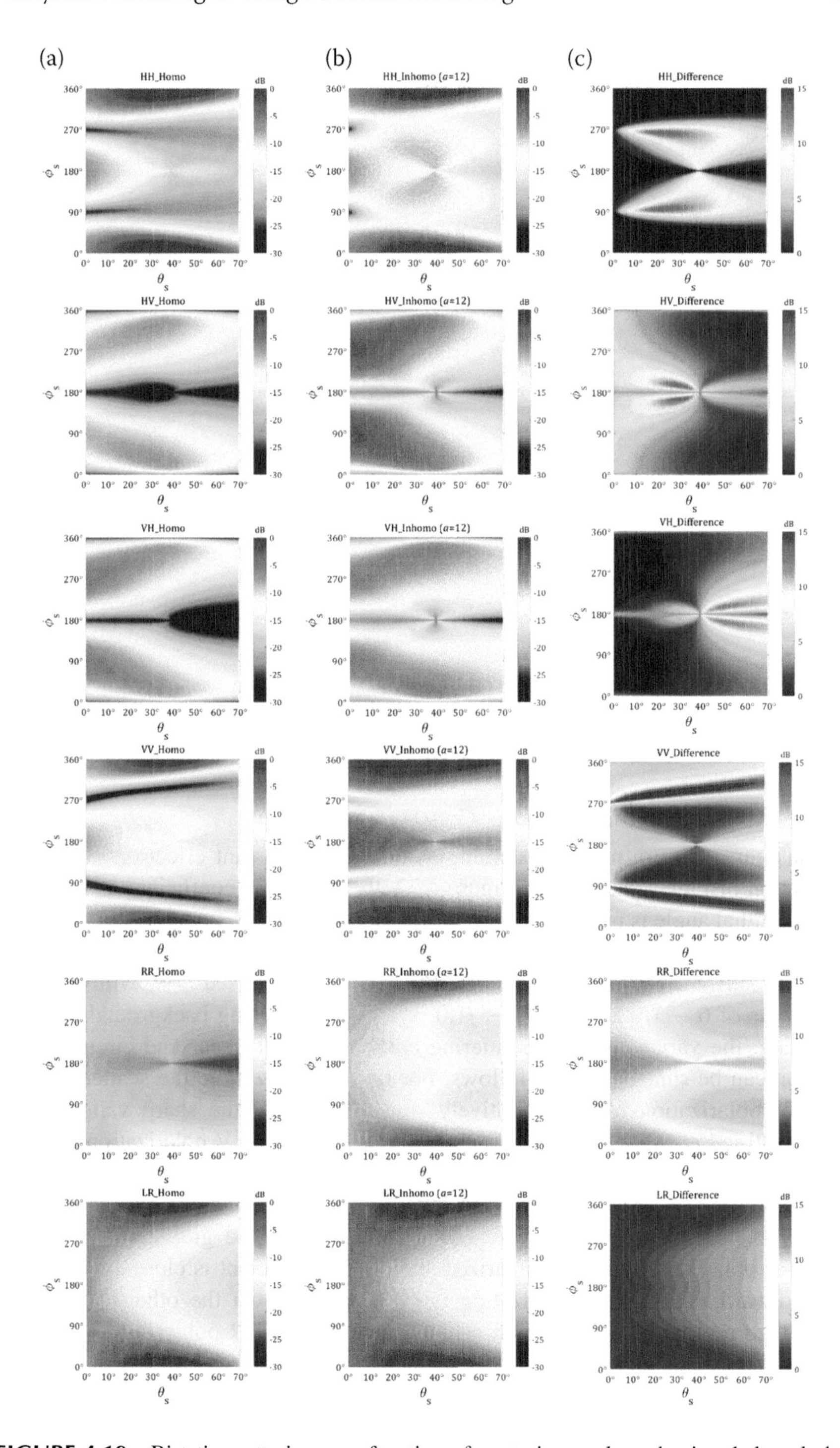

FIGURE 4.19 Bistatic scattering as a function of scattering angle and azimuthal angle between homogeneous and inhomogeneous rough surfaces with exponential correlated function for $kl = 5$, $k\sigma = 0.5$, $\varepsilon_r = 9 - j0.5$, $a = 12$, $\theta_i = 40°$: (a) homogenous case, (b) inhomogeneous case, (c) Differences between dB values of homogenous and inhomogeneous cases.

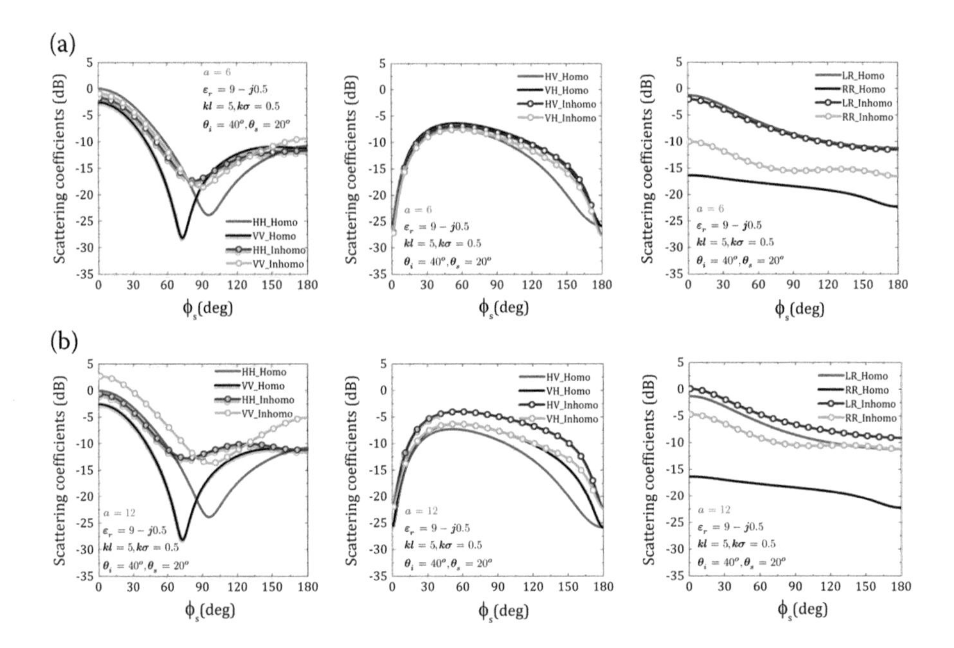

FIGURE 4.20 Full bistatic scattering difference between homogeneous and inhomogeneous rough surfaces with exponential correlated function for $kl = 5,\ \ k\sigma = 0.5$, $\varepsilon_r = 9\ -\ j0.5$, $\theta_i = 40°$, $\theta_s = 20°$. (a) $a = 6$, (a): $a = 12$.

4.5.1.2.2 Background Dielectric Constant Effects

In this part, we turn now to examine the dielectric constant effect on the full polarization bistatic scattering. In Figure 4.22, the scattering coefficient as a function of azimuthal angle is presented both for homogeneous and inhomogeneous surfaces. The incident and scattering angle are selected as 40° and 20°, respectively. The relative surface parameters are set as $kl = 5,\ \ k\sigma = 0.5,\ \ a = 8$, with dielectric constants of $6\ -\ j0.25$ and $15\ -\ j1.0$. With the increasing background dielectric constant, the variation of the scattering curves for both linear and circular polarization can be summarized as follows. For linear polarization, the scattering curve of VV polarization remains relatively smooth, and yet the sharp valley of HH polarization becomes more apparent. The small valley of VV polarization moves to the forward region, and that of HH polarization trends to the backward region. At the same time, the scattering curves of HV and VH polarizations over the azimuth direction become more symmetrical, and VH tends to be greater than HV. For circular polarization, the LR polarized scattering coefficient is close to the RR. In the forward region, LR polarization is less than RR. On the other hand, in the backward region, LR polarization remains greater than RR polarization.

As shown in Figure 4.23, the hemispherical plots show the effect of dielectric constant on the whole scattering plane. For the smaller dielectric constant (e.g., 6 – j0.25), the strong scattering focuses on the forward region both for linear and circular polarizations. As the background dielectric constant increases, some interesting features appearing in scattering patterns are depicted as follows. For linear

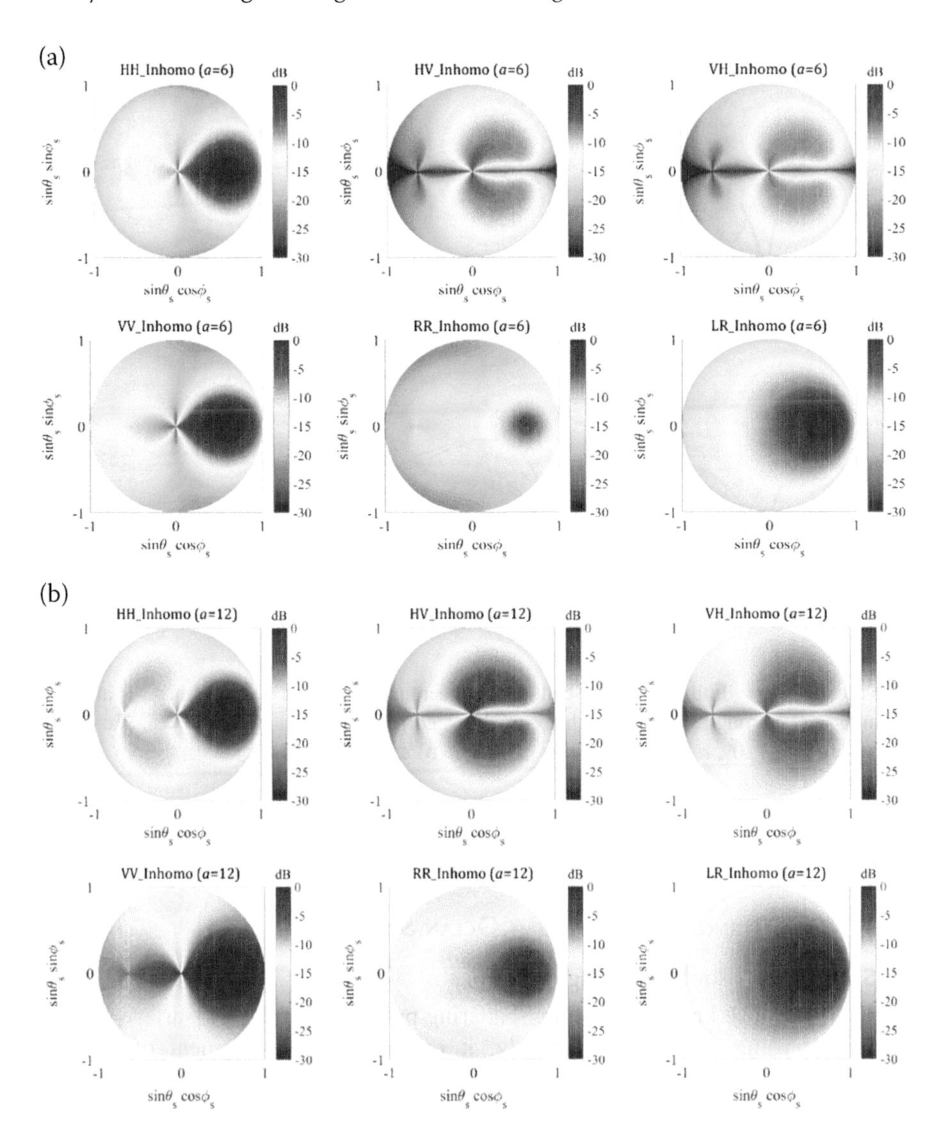

FIGURE 4.21 Comparison in bistatic scattering hemispherical plots for inhomogeneous rough surface with two transition rates, exponential correlated function $kl = 5$, $k\sigma = 0.5$, $\varepsilon_r = 9 - j0.5$, $\theta_i = 40°$: (a) $a = 6$, (b): $a = 12$.

polarization, the HH polarized scattering pattern exhibits a stronger angular dependence. The VV polarized scattering coefficient is reduced in the forward region but is enhanced notably in the backward region. The scattering coefficients for HV and VH polarizations are larger clustering on the cross-plane with a small scattering angle. For cross polarizations, HV polarization decreases, and, reversely, VH polarization increases. For circular polarization, the scattering coefficients both for LR

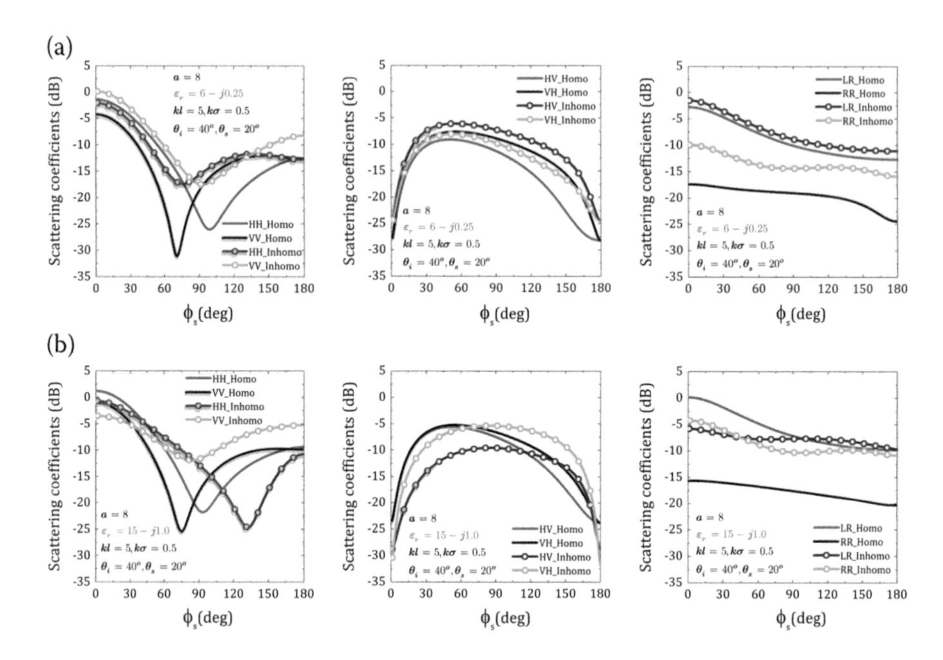

FIGURE 4.22 Full bistatic scattering difference between homogeneous and inhomogeneous rough surfaces with exponential correlated function for $kl = 5$, $k\sigma = 0.5$, $\theta_i = 40°$, $\theta_s = 20°$, $a = 8$. (a) $\varepsilon_r = 6 - j0.25$, (b) $\varepsilon_r = 15 - j1.0$.

and RR polarizations are enhanced in backward region. However, in the forward region, LR polarized scattering coefficient decreases, and yet RR increases.

4.5.2 Scattering Behaviors from Ocean Surface

4.5.2.1 Radar Backscattering Behaviors

To acquire more knowledge of the scattering process from the sea surface and use remote sensing technique effectively, it requires the development of accurate models including surface scattering models and wind wave spectrum models to predict the radar scattering cross-section (RCS) from the ocean surface [29]. Wind wave spectrum describes the quasi-periodic nature of the ocean surface oscillations and plays an indispensable role in the study of microwave electromagnetic scattering from the sea surface. Studies show that the effective roughness is selective from a portion of sea spectrum components $[k_L,\infty)$, with the lower limit k_L proportional to the effective wavenumber K determined by radar frequency and the incident angle ($K = 2k \sin \theta_i$) [30].

We now evaluate the performances of five common spectrum models (i.e., Fung spectrum, Durden–Vesecky spectrum, Apel spectrum, Elfouhaily spectrum, and the newest version of Hwang spectrum, H18) on the normalized radar backscattering cross-section (NRBCS) simulations based on advanced integral equation model (AIEM) at L-, C-, X-, and Ku-bands versus incidence angle, wind direction, and wind speed by comparing with the model and measured data for validation.

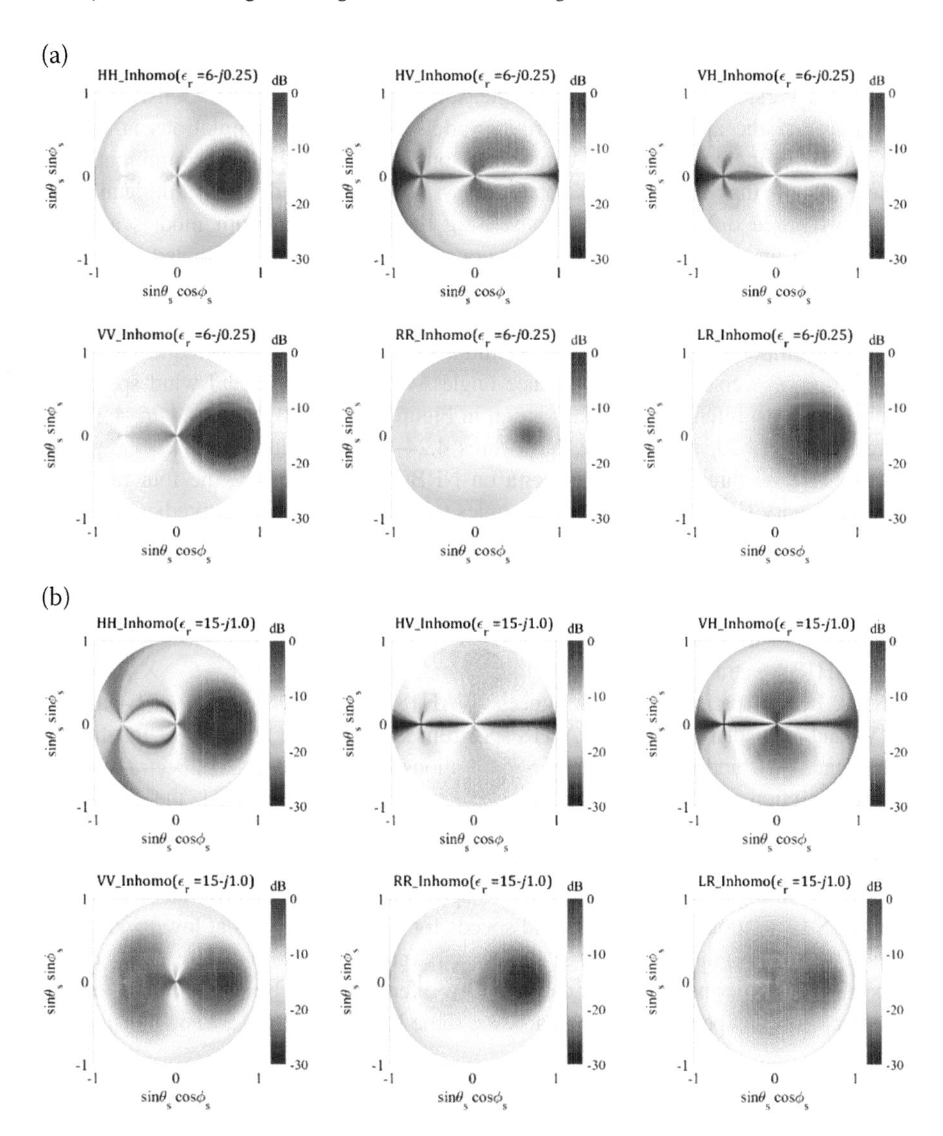

FIGURE 4.23 Comparison in bistatic scattering hemispherical plots for inhomogeneous rough surface with two dielectric constants, exponential correlated function $kl = 5$, $k\sigma = 0.5$, $a = 8$, $\theta_i = 40°$: (a) $\varepsilon_r = 6 - j0.25$, (b): $\varepsilon_r = 15 - j1.0$.

In [30], through comprehensive verification and evaluation of the performance of the five wind wave spectra, no one of them can obtain the satisfactory simulation results for all the frequencies, incidence angles, wind directions, and wind speeds. The better choice of the wave spectrum models for NRBCS simulations at different frequencies with the cases of incidence angle, wind direction, and wind speed are listed in Table 4.5. In Table 4.5, we can see that the A- and H18-spectra have the a wider applicability than the other spectrum models for different cases. This is maybe because radar observations participate in the modeling of the two wave spectra. In addition, the original DV spectrum seems performs worst, but the modified version by matching radar measurements was obtained. Considering the diversity of the appropriate spectrum models for NRBCS simulations at various frequencies, incidence angles, wind directions, and wind speeds, a combination of these spectra may have a wider applicability, which will be discussed in detail next.

Now, three different combinations using these wave spectrum models are constructed in an average way, i.e., the average of the five spectra (labeled "All"), the average of A-, E-, and H18-spectra (labeled "AEH18"), and the average of A- and H18-spectrum (labeled "AH18"). Similarly, we also evaluate the performances of the three average composite spectra on NRBCS simulations for different radar frequencies. Their performances versus various incidence angles, wind directions, and wind speeds are compared with reference data and shown in Figures 4.24–4.25, Figures 4.26–4.27, and Figures 4.28–4.29, respectively. In Figures 4.24–4.25, the differences of the performances of the three composite spectra on NRBCS simulations at the four radar frequencies vary slightly with incidence angles (about 2–3 dB), but the predictions of the AH18-spectrum are overall closer to the reference data especially at small incidence angles. For X-band, the reference data from XMOD2 at large incidence angles (e.g., greater than 45°) are not applicable for validation. In Figures 4.26–4.27, the maximum difference of the NRBCS predictions versus wind directions based on All-, AEH18-, and AH18-spectra are within 2 dB for L-band, and about 1 dB for C-, X-, and Ku-bands. By comparing the performances of AH18- and AEH18-spectra at L-band for VV- and HH-polarizations, the AH18 spectrum is recognized to be better for the four frequencies. In addition, the NRBCS simulations based on the three spectra in cross-wind direction at C- and Ku-bands are higher than reference data, the differences between the NRBCSs versus wind directions are determined by ASF. Overall, the AH18-spectrum is competent to simulate the NRBCSs versus wind directions. Similarly, in Figures 4.28–4.29, the AH18-spectrum also performs overall better on the NRBCS simulations at different wind speeds for L- and Ku-bands at low wind speeds (e.g., less than 20 m/s) but for high wind speeds (greater than 30 m/s) the results of All-spectrum at Ku-band are much closer to the reference data. Besides, for C-band, the three composite spectra produce almost the same level in NRBCS at low and moderate

TABLE 4.5
The Appropriate Wind Wave Spectrum Models for NRBCS Simulations at Different Frequencies with the Cases of Incidence Angles, Wind Directions, and Wind Speeds

Frequency Cases	L-Band	C-Band	X-Band	Ku-Band
Incidence angles	A and H18	A and H18	A	A, E, and H18
Wind directions	F, A, and H18	F, A, and H18	A	E and H18
Wind speeds	A and H18	F, A, and H18	F and A	A, E, and H18

F (Fung spectrum, 1982); **A** (Apel spectrum, 1994); **E** (Elfouhaily spectrum, 1997); **H18** (Hwang spectrum, 2018).

wind speeds (less than 30 m/s). There is a clear difference between All-spectrum and the other two combined spectra as the wind speed further increases to greater than 30 m/s. Their performances at wind speeds less than 10 m/s are in agreement with reference data, but deviate to a certain extent to 3 dB for VV-polarization and about 2 dB for HH-polarization at wind speed of 20 m/s and then close to the reference data again at wind speed greater than 20 m/s. At X-band, the VV-polarized performance predicted by the All-spectrum is consistent with XMOD2 but with an underestimate of 2 dB compared to the other two composite spectra; while performances of all the three spectra at HH-polarization are all underestimated by about 2 dB. As the wind speed rises (greater than 30 m/s), the extrapolated prediction by XMOD2 is consistent with the predictions of AEH18- and A-spectra.

Overall, the combination of wind-wave spectrum may be a feasible way to unify the various sea spectrum models for wider applications (e.g., in the cases of different

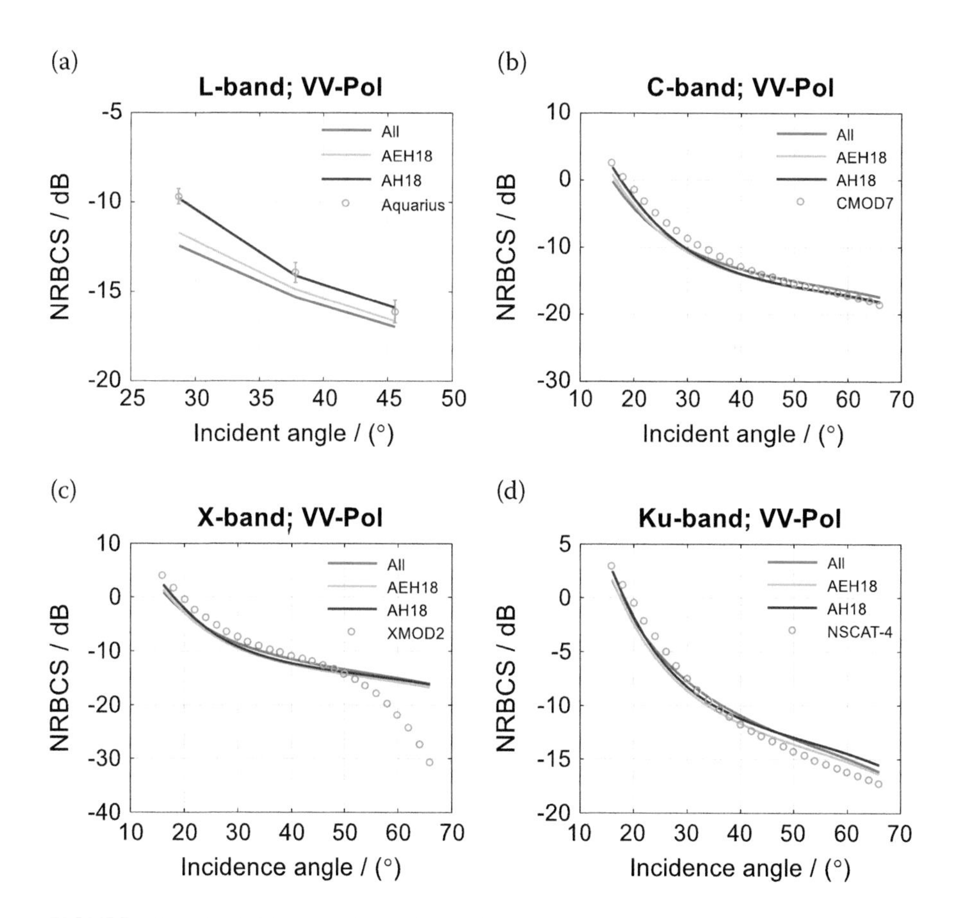

FIGURE 4.24 NRBCS simulations with the three composite wind wave spectra for upwind direction and wind speed of 10 m/s and comparison with validation data at L-, C-, X-, and Ku-bands versus incidence angles for VV polarization. (a) L-band, (b) C- band, (c) X-band, (d) Ku-band.

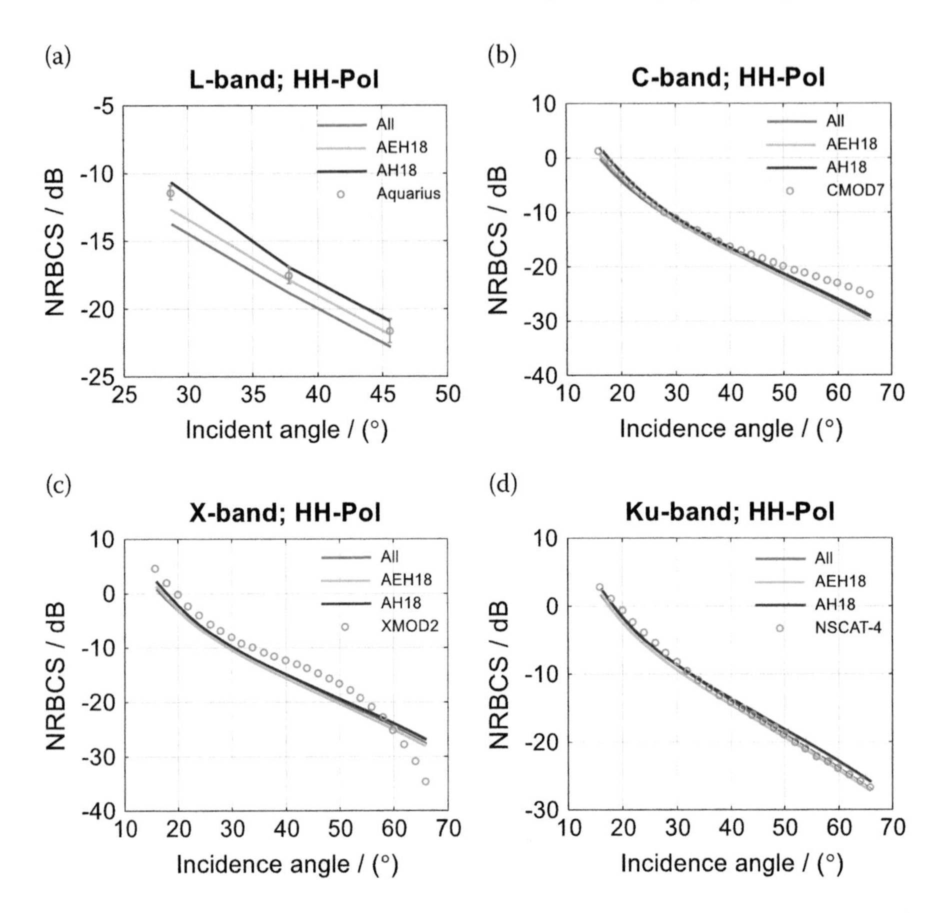

FIGURE 4.25 NRBCS simulations with the three composite wind wave spectra for upwind direction and wind speed of 10 m/s and comparison with validation data at L-, C-, X-, and Ku-bands versus incidence angles for HH polarization. (a) L- band, (b) C-band, (c) X- band, (b) Ku-band.

frequencies, incidence angles, wind directions, and wind speeds) in NRBCS simulations from sea surface. Although the performances from different combinations of sea spectra differ, the differences between them are usually about 2 dB, which is acceptable taking the complexity of the ocean into consideration. In general, the average combination of A- and H18-spectra in this study can be recognized to be the most of the three combinations suitable for the NRBCS simulations at different incidence angles, wind directions, and wind speeds for L-, C-, and X-bands. For high-frequency Ku-band, the AH18 is overall best for wind speed less than 30 m/s, otherwise, the All-spectrum performs satisfactorily.

The upwind-downwind asymmetry in normalized radar backscattering cross-section (NRBCS) from ocean surface is well-known; one acceptable and convincing reasoning is explained by the fact that the surface height distribution deviates from Gaussian one, which causes a non-zero skewness function, and consequently affects the radar

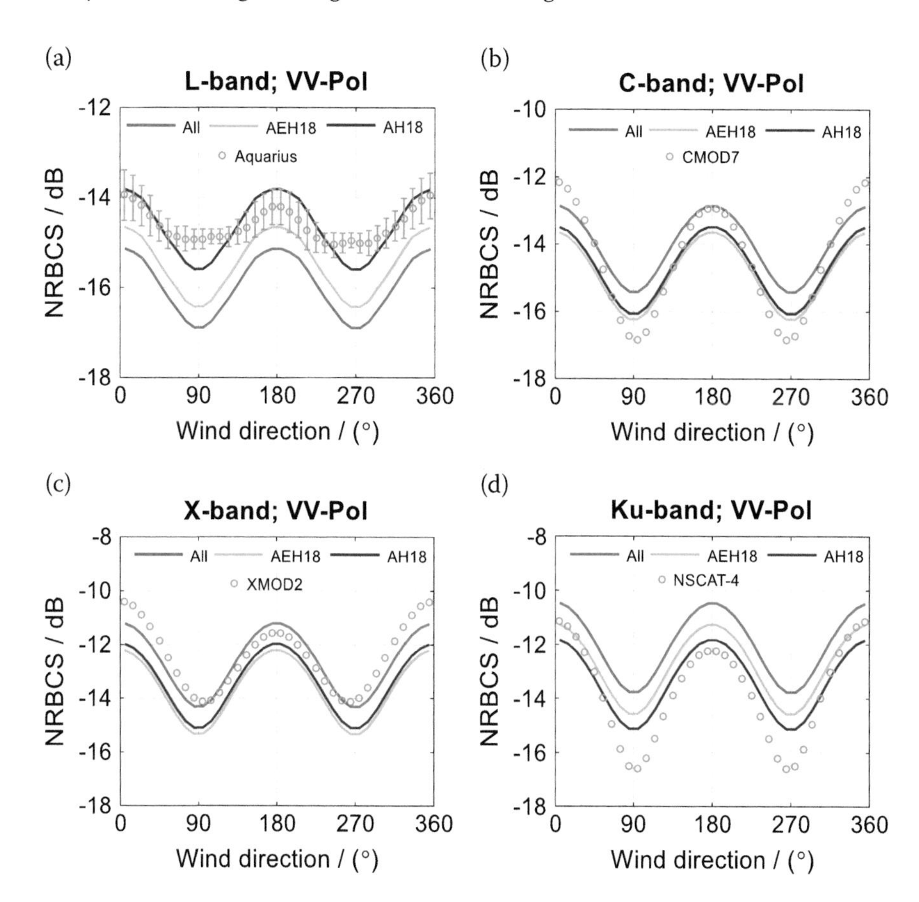

FIGURE 4.26 NRBCS simulations at L-, C-, X-, and Ku-bands and incidence angle of 38° with the three composite wind wave spectra for wind speed of 10 m/s and comparison with validation data versus wind directions for VV polarization. (a) L-band, (b) C-band, (c) X-band, (d) Ku-band.

cross-section in up- and downwind directions. In this regard, we examine the impact of Gaussian and exponential skewness functions on NRBCS. The simulated NBRCSs, with and without skewness contributions, are compared with measured data in upwind and downwind directions at L-, C-, and Ku-bands for different wind speeds.

As illustrated in Figure 4.30 at the incident angle of θ_t, the surface slope viewed along upwind direction is larger than that along downwind direction, giving rise to a higher NRBCS in upwind direction. Due to the scarcity of measured data for skewness function or bispectrum, it is difficult to model them with mathematical expressions like sea spectrum models. Until now, only some basic properties of skewness function were summarized and various expressions were proposed in [32,33]. The Gaussian- and exponential-like skewness functions are usually employed to explain the asymmetry of radar echoes from sea surface at upwind and downwind directions. Even if these functions belong to the same type (e.g.,

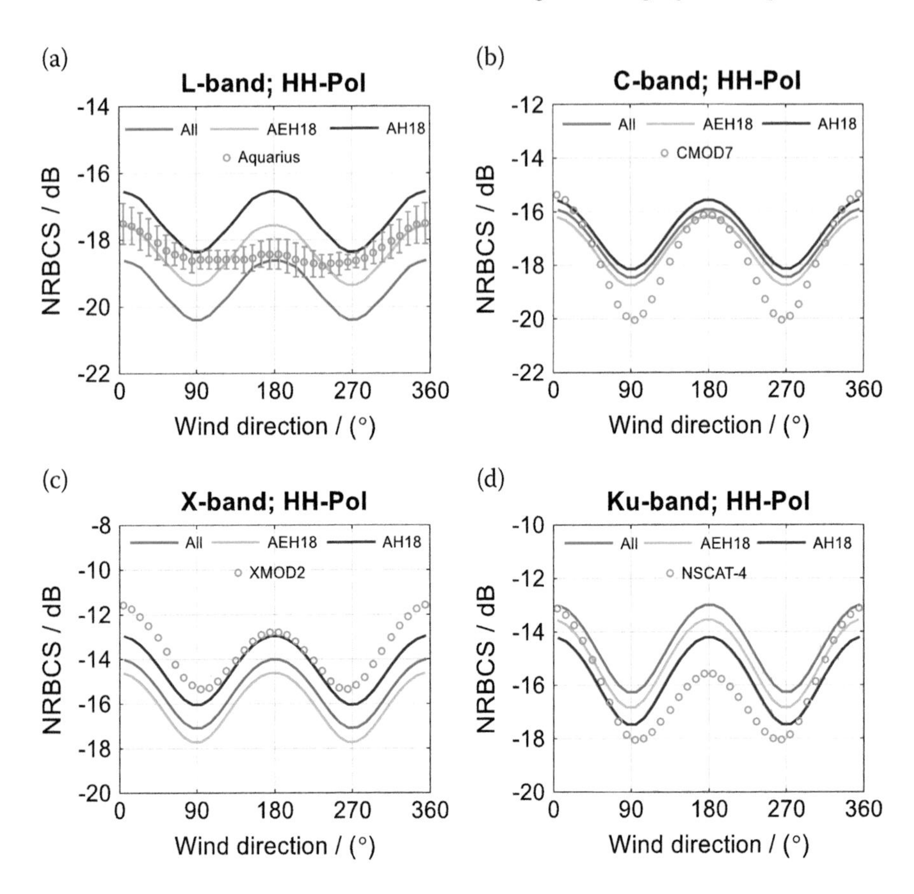

FIGURE 4.27 NRBCS simulations at L-, C-, X-, and Ku-bands and incidence angle of 38° with the three composite wind wave spectra for wind speed of 10 m/s and comparison with validation data versus wind directions for HH polarization (a) L- band, (b) C- band, (c) X- band, (d) Ku-band.

Gaussian type), their coefficients are also various [33,34]. However, the high order statistical moments contained in the slope distribution measured by Cox and Munk [35] are related to the skewness function. Therefore, the unknown coefficient can be derived from the slope distribution, which makes the derivation process of skewness function more well-founded. Based on the previous studies, we mainly aims at investigating the effect of the size of small scale waves on NRBCS contributed by skewness effect (skewness function or bispectrum) as well as the effect of Gaussian- and exponential-type skewness functions on NRBCS at different radar frequency, incidence angles, wind direction and wind speed.

The L-band sea surface backscattering measurements used are obtained from the scatterometry onboard the Aquarius satellite. This scatterometry operated at 1.26 GHz with incidence angles of 28.7°, 37.8°, and 45.6°. The comparisons of the simulated NRBCSs with and without the skewness contribution versus different wind directions with Aquarius data are shown in Figure 4.31. The solid lines

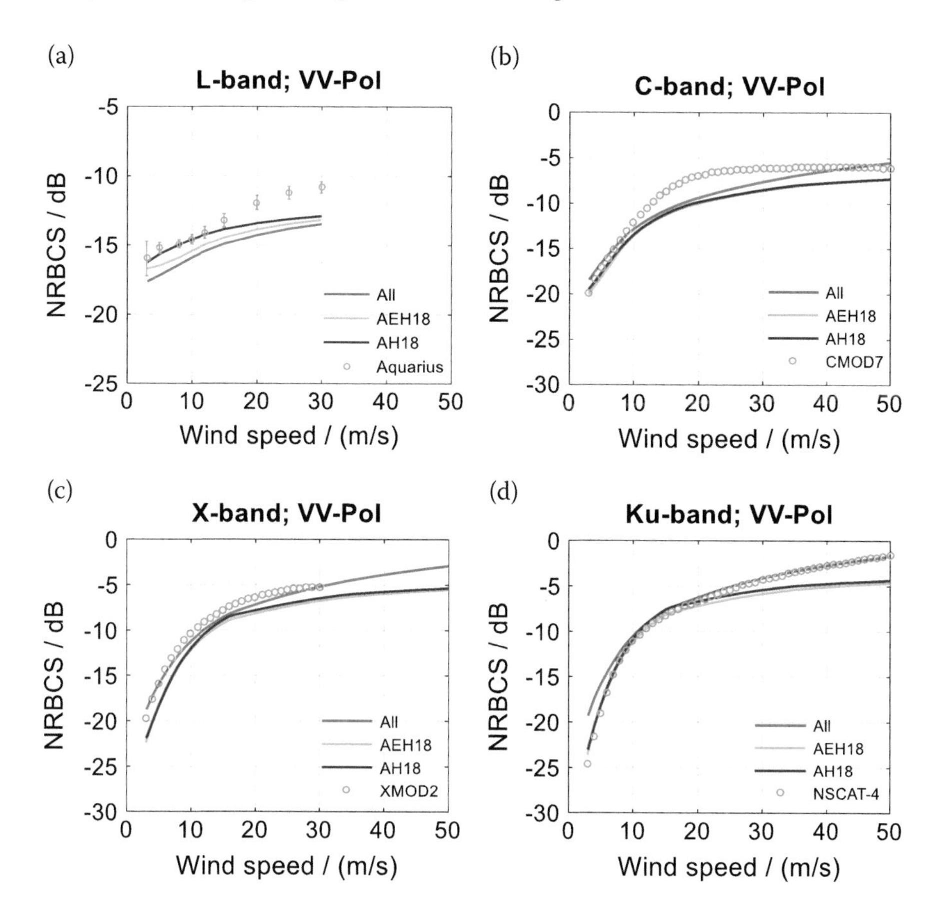

FIGURE 4.28 Simulated NRBCS at L-, C-, X-, and Ku-bands and incidence angle of 38° with the three composite wind wave spectra at upwind direction (except the azimuthally integrated NRBCS at L-band) and comparison with validation data versus wind speeds for the VV polarization. (a) L-band, (b) C-band, (c) X-band, (d) Ku-band.

(Simu-vv^b, hh^b) represent the simulations with the effect of Gaussian-type bispectrum based on Apel spectrum and without for the dotted line (Simu-vv, hh). It can be seen that the Gaussian-type bispectrum well represents the up/downwind asymmetry of radar echoes from non-Gaussian sea surface. The simulated NRBCSs with the skewness contribution agree well with the radar observations with a root mean square error (RMSE) about 1 dB. Besides, the skewness contribution also makes the minimum of NRBCS deviate from crosswind direction (i.e., $\phi = 90°$) and move to downwind direction (i.e., $\phi = 180°$) as well as the larger skewness contribution makes the minimum of NRBCS deviate from downwind direction more apparently, which is an effective feature for inferring wind directions from radar observations. For the determination of the c-value, the reader can refer to [36].

The validation data sources are obtained by mean of C-band (5.3 GHz) geophysical model functions (GMFs) CMOD2-I3 [37] and CMOD5 [38] $\theta_i = \{18°, 20°, \ldots, 58°\}$;

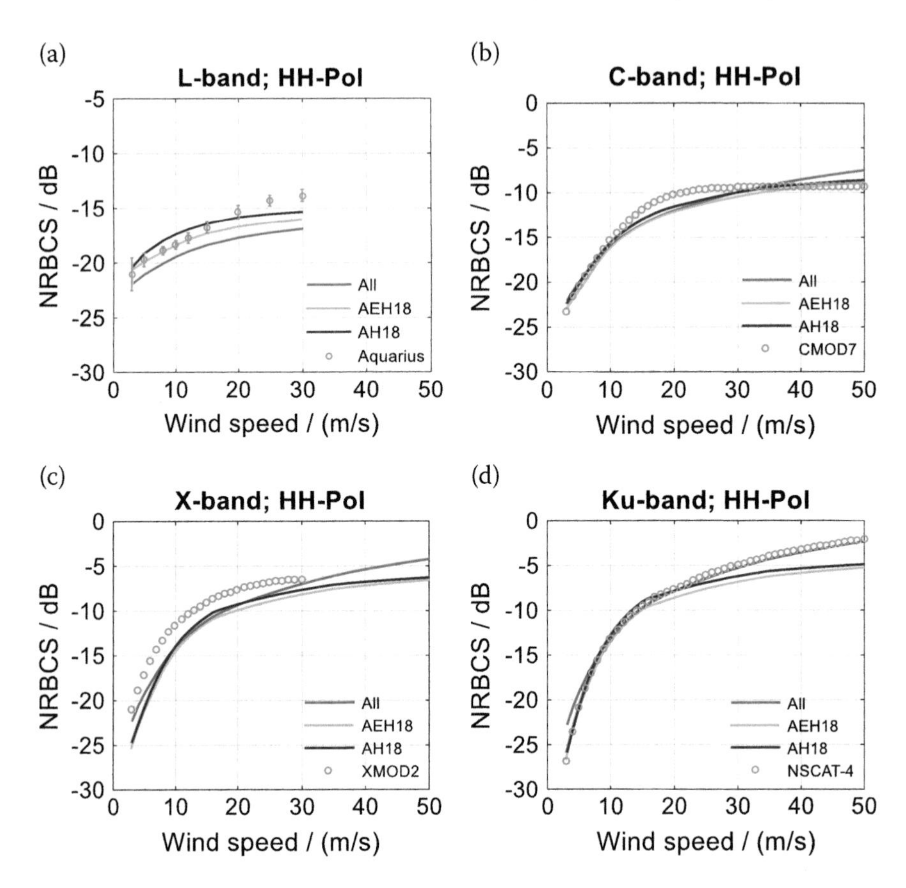

FIGURE 4.29 Simulated NRBCS at L-, C-, X-, and Ku-bands and incidence angle of 38° with the three composite wind wave spectra at upwind direction (except the azimuthally integrated NRBCS at L-band) and comparison with validation data versus wind speeds for the HH polarization. (a) L-band, (b) C-band, (c) X-band, (d) Ku-band.

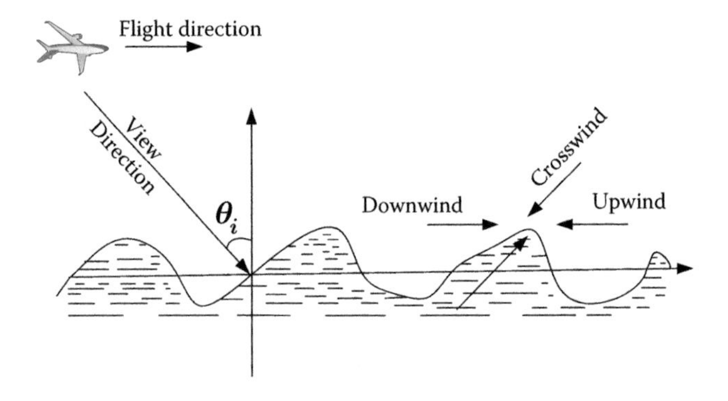

FIGURE 4.30 Geometry of radar observation (a case viewed from downwind direction) and skewed sea surface shaped by wind vector (Adapted from Chen et al., 1993 [32]).

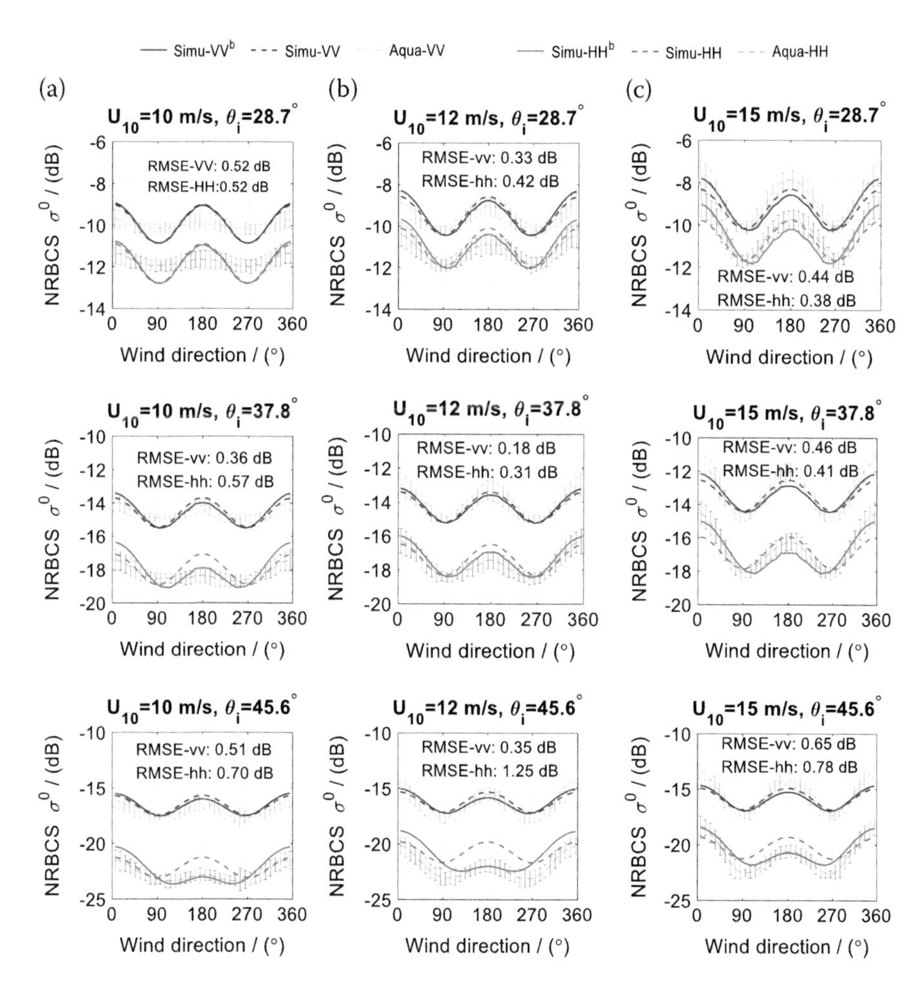

FIGURE 4.31 Comparison of the simulated NRBCS with Aquarius radar measured data at incidence angles of 28.7° (first row), 37.8° (second row), and 45.6° (third row). (a) U_{10} = 10 m/s, c = 0.35, RMS = 0.73 cm; (b) 12 m/s, c = 0.29, RMS = 0.89 cm; (c) 15 m/s, c = 0.25, RMS = 1.05 cm.

for Ku-band (14 GHz), $\theta_i = \{2°, 4°, \ldots, 60°\}$, they are obtained from the Seasat microwave scatterometer (SASS-II) model (the negative skewness contributions are forced to be zero by the authors) [39], and NASA scatterometer (NSCAT-4) model [40]. These empirical models are all derived from extensive experimental measurements and can be referred as the proxies of radar observations for the validation. Comparison results are shown in Figures 4.32–4.33 at 10 and 15 m/s, respectively. The simulations with Gaussian-type bispectrum contribution based on Elfouhaily spectrum in upwind (solid line) and downwind (dotted line) directions. We can see that the Gaussian-type skewness function also do well in predicting the NRBCS versus incidence angles at wind speeds of 15 m/s.

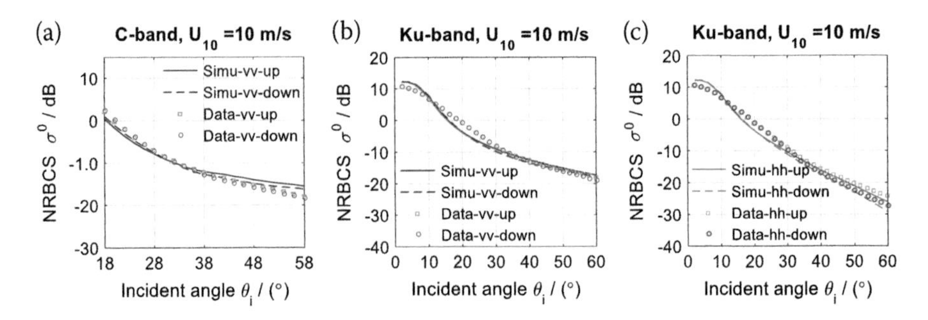

FIGURE 4.32 The comparison of simulated NRBCS with the reference data from CMOD2-I3 and CMOD5 for C-band and SASS-II and Ku-2011 for Ku-band at U_{10} = 10 m/s. (a) C-band for VV-polarization (c = 0.31, RMS = 0.16 cm); (b) and (c) Ku-band for VV- and HH-polarizations (c = 0.18, RMS = 0.11 cm).

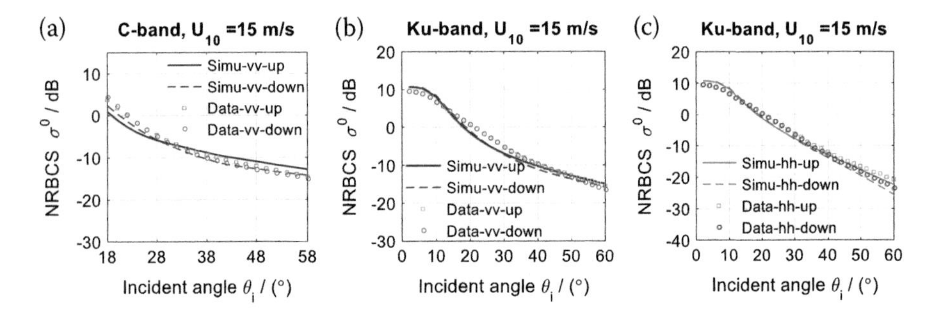

FIGURE 4.33 Similar to Figure 4.32 but at U_{10} = 15 m/s. (a) C-band for VV-polarization (c = 0.22, RMS = 0.26 cm); (b) and (c) Ku-band for VV- and HH-polarizations (c = 0.13, RMS = 0.19 cm).

In fact, sea surface essentially contains multiscale roughness with capillary waves of many sizes riding on large-scale waves, also of many sizes. It is instructive to exploit the effect of radar frequency and observation geometry on the effective roughness scales responsible for radar backscattering so that the scattering mechanism and the scattering source can be better understood and quantitated. Based on common sea spectra and a theoretical scattering model, attempt is made to attain the above objective. Model predictions, with selective roughness scales, are compared with wide validation data, including L-band radar observations, and predictions from C-band and Ku-band empirical models: geophysical model function (CMOD7) and NASA scatterometer (NSCAT-4) for C- and Ku-bands at different incident angles. For a single-scale rough surface, the sources of surface scattering for the classical geometric optics, Kirchhoff, and small perturbation or Bragg scattering models are the surface slope distribution, reflection and diffraction, and surface spectrum evaluated at Bragg wavenumbers, respectively [41]. They are valid, respectively, in the high-frequency limit, high-frequency region, and low-frequency region. For multiscale rough sea surface, it is instructive to know that only a part of roughness scales (effective roughness) are responsible for backscattering at a given view angle and exploring frequency [30,41]. The contributions to backscattering by the other roughness do not add up coherently along the backscattering direction and are filtered by effective sensing wavelength λ_e, determined by the incident wavelength and incident angle, i.e., $\lambda_e = \lambda/(2 \sin \theta_i)$, which is called "*wavelength filtering effect*" [41]. Therefore, the backscattering source for radar backscattering from multiscale sea surface is also the all orders of surface spectrum $\mathbf{W}^{(n)}(K, \phi)$ which attaches to a specific roughness determined by incident angle and frequency (i.e., effective roughness).

The comparisons with Aquarius data along various wind directions at several combinations of incident angle and wind speed are shown in Figures 4.34–4.36. We note that the simulated backscattering coefficients based on AIEM model and the Apel spectrum for L-band are overall in good agreement with the Aquarius radar measurements with the error of around 1 dB for HH- and VV-polarizations for the incidence angles of 28.7°, 37.8°, and 45.6° and the wind speeds of 3, 5, 8, 10, 12, and 15 m/s. The corresponding RMS height and correlation length for L-band radar backscattering and the effective wavelength are 3.2, 58, 19.4 cm for wind speed of 5 m/s, and 4.1, 63, 19.4 cm for wind speed of 15 m/s.

Similarly, for C- and Ku-bands, the simulated backscattering coefficients versus incident angles based on Apel spectrum for 5, 10, and 15 m/s also are in good

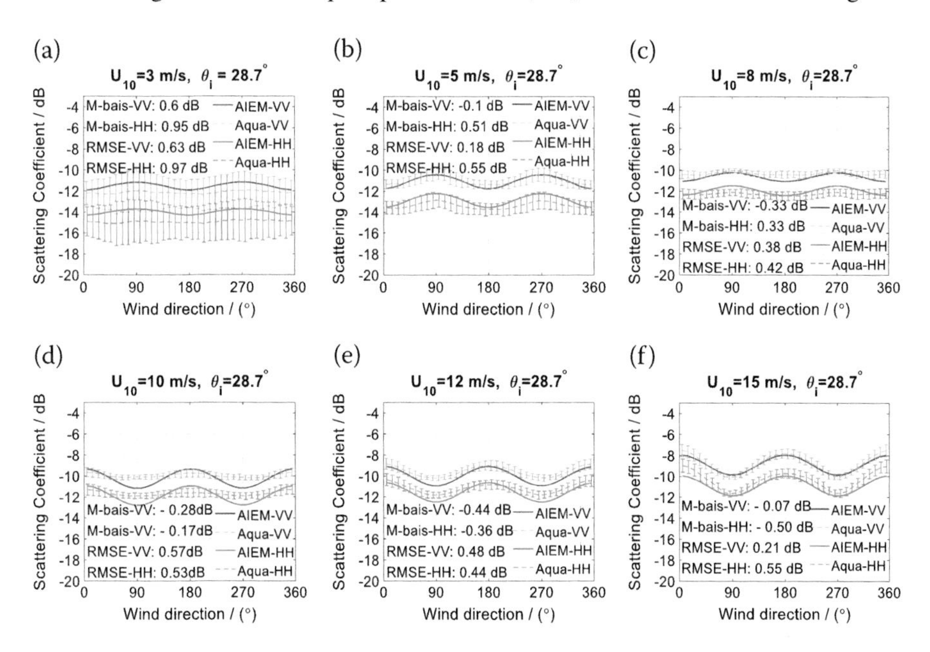

FIGURE 4.34 Comparison of simulated backscattering based on AIEM and Apel sea spectrum and Aquarius radar data at L-band for incidence angles of 28.7° and wind speeds of 3, 5, 8, 10, 12, and 15 m/s, respectively. (a) 3 m/s; (b) 5 m/s; (c) 8 m/s; (d) 10 m/s; (e) 12 m/s; (f) 15 m/s.

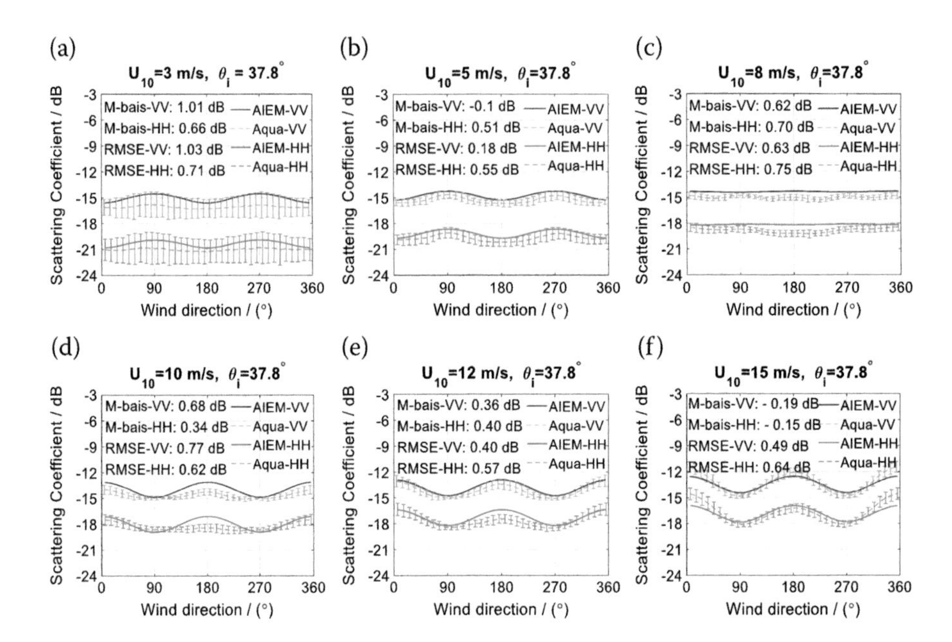

FIGURE 4.35 Comparison of simulated backscattering based on AIEM and Apel sea spectrum and Aquarius radar data at L-band for incidence angles of 37.8° and wind speeds of 3, 5, 8, 10, 12, and 15 m/s, respectively. (a) 3 m/s; (b) 5 m/s; (c) 8 m/s; (d) 10 m/s; (e) 12 m/s; (f) 15 m/s.

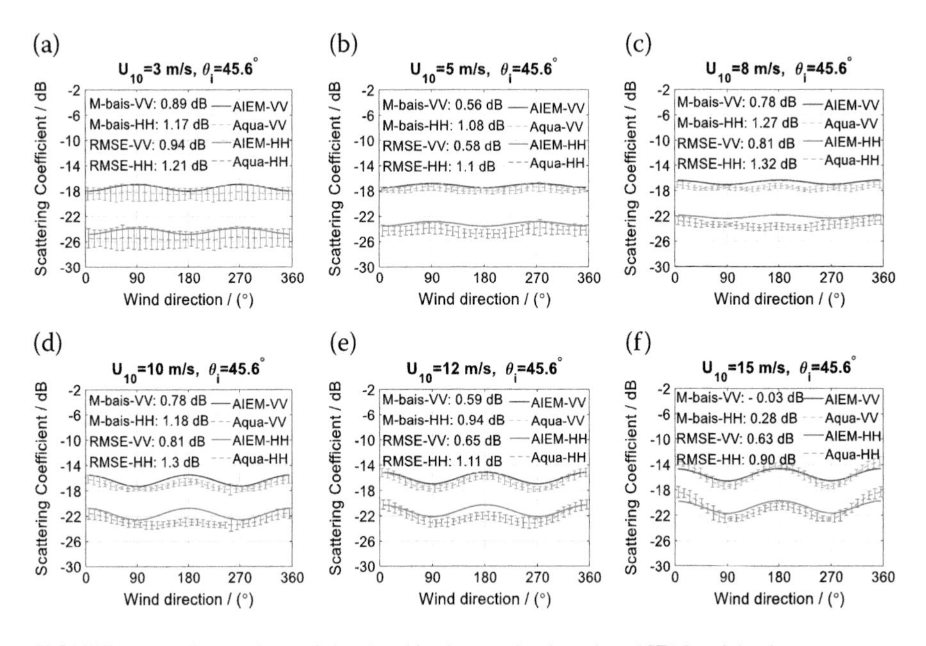

FIGURE 4.36 Comparison of simulated backscattering based on AIEM and Apel sea spectrum and Aquarius radar data at L-band for incidence angles 45.6° and wind speeds of 3, 5, 8, 10, 12, and 15 m/s, respectively. (a) 3 m/s; (b) 5 m/s; (c) 8 m/s; (d) 10 m/s; (e) 12 m/s; (f) 15 m/s.

agreement with the empirical CMOD7 (no HH polarization) and NSCAT-4 models except for Ku-band at 10 and 15 m/s. The related comparisons with the validation data were shown in Figures 4.37–4.38, respectively. For C-band, the simulated VV-polarized backscattering coefficients well close to the empirical CMOD7 model with the error of mean bias and root mean square error (RMSE) within 1 dB for the three wind speeds [see Figure 4.37]. As for Ku-band, the simulated backscattering versus incident angles based on Apel spectrum are consistent better with the NSCAT-4 model at small wind speed (i.e., 5 m/s, see Figure 4.38(a)). While at wind speeds of 10 m/s [see Figure 4.38(b)] and 15 m/s [see Figure 4.38(c)], the simulations are in agreement with the experimental data at small incidence angles ($\theta_i \leq 30°$), but substantially biased at medium to large incidence especially for VV-polarization. The overestimation of VV-polarization mentioned above may be given rise by the Apel spectrum itself and this behavior also was reported in [42]. This perhaps is because that the high-frequency spectral components in the Apel model is overestimated, which has been proved by Elfouhaily et al.

In Figure 4.39(a), the simulated VV polarization backscattering coefficients are closer to the experimental data, compared to simulated HH polarization, which is underestimated. Comparing with Figures 4.38 and 4.39, we note that the simulated backscattering based on the Apel spectrum is higher than that based on the

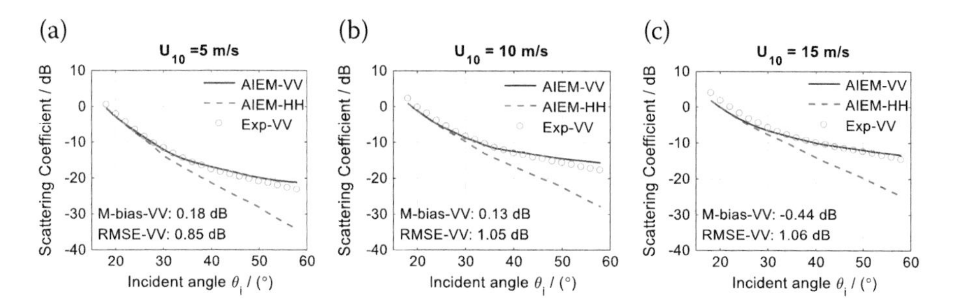

FIGURE 4.37 Comparisons of the simulated backscattering based on Apel spectrum for C-band with the CMOD7 data (no HH polarization) at the wind speeds of 5–15 m/s, the first row: (a) 5 m/s; (b) 10 m/s; (c) 15 m/s.

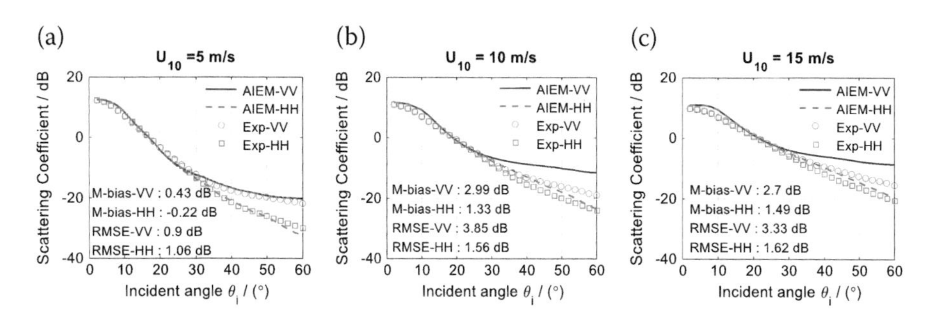

FIGURE 4.38 Comparisons of the simulated backscattering based on Apel spectrum for Ku-band with NSCAT-4 model at the wind speeds of (a) 5 m/s, (b) 10 m/s, and (c) 15 m/s.

Elfouhaily spectrum at large incident angles. To further verify the deviation for Ku-band at high wind speeds impacted by sea spectrum itself, the simulated scattering coefficients based on the Elfouhaily spectrum by the first order SSA-1 scattering model are calculated and shown in Figure 4.39(a). They are in better agreement with each other except for slight deviation in the incidence region ranging from 10° to 30°. For the simulations for VV-polarization at C-band based on Elfouhaily spectrum, they were verified with empirical model in [39] and the simulated results based on SSA-1 and Elfouhaily spectrum coincide with the simulations using the Apel spectrum here.

Besides, in order to investigate the effect of sea spectrum on the simulations at low frequency, the simulation results at L-band in the upwind direction based on the Elfouhaily and Apel spectra were also compared in Figure 4.40. It is clearly seen that the simulations (red symbols) based on the Apel spectrum are closer to the radar measurements (black symbols) especially for 3–12 m/s while the simulations (blue symbols) based on the Elfouhaily spectrum are underestimated significantly compared with the measured data. Therefore, for the low-frequency L-band, the Apel spectrum is more suitable for simulating radar backscattering. This may be because the related parameters used for building the Apel spectrum model are determined by radar data, making the low-frequency spectral components (gravity and capillary-gravity waves) more accurate [43]. While, for Elfouhaily spectrum, only the measurements in the tank were involved but the RMS slope of the spectrum, mainly determined by the high-frequency spectral components, is more consistent with the Cox and Munk measurements [35]. This perhaps explains why the Elfouhaily spectrum is more applicable to high-frequency Ku-band at high wind speed (e.g., greater than 10 m/s) and large incident angles (e.g., greater than 30°). Overall, the roughness generated by using the spectral components from the classical Apel and Elfouhaily spectra is responsible for radar backscattering.

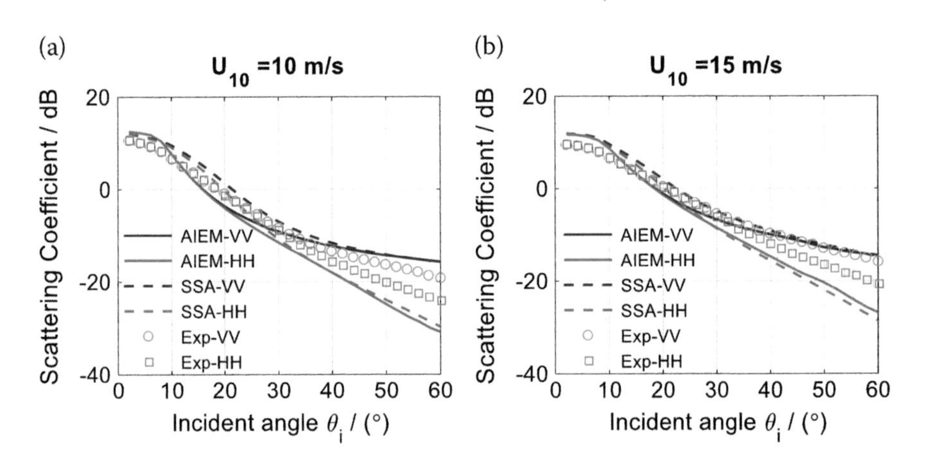

FIGURE 4.39 Comparisons of simulated backscattering, predicted by AIEM and SSA-1 model, versus incident angles based on Elfouhaily spectrum for Ku band with NSCAT-4 model at the wind speeds of (a) 10 m/s and (b) 15 m/s.

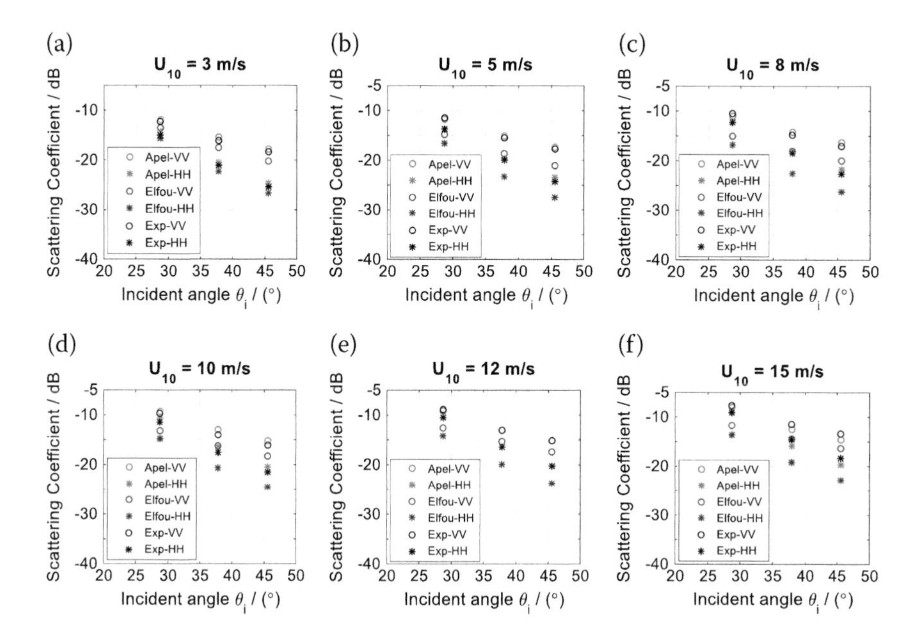

FIGURE 4.40 Comparisons of simulated backscattering based on Apel and Elfouhaily spectra with Aquarius data in the upwind direction at incident angles of 28.7°, 37.8°, and 45.6° for L-band at wind speeds of 3–15 m/s, respectively. (a) 3 m/s; (b) 5 m/s; (c) 8 m/s; (d) 10 m/s; (e) 12 m/s; (f) 15 m/s.

4.5.2.2 Radar Bistatic Scattering Behaviors

In Figures 4.41 and 4.42, we present the hemispherical plots of bistatic scattering on the whole scattering plane at wind speeds of 6.5 m/s and 25 m/s in the upwind direction, respectively. The incident angle was fixed at 33° for wind speed of 6.5 m/s, and 49° for wind speed of 25 m/s. The scattering polar and azimuthal angles each varies with step of 1°. Recall that the dependence pattern of bistatic scattering on wind speed is altered by scattering polar and azimuthal angles. For co-polarization, the strong scattering region lies in around the specular direction. Otherwise, the scattering behaviors between HH and VV polarizations are quite different, with VV being more diffuse on the scattering polar direction. The deep scattering valley across the scattering polar angle appears at the cross-plane ($\phi_s = \pm 90°$) for HH polarization, whereas for VV polarization, it is not so strict at the cross direction but instead tends to be in forward region ($\phi_s = 60°, 300°$). For HV and VH polarizations, generally they looks similar at first glance; however, looking into more detail, we find the VH polarization is more diffuse in the scattering polar direction. The reflection symmetry is not applicable here, as expected. The scattering valley of HV and VH polarizations occurs in the incident plane ($\phi_s = 0°$), especially in the large scattering polar angle region. Namely, the depolarization effect is relatively weak in the incident plane. By comparing the scattering patterns of Figures 4.41 and 4.42, it is clear that as the wind speed increases, we assert that (1) bistatic scattering strengthen except around the specular direction. It should be noted that the scattering here includes only the incoherent scattering; (2) the dynamic range of bistatic scattering strength reduces on the whole scattering plane; (3) the scattering tends to be more omnidirectional because the large surface roughness, and the difference between forward and backward scattering region decreases; (4) for HV and VH polarizations, the scattering strength in backward region increases due to stronger multiple scattering at high wind speeds. In addition, at low wind speeds, the scattering coefficients is quite small in backscattering, but not so in other scattering directions.

To inspect the impact of the wind direction on bistatic scattering, we plot the difference of co- and cross-polarized scattering between upwind and crosswind directions for the wind speed of 15 m/s, as shown in Figure 4.43. The incident angle was fixed at 33°. We observe that the bistatic scattering responds to wind direction on the whole scattering plane. For backscattering, the scattering strength in the upwind direction is always greater than that in the crosswind direction. However, this phenomenon not necessary, perhaps not always, occurs for bistatic scattering. Note that the difference of the scattering coefficients between the upwind and crosswind directions is greater than zero except around the specular direction, in which scattering in the crosswind direction is stronger than in the upwind direction. In this simulation case, as large as 8 dB of scattering strength from upwind direction

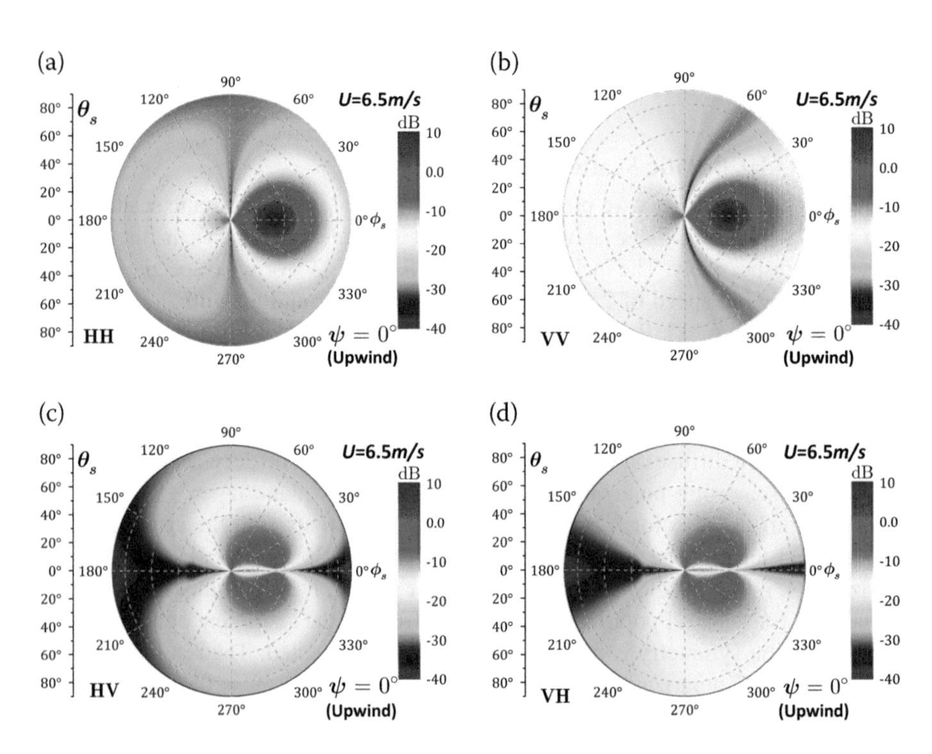

FIGURE 4.41 The bistatic scattering hemispherical plots at windspeed of 6.5 m/s in the upwind direction $\theta_i = 33°$ (a) HH polarization, (b) VV polarization, (c) HV polarization, (d) VH polarization.

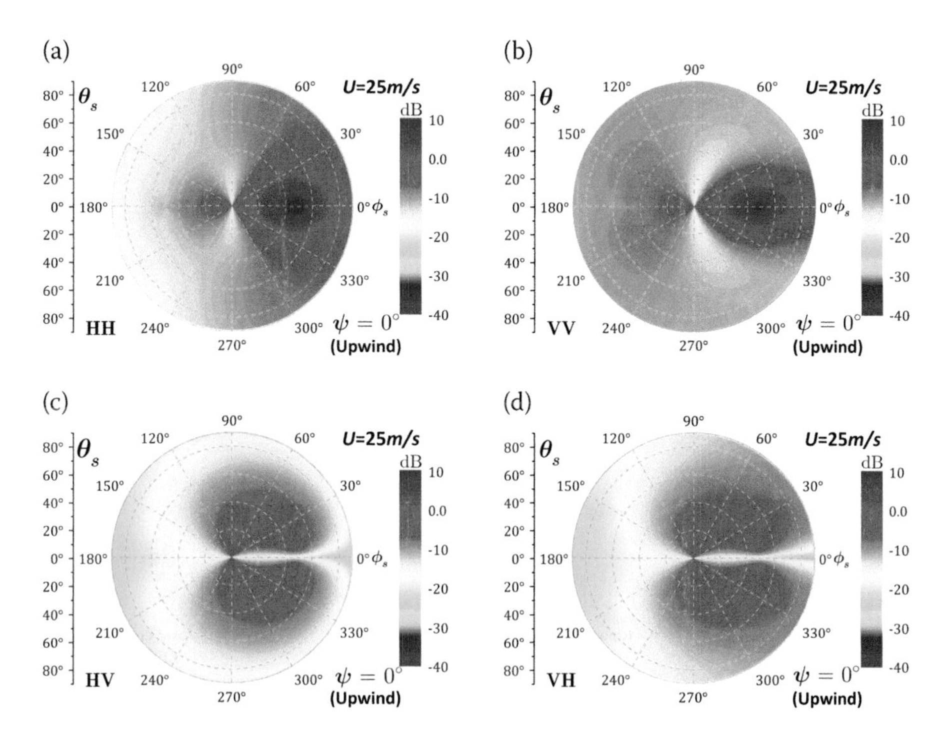

FIGURE 4.42 The bistatic scattering hemispherical plots at windspeed of 25 m/s in the upwind direction $\theta_i = 49°$ (a) HH polarization, (b) VV polarization, (c) HV polarization, (d) VH polarization.

and crosswind direction in the cross plane for HH polarization and in forward region for VV polarization. It implies that there exist scattering direction diversity along with polarization diversity to sense the wind direction. This difference is obvious for HV and VH polarizations at the incident plane. In terms of sensitivity to wind direction, HH polarization is higher in the cross-plane, and VV polarization is more so in the forward region. The HV and VH polarizations are more sensitive to wind direction in the incident plane. Under the bistatic configuration, these highly sensitive region for various polarizations is potentially vital to determine wind direction.

To catch a more complete picture how the bistatic scattering pattern depends on wind speed, Figures 4.44 and 4.45 display hemispherical plots for the upwind direction. The difference of co-and cross-polarized scattering coefficient between the low wind speed of 6.5 m/s and medium wind speed of 15 m/s are presented in Figure 4.44. In this upwind case, the incident angle is 33°. In order to analyze the effect of wind speed in more detail, we also explore the difference between the strong wind speeds of 25 and 40 m/s in Figure 4.45 at 49° incidence. In general, the similar observations as above the effect of wind direction are applicable. By comparing with Figures 4.44 and 4.45, we might argue the following points: (1) the feature of scattering difference is similar, but there is a noticeable narrowing of dynamic range

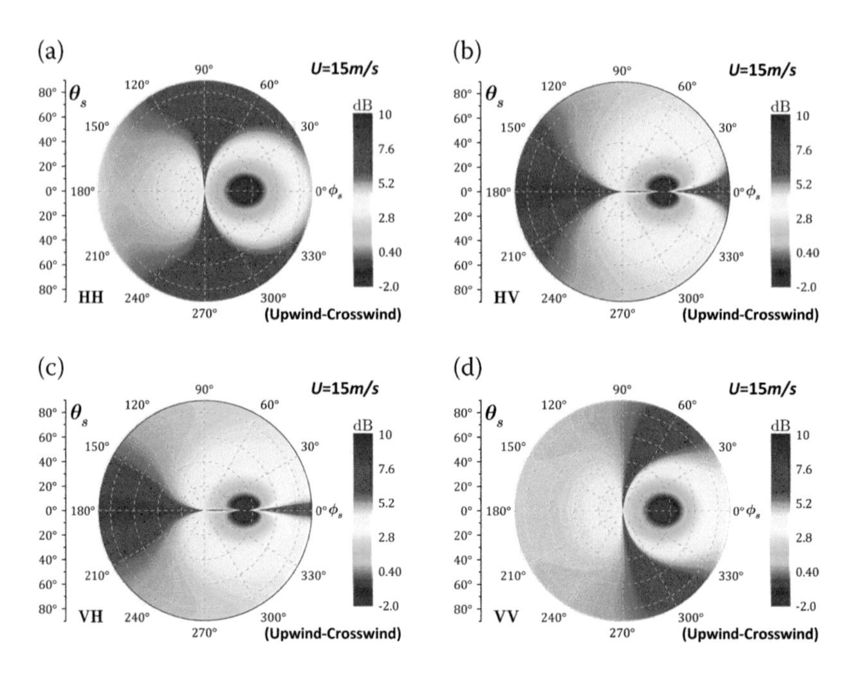

FIGURE 4.43 Effect of wind direction on bistatic scattering in the scattering plane with 33° of incidence at wind speeds of 15 m/s. Difference of scattering coefficients in the upwind and crosswind directions. (a) HH, (b) HV, (c) VH, (d) VV.

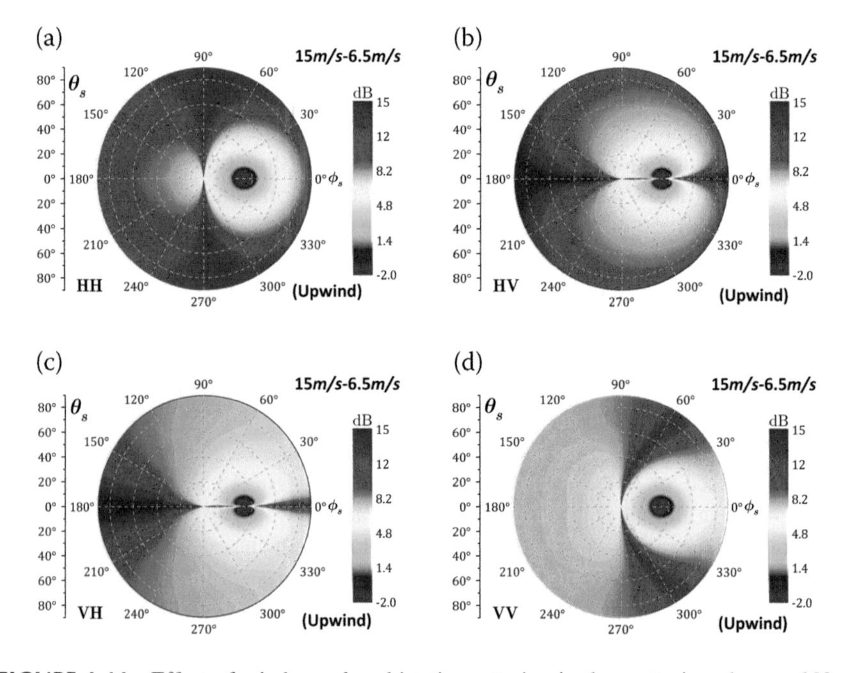

FIGURE 4.44 Effect of wind speed on bistatic scattering in the scattering plane at 33° of incidence in the upwind direction. Difference of scattering coefficients at 15 and 6.5 m/s. (a) HH, (b) HV, (c) VH, (d) VV.

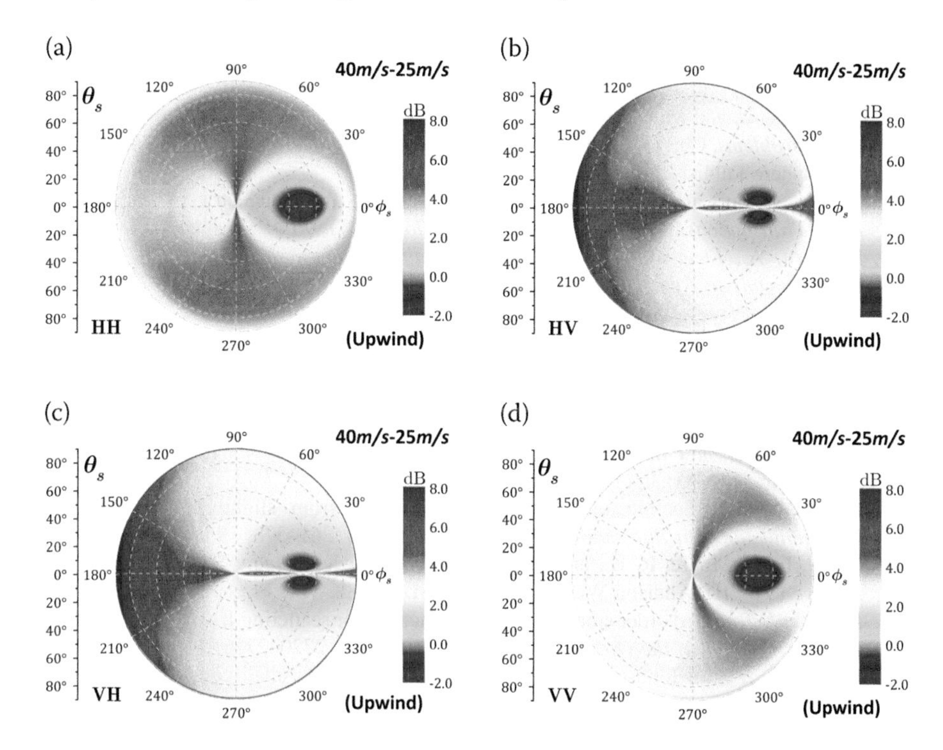

FIGURE 4.45 Effect of wind speed on bistatic scattering in the scattering plane at 49° incidence for upwind case. Difference of scattering coefficient at 40 and 25 m/s. (a) HH, (b) HV, (c) VH, (d) VV.

on the whole scattering plane for the strong wind speed; (2) the area of negative differences expands and its location appears around the specular direction. This is because that the specular scattering decreases as the wind speed increases; (3) for high wind speeds, in bistatic configuration, the difference in scattering coefficient between different wind speeds is obvious (~6 dB), especially in the cross plane for HH polarization, in the forward region for VV polarization, and in the incident plane for both HV and VH polarizations. Compared to backscattering, the saturation and damping of co-polarization at high wind speeds are not shown, implying a great advantage of bistatic scattering in inversion of high wind speed.

REFERENCES

1. Stratton, J. A., *Electromagnetic Theory*, McGraw-Hill, New York, 1941.
2. Mueller, C., *Foundations of the Mathematical Theory of Electromagnetic Waves*, Springer, Berlin, 1969.
3. Colton, D., and Kress, R., *Integral Equation Methods in Scattering Theory*, Wiley, New York, 1983.
4. Ulaby, F. T., Moore, R. K., and Fung, A. K., *Microwave Remote Sensing*, Addison-Wesley, Reading, MA, 1982.

5. Tsang, L., Kong, J. A., and Shin, R. T., *Theory of Microwave Remote Sensing*, John Wiley, New York, 1985.
6. Poggio, A. J., and Miller, E. K., *Integral equation solutions of three-dimensional scattering problems*, Chapter 4, in *Computer Techniques for Electromagnetics*, R. Mittra ed., Hemisphere Publishing, New York, 1987.
7. Iodice, A., Forward–backward method for scattering from dielectric rough surfaces, *IEEE Transactions on Antennas and Propagation*, 50(7), 901–911, July 2002.
8. Ishimaru, A., *Electromagnetic Wave Propagation, Radiation, and Scattering: From Fundamentals to Applications*, Prentic Hall, Englewood Cliffs, NJ, 1991, 2nd Edition reprinted by Wiley-IEEE Press, New York, 2017.
9. Tsang, L., Ding, K. H., Huang, S. W., and Xu, X. L., Electromagnetic computation in scattering of electromagnetic waves by random rough surface and dense media in microwave remote sensing of land surfaces, *Proceedings of the IEEE*, 101(2), 255–279, February 2013.
10. Beckmann, P., and Spizzichino, A., *The Scattering of Electromagnetic Waves from Rough Surfaces*, Pergamon Press, Oxford, England, 1963.
11. Rice, S. O., Reflection of electromagnetic waves from slightly rough surfaces, *Communications on Pure and Applied Mathamatics*, **4**, 361–378, 1951.
12. Ulaby, F. T., Moore, R. K., and Fung, A. K., *Microwave Remote Sensing: Active and Passive* (Vol. II). Addison-Wesley, Reading, MA, 1982.
13. Johnson, J. T., Third order small perturbation method for scattering from dielectric rough surfaces, *Journal of the Optical Society of America*, 16, 2720–2726, 1999.
14. Demir, M. A., and Johnson, J. T., Fourth- and higher-order small-perturbation solution for scattering from dielectric rough surfaces, *The Journal of the Optical Society of America A: Optics Image Science and Vision*, 20(12), 2330–2337, 2003.
15. Voronovich, A. G., Small-slope approximation in wave scattering by rough surfaces, *Soviet Physics–JETP*, 62, 65–70. 1985.
16. Voronovich, A. G., *Wave Scattering from Rough Surfaces*, 2nd Edition, Springer Series on Wave Phenomena. Springer-Verlag, Berlin, Germany, 1999.
17. Gilbert, M. S., and Johnson, J. T., A study of the high order small-slope approximation for scattering from a Gaussian rough surface, *Waves Random Media*, 13, 137–149, 2003.
18. Fung, A. K., Li, Q., and Chen, K. S., Backscattering from a randomly rough dielectric surface, *IEEE Transactions on Geoscience and Remote Sensing*, 30(2), 356–369, March 1992.
19. Fung, A. K., *Microwave Scattering and Emission Models and Their Applications*, Artech House, Norwood, MA, 1994.
20. Fung, A. K., and Chen, K. S., *Microwave Scattering and Emission Models for Users*, Artech House, Norwood, MA, 2010.
21. Chen, K. S., Wu, T. D., Tsang, L., Li, Q., Shi, J. C., and Fung, A. K., Emission of rough surfaces calculated by the integral equation method with comparison to three-dimensional moment method simulations, *IEEE Transactions on Geoscience and Remote Sensing*, 41(1), 90–101, January 2003.
22. Oh, Y., Sarabandi, K., and Ulaby, F. T., An empirical model and an inversion technique for radar scattering from bare soil surfaces, *IEEE Transactions on Geoscience and Remote Sensing*, 30(2), 370–381, February 1992.
23. Chen, K. S., Wu, T. D., and Fung, A. K., A note on the multiple scattering in an IEM model, *IEEE Transactions on Geoscience and Remote Sensing*, 38(1), 249–256, January 2000.
24. Guerin, C. A., Soriano, G., and Elfouhaily, T., Weighted curvature approximation: Numerical tests for 2D dielectric surfaces, *Waves Random Media*, 14(3), 349–363, July 2004.
25. Oh, Y., Sarabandi, K., and Ulaby, F. T., Semi-empirical model of the ensemble-averaged differential muller matrix for microwave backscattering from bare soil surface, *IEEE Transactions on Geoscience and Remote Sensing*, 40(6),1348–1355, June 2002.

26. Long, D. G., and Drinkwater, M. R., Azimuth variation in microwave scatterometer and radiometer data over Antarctica, *IEEE Transactions on Geoscience and Remote Sensing*, 38(4), 1857–1870, July 2000.
27. Brekhovskikh, L. M., *Waves in Layered Media*, 2nd Edition, Translated by R. T. Beyer, Academic Press, New York, 1980.
28. Fung, A. K., Dawson, M. S., Chen, K. S., Hsu, A. Y., Engman, E.T., O'Neill, P. O., and Wang, J., A modified IEM model for scattering from soil surface with application to soil moisture sensing, *IEEE IGARSS*, 2(2), 1297–1299, 1996.
29. Valenzuela, G. R., Theories for the interaction of electromagnetic and oceanic waves—A review, *Boundary-Layer Meteorology*, 13(1–4), 61–85, January 1978.
30. Xie, D., Chen, K.-S., & Zeng, J. The frequency selective effect of radar backscattering from multiscale sea surface. *Remote Sensing*, 11(2), 160, 2019. doi:10.3390/rs11020160.
31. Xie, D. F., Chen, K. S., Yang, X. F., Effects of wind wave spectra on radar backscatter from sea surface at different microwave bands: a numerical study, *IEEE Transactions on Geoscience and Remote Sensing*, 57(9), 6325–6334, September 2019.
32. Chen, K. S., Fung, A. K., and Amar, F., An empirical bispectrum model for sea surface scattering, *IEEE Transactions on Geoscience and Remote Sensing*, 31(4), 830–835, July 1993.
33. Fung, A. K., and Chen, K. S., Kirchhoff model for a skewed random surface, *Journal of Electromagnetic Waves and Applications*, 5(2), 205–216, 1991.
34. Chen, K. S., Fung, A. K., and Weissman, D. E., A backscattering model for ocean surfaces, *IEEE Transactions on Geoscience and Remote Sensing*, 30(4), 811–817, July 1992.
35. Cox, C., and Munk, W., Statistics of the sea surface derived from sun glitter, *Journal of Marine Research*, 13, 198–227, 1954.
36. Xie, D. F., Chen, K. S., Yang, X. F., Effect of bispectrum on radar backscattering from non-Gaussian sea surface, *IEEE Journal of Selected Topics in Applied Earth Observation and Remote Sensing*, 12(1), 4367–4378, October 2019.
37. Bentamy, A., Quilfen, Y., Queuffeulou, P., *Calibration of the ERS-1 Scatterometer C-Band Model*, Institute Française de Recherche pour l'Exploitation de la Mer, 1994.
38. Hersbach, H., Stoffelen, A., de Haan, S., An improved C-band scatterometer ocean geophysical mode function: CMOD5, *Journal of Geophysical Research*, 112(C3), C03006, March 2007.
39. Voronovich, A. G., Zavorotny, V. U., Theoretical model for scattering of radar signals in Ku- and C-bands from a rough sea surface with breaking waves, *Waves Random Media*, 11, 247–269, March 2001.
40. OSI SAF, *Algorithm Theoretical Basis Document for the OSI SAF wind products, SAF/OSI/CDOP2/KNMI/SCI/MA/197*, and "NSCAT-4 Geophysical Model Function," EUMETSAT, Darmstadt, Germany, 2014 [Online]. Available: http://projects.knmi.nl/scatterometer/nscatgmf.
41. Fung, A. K., *Backscattering from Multiscale Rough Surfaces with Application to Wind Scatterometry*, Artech House, Norwood, MA, 2015.
42. Apel, J. R., An improved model of the ocean surface wave vector spectrum and its effects on radar backscatter, *Journal of Geophysical Research: Oceans*, 99(C8), 16269–16291, 1994.
43. Yang, Y., Chen, K. S., Tsang, L., and Liu, Y., Depolarized backscattering of rough surface by AIEM model. *IEEE Journal of Selected Topics in Applied Earth Observations and Remote Sensing*, 10(11), 4740–4752, 2017.

APPENDIX 4A

Expressions of Factors g^{kc_l}, g^{c_i} [43]

The explicit expressions of the factors g^{kc_l}, g^{c_i}, g^{c_j} appearing in Equations 4.61 and 4.63 up to double bounce, are summarized below:

$$g^{kc_1}(u,v,q) = e^{-\sigma^2\left(k_{sz}^2+k_z^2+k_{sz}k_z+q^2-qk_{sz}+qk_z\right)} \sum_{m=1}^{\infty}\frac{[\sigma^2(k_z+q)(k_{sz}+k_z)]^m}{m!}\mathbf{W}^{(m)}\left(k_x+u,\,k_y+v\right)$$
$$\sum_{n=1}^{\infty}\frac{[\sigma^2(k_{sz}-q)(k_{sz}+k_z)]^n}{n!}\mathbf{W}^{(n)}\left(k_{sx}+u,\,k_{sy}+v\right) \tag{4A.1}$$

$$g^{kc_2}(u,v,q) = e^{-\sigma^2\left(k_{sz}^2+k_z^2+k_{sz}k_z+q^2-qk_{sz}+qk_z\right)} \sum_{m=1}^{\infty}\frac{[\sigma^2(k_z+q)(k_{sz}+k_z)]^m}{m!}\mathbf{W}^{(m)}\left(k_x-k_{sx},\,k_y-k_{sy}\right)$$
$$\sum_{n=1}^{\infty}\frac{[-\sigma^2(k_{sz}-q)(k_z+q)]^n}{n!}\mathbf{W}^{(n)}\left(k_{sx}+u,\,k_{sy}+v\right) \tag{4A.2}$$

$$g^{kc_3}(u,v,q) = e^{-\sigma^2\left(k_{sz}^2+k_z^2+k_{sz}k_z+q^2-qk_{sz}+qk_z\right)} \sum_{m=1}^{\infty}\frac{[-\sigma^2(k_{sz}-q)(k_z+q)]^m}{m!}\mathbf{W}^{(m)}\left(k_x+u,\,k_y+v\right)$$
$$\sum_{n=1}^{\infty}\frac{[\sigma^2(k_{sz}-q)(k_{sz}+k_z)]^n}{n!}\mathbf{W}^{(n)}\left(k_x-k_{sx},\,k_y-k_{sy}\right) \tag{4A.3}$$

$$g^{c_1}(u,v,q,q') = e^{-\sigma^2\left(k_{sz}^2+k_z^2+q^2+q'^2-(k_{sz}-k_z)(q+q')\right)} \sum_{n=1}^{\infty}\frac{[\sigma^2(k_{sz}-q)(k_{sz}-q')]^n}{n!}$$
$$\mathbf{W}^{(n)}\left(k_{sx}+u,\,k_{sy}+v\right)\sum_{n=1}^{\infty}\frac{[\sigma^2(k_z+q)(k_z+q')]^m}{m!}\mathbf{W}^{(m)}\left(k_x+u,\,k_y+v\right) \tag{4A.4}$$

$$g^{c_2}(u,v,q,q') = e^{-\sigma^2\left(k_{sz}^2+k_z^2+q^2+q'^2-(k_{sz}-k_z)(q+q')\right)} \sum_{n=1}^{\infty}\frac{[\sigma^2(k_{sz}-q)(k_z+q')]^n}{n!}$$
$$\mathbf{W}^{(n)}\left(k_{sx}+u,\,k_{sy}+v\right)\sum_{n=1}^{\infty}\frac{[\sigma^2(k_z+q)(k_{sz}-q')]^m}{m!}\mathbf{W}^{(m)}\left(k_x+u,\,k_y+v\right) \tag{4A.5}$$

$$g^{c_3}(u,v,q,q') = e^{-\sigma^2\left(k_{sz}^2+k_z^2+q^2+q'^2-(k_{sz}-k_z)(q+q')\right)} \sum_{n=1}^{\infty}\frac{[\sigma^2(k_{sz}-q)(k_z+q')]^n}{n!}$$
$$\mathbf{W}^{(n)}\left(k_{sx}+u,\,k_{sy}+v\right)\sum_{m=1}^{\infty}\frac{[\sigma^2(k_z+q)(k_z+q')]^m}{m!}\mathbf{W}^{(m)}\left(k_x+u,\,k_y+v\right) \tag{4A.6}$$

$$g^{c_4}(u, v, q, q') = e^{-\sigma^2\left(k_{sz}^2+k_z^2+q^2+q'^2 - (k_{sz} - k_z)(q+q')\right)} \sum_{n=1}^{\infty} \frac{[\sigma^2(k_{sz} - q)(k_z + q')]^n}{n!}$$
$$\mathbf{W}^{(n)}\left(k_x - k_{sx}, k_y - k_{sy}\right) \sum_{m=1}^{\infty} \frac{[-\sigma^2(k_{sz} - q)(k_z + q)]^m}{m!}\mathbf{W}^{(m)}\left(k_x + u, k_y + v\right) \tag{4A.7}$$

$$g^{c_5}(u, v, q, q') = e^{-\sigma^2\left(k_{sz}^2+k_z^2+q^2+q'^2 - (k_{sz} - k_z)(q+q')\right)} \sum_{n=1}^{\infty} \frac{[\sigma^2(k_z + q)(k_z + q')]^n}{n!}$$
$$\mathbf{W}^{(n)}\left(k_x - k_{sx}, k_y - k_{sy}\right) \sum_{m=1}^{\infty} \frac{[-\sigma^2(k_{sz} - q)(k_z + q)]^m}{m!}\mathbf{W}^{(m)}\left(k_{sx} + u, k_{sy} + v\right) \tag{4A.8}$$

$$g^{c_8}(u, v, q, q') = e^{-\sigma^2\left(k_{sz}^2+k_z^2+q^2+q'^2 - (k_{sz} - k_z)(q+q')\right)} \sum_{n=1}^{\infty} \frac{[\sigma^2(k_z + q)(k_{sz} - q')]^n}{n!}$$
$$\mathbf{W}^{(n)}\left(k_x - k_{sx}, k_y - k_{sy}\right) \sum_{m=1}^{\infty} \frac{[-\sigma^2(k_{sz} - q)(k_z + q)]^m}{m!}\mathbf{W}^{(m)}\left(k_{sx} + u, k_{sy} + v\right) \tag{4A.9}$$

$$g^{c_9}(u', v', q, q') = e^{-\sigma^2\left(k_{sz}^2+k_z^2+q^2+q'^2 - (k_{sz} - k_z)(q+q')\right)} \sum_{n=1}^{\infty} \frac{[\sigma^2(k_z + q)(k_{sz} - q')]^n}{n!}$$
$$\mathbf{W}^{(n)}\left(k_{sx} + u', k_{sy} + v'\right) \sum_{m=1}^{\infty} \frac{[\sigma^2(k_z + q)(k_z + q')]^m}{m!}\mathbf{W}^{(m)}\left(k_x + u', k_y + v'\right) \tag{4A.10}$$

$$g^{c_{10}}(u', v', q, q') = e^{-\sigma^2\left(k_{sz}^2+k_z^2+q^2+q'^2 - (k_{sz} - k_z)(q+q')\right)} \sum_{n=1}^{\infty} \frac{[\sigma^2(k_z + q)(k_{sz} - q')]^n}{n!}$$
$$\mathbf{W}^{(n)}\left(k_x - k_{sx}, k_y - k_{sy}\right) \sum_{m=1}^{\infty} \frac{[-\sigma^2(k_{sz} - q')(k_z + q')]^m}{m!}\mathbf{W}^{(m)}\left(k_x + u', k_y + v'\right) \tag{4A.11}$$

$$g^{c_{11}}(u', v', q, q') = e^{-\sigma^2\left(k_{sz}^2+k_z^2+q^2+q'^2 - (k_{sz} - k_z)(q+q')\right)} \sum_{n=1}^{\infty} \frac{[\sigma^2(k_z + q)(k_z + q')]^n}{n!}$$
$$\mathbf{W}^{(n)}\left(k_x - k_{sx}, k_y - k_{sy}\right) \sum_{m=1}^{\infty} \frac{[-\sigma^2(k_{sz} - q')(k_z + q')]^m}{m!}\mathbf{W}^{(m)}\left(k_{sx} + u', k_{sy} + v'\right) \tag{4A.12}$$

$$g^{c_{12}}(u', v', q, q') = e^{-\sigma^2\left(k_{sz}^2+k_z^2+q^2+q'^2 - (k_{sz} - k_z)(q+q')\right)} \sum_{n=1}^{\infty} \frac{[\sigma^2(k_{sz} - q)(k_{sz} - q')]^n}{n!}$$
$$\mathbf{W}^{(n)}\left(k_{sx} + u', k_{sy} + v'\right) \sum_{m=1}^{\infty} \frac{[\sigma^2(k_{sz} - q)(k_z + q')]^m}{m!}\mathbf{W}^{(m)}\left(k_x + u', k_y + v'\right) \tag{4A.13}$$

$$g^{c_{13}}(u', v', q, q') = e^{-\sigma^2\left(k_{sz}^2+k_z^2+q^2+q'^2 - (k_{sz} - k_z)(q+q')\right)} \sum_{n=1}^{\infty} \frac{[\sigma^2(k_{sz} - q)(k_{sz} - q')]^n}{n!}$$
$$\mathbf{W}^{(n)}\left(k_{sx} - k_x, k_{sy} - k_y\right) \sum_{m=1}^{\infty} \frac{[-\sigma^2(k_{sz} - q')(k_z + q')]^m}{m!}\mathbf{W}^{(m)}\left(k_x + u', k_y + v'\right) \tag{4A.14}$$

$$g^{c_{14}}(u', v', q, q') = e^{-\sigma^2\left(k_{sz}^2+k_z^2+q^2+q'^2 - (k_{sz} - k_z)(q+q')\right)} \sum_{n=1}^{\infty} \frac{[\sigma^2(k_{sz} - q)(k_z + q')]^n}{n!}$$
$$\mathbf{W}^{(n)}\left(k_{sx} - k_x, k_{sy} - k_y\right) \sum_{m=1}^{\infty} \frac{[-\sigma^2(k_{sz} - q')(k_z + q')]^m}{m!}\mathbf{W}^{(m)}\left(k_{sx} + u', k_{sy} + v'\right) \tag{4A.15}$$

APPENDIX 4B

The Complementary Field Coefficients [43]

We give explicit expressions of the upward and downward reradiation coefficients appearing in Equation 4.63.

$$F_{vv}^{+} = -\left(\frac{1 - R_v}{q_1}\right)(1 + R_v)C_1 + \left(\frac{1 - R_v}{q_1}\right)(1 - R_v)C_2 + \left(\frac{1 - R_v}{q_1}\right)(1 + R_v)C_3$$
$$+ \left(\frac{1 + R_v}{q_1}\right)(1 - R_v)C_4 + \left(\frac{1 + R_v}{q_1}\right)(1 + R_v)C_5 + \left(\frac{1 + R_v}{q_1}\right)(1 - R_v)C_6 \tag{4B.1}$$

$$G_{vv}^{+} = \left(\frac{(1 + R_v)u_r}{q_2}\right)(1 + R_v)C_{1t} - \left(\frac{1 + R_v}{q_2}\right)(1 - R_v)C_{2t} - \left(\frac{1 + R_v}{q_2\varepsilon_r}\right)(1 + R_v)C_{3t}$$
$$- \left(\frac{(1 - R_v)\varepsilon_r}{q_2}\right)(1 - R_v)C_{4t} - \left(\frac{1 - R_v}{q_2}\right)(1 + R_v)C_{5t} - \left(\frac{1 - R_v}{q_2 u_r}\right)(1 - R_v)C_{6t} \tag{4B.2}$$

$$F_{hh}^{+} = \left(\frac{1 - R_h}{q_1}\right)(1 + R_h)C_1 - \left(\frac{1 - R_h}{q_1}\right)(1 - R_h)C_2 - \left(\frac{1 - R_h}{q_1}\right)(1 + R_h)C_3$$
$$- \left(\frac{1 + R_h}{q_1}\right)(1 - R_h)C_4 - \left(\frac{1 + R_h}{q_1}\right)(1 + R_h)C_5 - \left(\frac{1 + R_h}{q_1}\right)(1 - R_h)C_6 \tag{4B.3}$$

$$G_{hh}^{+} = -\left(\frac{(1 + R_h)\varepsilon_r}{q_2}\right)(1 + R_h)C_{1t} + \left(\frac{1 + R_h}{q_2}\right)(1 - R_h)C_{2t} + \left(\frac{1 + R_h}{q_2 u_r}\right)(1 + R_h)C_{3t}$$
$$+ \left(\frac{(1 - R_h)u_r}{q_2}\right)(1 - R_h)C_{4t} + \left(\frac{1 - R_h}{q_2}\right)(1 + R_h)C_{5t} + \left(\frac{1 - R_h}{q_2\varepsilon_r}\right)(1 - R_h)C_{6t} \tag{4B.4}$$

$$F_{hv}^{+} = \left(\frac{1-R}{q_1}\right)(1+R)B_1 - \left(\frac{1-R}{q_1}\right)(1-R)B_2 - \left(\frac{1-R}{q_1}\right)(1+R)B_3 + \left(\frac{1+R}{q_1}\right)(1-R)B_4 + \left(\frac{1+R}{q_1}\right)(1+R)B_5 + \left(\frac{1+R}{q_1}\right)(1-R)B_6 \quad (4B.5)$$

$$G_{hv}^{+} = -\left(\frac{(1+R)\mu_r}{q_2}\right)(1+R)B_{1t} + \left(\frac{1+R}{q_2}\right)(1-R)B_{2t} + \left(\frac{1+R}{q_2\varepsilon_r}\right)(1+R)B_{3t} - \left(\frac{(1-R)\varepsilon_r}{q_2}\right)(1-R)B_{4t} - \left(\frac{1-R}{q_2}\right)(1+R)B_{5t} - \left(\frac{1-R}{q_2\mu_r}\right)(1-R)B_{6t} \quad (4B.6)$$

$$F_{vh}^{+} = \left(\frac{1+R}{q_1}\right)(1-R)B_1 - \left(\frac{1+R}{q_1}\right)(1+R)B_2 - \left(\frac{1+R}{q_1}\right)(1-R)B_3 + \left(\frac{1-R}{q_1}\right)(1+R)B_4 + \left(\frac{1-R}{q_1}\right)(1-R)B_5 + \left(\frac{1-R}{q_1}\right)(1+R)B_6 \quad (4B.7)$$

$$G_{vh}^{+} = -\left(\frac{(1-R)\varepsilon_r}{q_2}\right)(1-R)B_{1t} + \left(\frac{1-R}{q_2}\right)(1+R)B_{2t} + \left(\frac{1-R}{q_2\mu_r}\right)(1-R)B_{3t} - \left(\frac{(1+R)\mu_r}{q_2}\right)(1+R)B_{4t} - \left(\frac{1+R}{q_2}\right)(1-R)B_{5t} - \left(\frac{1+R}{q_2\varepsilon_r}\right)(1+R)B_{6t} \quad (4B.8)$$

APPENDIX 4C

The B and C Coefficients [43]

The B and C coefficients with $\phi_i \neq 0$ in Equations 4B.1 to 4B.8 are given below.

$$C_1 = -\cos\phi_s(-\cos\phi - z_x z'_x \cos\phi - z_x z'_y \sin\phi) + \sin\phi_s(\sin\phi + z'_x z_y \cos\phi + z_y z'_y \sin\phi) \quad (4C.1)$$

$$C_2 = -\cos\phi_s(-q_1\cos\theta_i\cos\phi - u z_x\cos\theta\cos\theta - v z'_y\cos\theta\cos\phi - q_1 z'_x\sin\theta - u z_x z'_x\sin\theta_i - v z_x z'_y\sin\theta - v z_x\cos\theta\sin\theta + v z'_x\cos\theta\sin\phi) + \sin\phi_s(u z_y\cos\theta\cos\phi - u z'_y\cos\theta\cos\theta + u z'_x z_y\sin\theta + q_1 z'_y\sin\theta + v z_y z'_y\sin\theta_i + q_1\cos\theta\sin\phi + u z'_x\cos\theta\sin\phi + v z_y\cos\theta\sin\phi) \quad (4C.2)$$

$$C_3 = \cos\phi_s(uz'_x\cos\theta\cos\phi - q_1z_xz'_x\cos\theta\cos\phi - u\sin\theta + q_1z_x\sin\theta + uz'_y\cos\theta\sin\phi - q_1z_xz'_y\cos\theta\sin\phi) + \sin\phi_s(vz'_x\cos\theta\cos\phi - q_1z'_xz_y\cos\theta\cos\phi - v\sin\theta + q_1z_y\sin\theta + vz'_y\cos\theta\sin\phi - q_1z_yz'_y\cos\theta\sin\phi) \quad (4C.3)$$

$$C_4 = \sin\theta_s(-z_x\cos\theta\cos\phi - z_xz'_x\sin\theta - z_yz'_y\sin\theta - z_y\cos\theta\sin\phi) - \cos\theta_s\cos\phi_s(-\cos\theta\cos\phi - z_yz'_y\cos\theta\cos\varphi - z'_x\sin\theta + z'_xz_y\cos\theta\sin\phi) - \cos\theta_s\sin\phi_s(z_xz'_y\cos\theta\cos\phi - z'_y\sin\theta - \cos\theta\sin\phi - z_xz'_x\cos\theta\sin\phi) \quad (4C.4)$$

$$C_5 = \sin\theta_s(q_1z_x\cos\phi + uz_xz'_x\cos\phi + vz'_xz_y\cos\phi + q_1z_y\sin\phi + uz_xz'_y\sin\phi + vz_yz'_y\sin\phi) - \cos\theta_s\cos\phi_s(q_1\cos\phi + uz'_x\cos\phi + vz_y\cos\phi - uz_y\sin\phi + uz'_y\sin\phi) - \cos\theta_s\sin\phi_s(-vz_x\cos\phi + vz'_x\cos\phi + q_1\sin\phi + uz_x\sin\phi + vz'_y\sin\phi) \quad (4C.5)$$

$$C_6 = \sin\theta_s(vz_xz'_y\cos\phi - uz_yz'_y\cos\phi - vz_xz'_x\sin\phi + uz'_xz_y\sin\phi) + \cos\theta_s\cos\phi_s(-vz'_y\cos\phi + q_1z_yz'_y\cos\phi + vz'_x\sin\phi - q_1z'_xz_y\sin\phi) + \cos\theta_s\sin\phi_s(uz'_y\cos\phi - q_1z_xz'_y\cos\phi - uz'_x\sin\phi + q_1z_xz'_x\sin\phi) \quad (4C.6)$$

$$B_1 = \sin\theta_s(-z_y\cos\phi + z_x\sin\phi) - \cos\theta_s\cos\phi_s(z'_xz_y\cos\phi + \sin\phi + z_yz'_y\sin\phi) - \cos\theta_s\sin\phi_s(-\cos\phi - z_xz'_x\cos\phi - z_xz'_y\sin\phi) \quad (4C.7)$$

$$B_2 = \sin\theta_s(-q_1z_y\cos\theta\cos\phi - uz_xz'_y\cos\theta\cos\phi - vz_yz'_y\cos\theta\cos\phi - q_1z'_xz_y\sin\theta + q_1z_xz'_y\sin\theta + q_1z_x\cos\theta\sin\phi + uz_xz'_x\cos\theta\sin\phi + vz'_xz_y\cos\theta\sin\phi) - \cos\theta_s\cos\phi_s(uz_y\cos\theta\cos\phi - uz'_y\cos\theta\cos\phi + uz'_xz_y\sin\theta + q_1z'_y\sin\theta + vz_yz'_y\sin\theta + q_1\cos\theta\sin\phi + uz'_x\cos\theta\sin\phi + vz_y\cos\theta\sin\phi) - \cos\theta_s\sin\phi_s(-q_1\cos\theta\cos\phi - uz_x\cos\theta\cos\phi - vz'_y\cos\theta\cos\phi - q_1z'_x\sin\theta - uz_xz'_x\sin\theta - vz_xz'_y\sin\theta - vz_x\cos\theta\sin\phi + vz'_x\cos\theta\sin\phi) \quad (4C.8)$$

$$B_3 = \sin\theta_s(vz_xz'_x\cos\theta\cos\phi - uz_yz'_x\cos\theta\cos\phi - vz_x\sin\theta + uz_y\sin\theta + vz_xz'_y\cos\theta\sin\phi - uz_yz'_y\cos\theta\sin\phi) + \cos\theta_s\cos\phi_s(-vz'_x\cos\theta\cos\phi + q_1z'_xz_y\cos\theta\cos\phi + v\sin\theta - q_1z_y\sin\theta - vz'_y\cos\theta\sin\phi + q_1z_yz'_y\cos\theta\sin\phi) + \cos\theta_s\sin\phi_s(uz'_x\cos\theta\cos\phi - q_1z_xz'_x\cos\theta\cos\phi - u\sin\theta + q_1z_x\sin\theta + uz'_y\cos\theta\sin\phi - q_1z_xz'_y\cos\theta\sin\phi) \quad (4C.9)$$

$$B_4 = -\cos\phi_s(z_x z'_y \cos\theta\cos\phi - z'_y \sin\theta - \cos\theta\sin\phi - z_x z'_x \cos\theta\sin\phi) + \sin\phi_s(-\cos\theta\cos\phi - z_y z'_y \cos\theta\cos\phi - z'_x \sin\theta + z'_x z_y \cos\theta\sin\phi) \tag{4C.10}$$

$$B_5 = -\cos\phi_s(-vz_x \cos\phi + vz'_x \cos\phi + q_1 \sin\phi + uz_x \sin\phi - vz'_y \sin\phi) + \sin\phi_s(q_1 \cos\phi + uz'_x \cos\phi + vz_y \cos\phi - uz_y \sin\phi + uz'_y \sin\phi) \tag{4C.11}$$

$$\begin{aligned} B_6 = {} & \cos\phi_s(uz'_y \cos\phi - q_1 z_x z'_y \cos\phi - uz'_x \sin\phi + q_1 z_x z'_x \sin\phi) \\ & + \sin\phi_s(vz'_y \cos\phi - q_1 z_y z'_y \cos\phi - vz'_x \sin\phi + q_1 z'_x z_y \sin\phi) \end{aligned} \tag{4C.12}$$

Notice that the coefficients for the lower medium B_{1t}, B_{2t}, B_{3t}, B_{4t}, B_{5t}, B_{6t} and C_{1t}, C_{2t}, C_{3t}, C_{4t}, C_{5t}, C_{6t} can be obtained by replacing q_1 by q_2 in Equations 4C.1 through 4C.12.

5 Sensitivity Analysis of Radar Scattering of Rough Surface

Sensitivity analysis (SA) of electromagnetic waves scattering from a randomly rough surface is pivotal in the field of remote sensing of surface geometrical and dielectric parameters [1–3]. It has been routine to retrieve the geophysical parameters of interest, such as soil moisture and roughness, from the radar scattering measurements [4–7], based on analyzing the sensitivity of the scattering behavior and mechanisms. By capturing the scattering patterns, one could effectively avoid undesired parameters, while devising a means to retrieve the parameters of interest. Knowing how the variation in the model outputs attributes to the variation of the inputs is critical [8]. SA can be on local and global scales [9]. Local SA is a gradient-based method that analyzes only one parameter in a model at a time while fixing all of the remaining parameters to their nominal values [9]. Local SA is simple and computationally efficient. However, it fails to quantify the interactions among all the parameters of interest; hence it may not be a good choice for the nonlinear and non-monotonic model [10]. Global SA, on the other hand, considers the mutual coupling among all the parameters within their ranges. That is, all the parameters in a model are simultaneously analyzed in global SA.

Among the global SA methods, the extended Fourier amplitude sensitivity test (EFAST) [11–16] and entropy-based sensitivity analysis (EBSA) [17], are two common methods. EFAST is one of the most popular methods for SA and is applicable to nonlinear and non-monotonic models, and can quantify the parameter interaction effects as well. Besides, to preserve the maximum information content of the parameters of interest, which are transferred from inputs to outputs of the radar response, EBSA has been proposed to characterize the information content in terms of communications theory, to quantitatively and objectively evaluate the information of the sensor configurations, and to evaluate observational uncertainty that is associated with parameter sampling [18]. In this chapter, we will focus on these two methods and show the numerical results.

5.1 EXTENDED FOURIER AMPLITUDE SENSITIVITY TEST (EFAST)

EFAST, a variance-based, is developed based on FAST and Sobol [19]. The FAST method offers a high-efficiency sampling method that is based on a suitably defined search curve. However, it can only calculate the "main effect" contribution of each input parameter to the model outputs. Thus, it cannot identify the parameter interaction effects on the model outputs. The Sobol method is capable of computing higher interaction terms but is computationally expensive since it adopts Monte Carlo sampling. EFAST fuses the strengths of FAST and Sobol to obtain the total and the main sensitivity indices (SIs) of each input parameter.

The EFAST algorithm mainly includes two steps: sampling and SIs computation. The random sample is first generated by converting the multidimensional parameter space into a 1-D space via a transformation function. The sample size N_s is calculated by using the largest sampling frequency $\omega_{\max}$ and the number of exploring curves N_r as

$$N_s = (2M\omega_{\max} + 1)N_r \tag{5.1}$$

where M is a interference factor. To ensure a reasonable sampling density, Saltelli et al. [11] suggested some constraints for the parameters: $M = 4$, $\omega_{\max} \geq 8$ and $16 \leq \omega_{\max}/N_r \leq 64$. Once $\omega_{\max}$ and N_r are fixed, the number of random samples can be determined.

The second step is to calculate the SIs for each model input, which is expressed by two indices: the main sensitivity index (MSI) and the total sensitivity index (TSI), as

$$MSI_i = \widehat{Var_i}(Y)/\widehat{Var}(Y) \tag{5.2}$$

$$TSI_i = 1 - \widehat{Var_{(-i)}}(Y)/\widehat{Var}(Y) \tag{5.3}$$

$$\sum_{j=1,\, j\neq i}^{n} S_{i,j} = TSI_i - MSI_i \tag{5.4}$$

where $\widehat{Var_i}(Y)$ is the estimated conditional variance of the ith factor, $\widehat{Var_{(-i)}}(Y)$ is the estimated conditional variance except for the ith factor, and $\widehat{Var}(Y)$ is the variance of output Y. MSI_i is the main (or first-order) sensitivity index of the ith factor, which represents the main contributions of each input parameter to the variance of the model output, and TSI_i is the total (including higher-order effects) sensitivity index of the ith factor, which considers the interactions among parameters. If MSI_i differs from TSI_i, interactions between parameters exist. Both MSI_i and TSI_i range between 0 and 1; higher MSI_i and TSI_i values suggest more important effects of the ith factor on the output. In Equation (5.4), $S_{i,j}$ represents the parameter interaction. For a detailed description and mathematical derivations of EFAST, readers are referred to [9], [11].

5.2 ENTROPY-BASED SENSITIVITY ANALYSIS (EBSA)

For a random rough surface, the scattering response varies nonlinearly with the surface conditions. There exists a well-known problem of uniqueness in the radar response measurement, because of the complex wave-target interactions, as well as uncertainty sources, such as the random characteristics of the rough surface and noise-corrupted radar echoes [20, 21]. It has been proven that, for the backscattering from a randomly rough surface, the Shannon entropy is a good indicator of the sensitivity of radar response to surface parameters because it not only reflects the probabilistic distribution of scattering coefficient but also the deviation [18]. In EBSA, more information on scattering behavior can be preserved through evaluating detailed parameter sensitivities, predicting the scattering signal saturation, estimating the advantage of multi-polarization and multi-angle, and identifying less significant variables. It is convinced that EBSA offers richer details of scattering sensitivity. The procedure EBSA composes of three steps: sampling input data, computing Shannon entropy (*SE*) and performing sensitivity analysis.

Sampling input data: generating sufficiently ample normally distributed random samples of surface parameters and bistatic configurations. The range of parameters includes surface roughness, dielectric constant, and bistatic configuration.
Computing Shannon entropy: feeding the generated samples into the AIEM model to compute the scattering coefficients and the corresponding Shannon entropy *SE* [22]:

$$SE = \mathrm{E}[\mathbf{H}(\mathbf{b})] = \mathrm{E}[-\ln(P(\mathbf{b}))] \tag{5.5}$$

where E is an expected value operator, and $\mathbf{H}$ is the Hartley's information content of observation $\boldsymbol{b}$, and P is the probability density function (pdf), which can be estimated by kernel density method with a Gaussian kernel [23].
Performing sensitivity analysis: tackling the signal changes in the bistatic scattering pattern by a varying one single parameter with the other parameters remaining unchanged to analyze the radar signal sensitivity to the surface roughness, moisture content, and bistatic configuration at both co- and cross-polarizations are examined.

To assess the performance of SE as well as scattering behavior, we may look into some distribution parameters, including skewness γ, kurtosis κ, and standard deviation σ_{std}.

$$\gamma = \frac{\mathrm{E}[(x - \mu)^3]}{{\sigma_{std}}^3} \tag{5.6}$$

$$\kappa = \frac{\mathrm{E}[(x - \mu)^4]}{{\sigma_{std}}^4} - 3 \tag{5.7}$$

$$\sigma_{std} = \sqrt{\left(\sum_{i=1}^{n} (x_i - \bar{x})^2\right)/n} \tag{5.8}$$

where μ is the mean of the random variable x, and σ_{std} measures statistical dispersion, and γ measures the asymmetry of the data around the sample mean, and

κ indicates how outlier-prone a distribution would be. Thus, γ and κ, equal to zero for a normal distribution. If γ is negative-valued, the data are spread out more to the left of the mean than to the right. Distributions that are more outlier-prone than the normal distribution have a positive value of κ, with a negative value of κ indicating distributions that are less outlier-prone than the normal distribution.

5.3 MONOSTATIC vs. BISTATIC SCATTERING PATTERNS

5.3.1 Monostatic Scattering Patterns

5.3.1.1 Distribution Response of the Backscattering Coefficient

Figure 5.1 is a scattering coefficient distribution with the normal disturbed rough surface parameter as inputs. The exponential correlation function is used at an incident angle of 45°. In the figure, we can see that, for dry and smooth surface, the scattering coefficient remains normal distributed. With increasing surface roughness and moisture content, the behavior of the rough surface scattering is not normally distributed, as is frequently assumed. It is more likely to be Beta distribution, as shown in Figure 5.2. This disagreement can be described by the departures of the γ and κ of returns from normal distribution counterparts. Moreover, we can see that, compared to the surface with varying kl and m_v, the scattering response deviation is more obvious when varying $k\sigma$. Finally, it can be observed that the HH-polarized distribution response is generally similar, but not exactly the same, as the VV-polarized ones.

5.3.1.2 SA by Information Entropy

Figure 5.3 is a volumetric slice of SE with three key surface parameters, namely, $k\sigma$, kl, and m_v in the x, y, z directions, respectively. In the figure, we observe that the entropy related to the HH polarization is generally similar to that of VV polarization. Moreover, the backscattering signals were found to be more sensitive to $k\sigma$ than to m_v and kl. Those results are highly consistent with those by other methods, such as EFAST methods. The consistency between the different models indicates again the feasibility of the proposed entropy-based method in SA.

Note that, compared to other SA methods, the entropy-based method provide more specific and detailed parameter sensitivity under various rough surface conditions, while there is always only single SA data for global SA or limited data for local SA. Therefore, the proposed method can give us more insight into the scattering behaviors of the rough surface and, thus, ensure better use of microwave scattering data in surface monitoring. Another interesting finding is that it is easy to determine the saturation point of backscattering signals in the form of rapidly decreasing entropy. For example, it can be seen that backscattering signals tend to saturate for a rougher surface, and wet soil prompts the saturation.

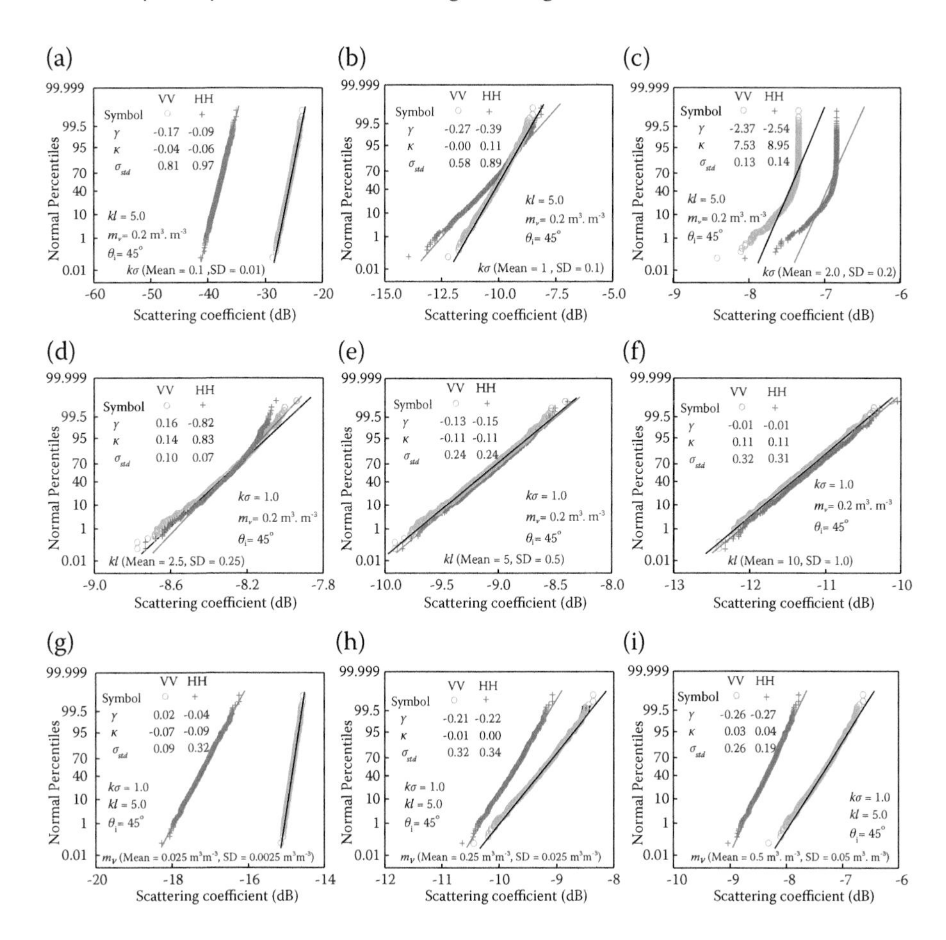

FIGURE 5.1 Probability of the scattering coefficient with normal distributed input surface parameters of **(a–c)** $k\sigma$; **(d–f)** kl; and **(g–i)** m_v, at the incident angle of 45°. The solid line is the reference line for a normal distribution fit. (a, d, g) small $k\sigma$, kl, and m_v; (b, e, h) medium $k\sigma$, kl, and m_v; and (c, f, i) large $k\sigma$, kl, and m_v . The parameters on upper left of the figures show the distribution of the scattering coefficients, while on the bottom are the input parameters.

5.3.2 Bistatic Scattering Patterns

To explore the distribution response of the bistatic scattering coefficient, we start with an intermedium roughness ($k\sigma = 1.0$), with the remaining parameters being fixed to ($kl = 5.0$, $m_v = 0.2\ m^3 m^{-3}$). Figure 5.4 shows the hemispherical plots of the standard deviation σ_{std} and their *SE* values at an incident angle of 50°, for both cross- and co-polarizations. In this plot, the left and right halves correspond to the backscattering and forward scattering direction, respectively, and the horizontal and vertical axes represent the incident and cross planes, respectively. It must be noted that, the exponential correlation function is used because of its good representation of natural surface conditions. Regarding σ_{std}, a stronger co-polarization σ_{std} closes

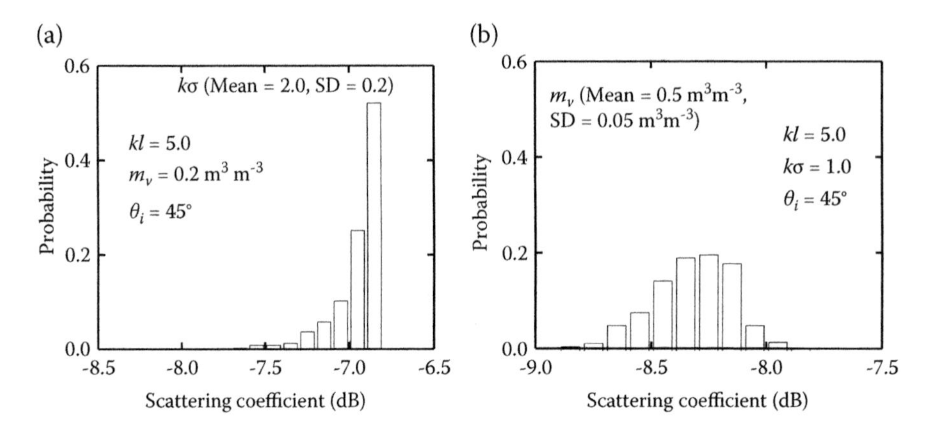

FIGURE 5.2 Probability distribution response of backscattering for rough surface with (**a**) great roughness; and (**b**) high water content.

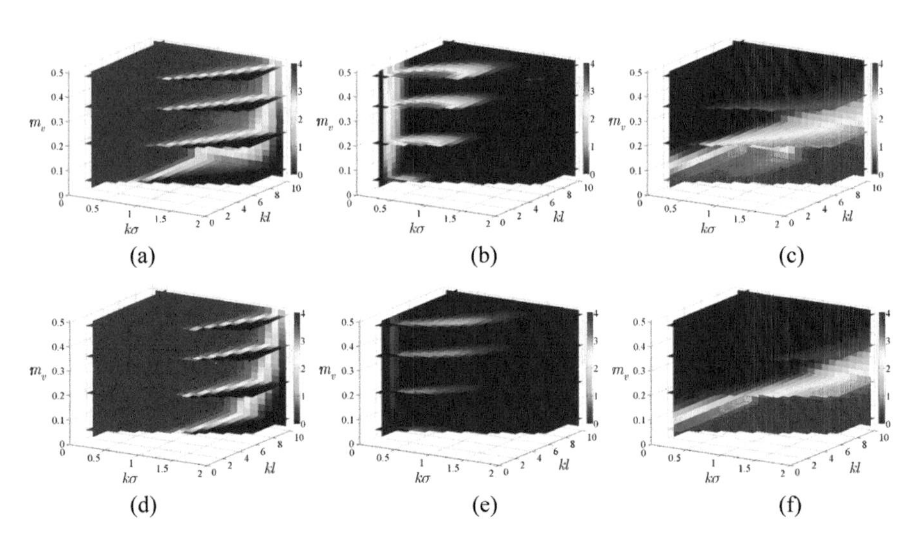

FIGURE 5.3 SE response to rough surface when varying: (**a, d**) $k\sigma$; (**b, e**); and (**c, f**) m_v at a polarization of: (**a–c**) VV; and (**d–f**) HH. The incident angle is 45° and the frequency is 1.26 GHz. Normal distribution disturbances were added to the input parameters, with σ equal to 1% of the maximum of the corresponding surface parameters. As an example, SD is set to 0.02, 0.1, and 0.005 $m^3{\cdot}m^{-3}$ when varying: (**a, d**) $k\sigma$; (**b, e**) kl; and (**c, f**) m_v, respectively.

to the specular region, a smaller σ_{std} in the backward region, and the appearance of a minimum HH-polarized σ_{std} around the cross-plane (the vertical axis), the apparent locations of the VV minima along the arc, and a nearly disappearing cross-polarized σ_{std} in the incidence plane (the horizontal axis). Accordingly, a high SE always corresponds to small deviations from a normal distribution and a large standard deviation σ_{std}. Moreover, the overall cross- and co-polarized distribution responses are generally similar, despite not being exactly the same. It is interesting to find that

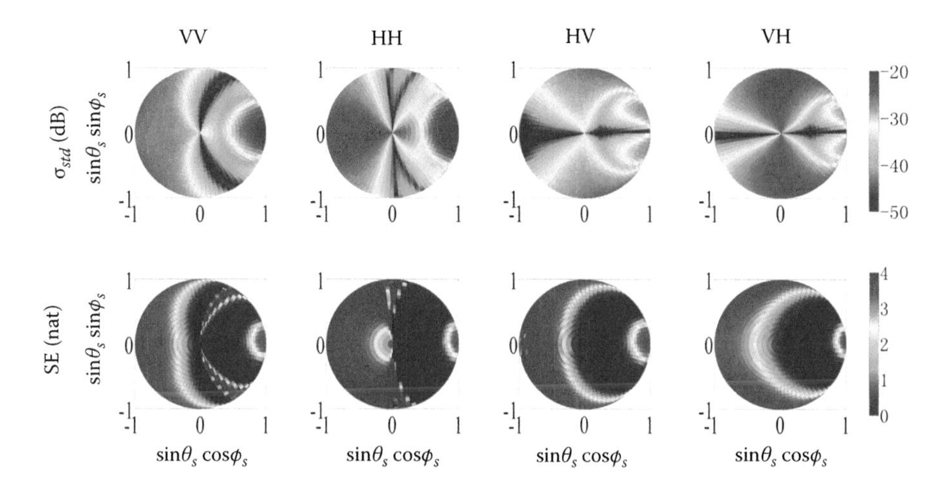

FIGURE 5.4. Hemispherical plots of the standard deviation σ_{std} and the corresponding Shannon entropy (SE) with the normally disturbed $k\sigma$ as the input: VV (first column); HH (second column); HV (third column); VH (fourth column), at $\theta_I = 50°$, with $kl = 5.0$, and $m_v = 0.2\ m^3 m^{-3}$.

the greatest scattering sensitivity to $k\sigma$ is found in the backward direction, and the scattering signals always saturate in the forward direction.

Figure 5.5 shows the hemispherical plots of the standard deviation σ_{std}, and their SE values at an incident angle of 50°, for both cross- and co-polarizations. The scattering response to kl is found to be quite different from that of $k\sigma$. A number of significant SE values are found at large scattering angles in the forward and backward directions, for both cross- and co-polarizations, and around the perpendicular planes for co-polarizations. Thus, the effect of correlation length at those directions should be carefully considered in bistatic observation, whereas it has almost no effect on mono-static observation for most rough surface conditions. Moreover, there appears to be a new, nonsensitive scattering response region in the backward direction with small σ_{std}, such as regions at small scattering angles for VV polarization and around the perpendicular scattering plane for HH polarization.

Figure 5.6 shows hemispherical plots of the bistatic scattering coefficient for surface with normally disturbed water content. In general, the scattering sensitivities to m_v are different for all four polarizations. For VV polarization, scattering in the forward direction is more sensitive than in the backward direction. For the HH polarization, the behavior is the opposite.

5.4 DEPENDENCES ON RADAR PARAMETERS

5.4.1 Sensitivity to Incident Angle

Figure 5.7 shows the entropy as a function of the incident angles. The SE decreases first and then increases, and the smallest entropy is always found around the small

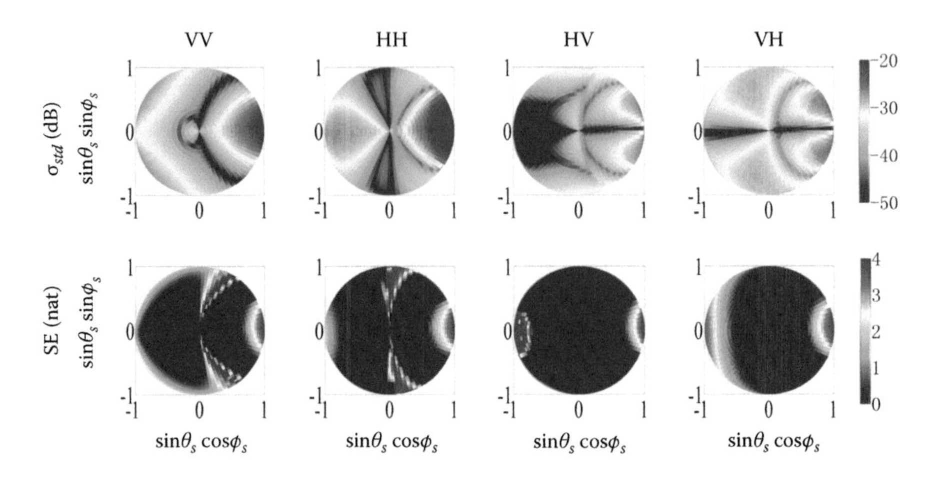

FIGURE 5.5 Hemispherical plots of the standard deviation σ_{std} and the corresponding Shannon entropy (SE) with the normally disturbed kl as the input: VV (first column); HH (second column); HV (third column); VH (fourth column), at $\theta_i = 50°$, with $k\sigma = 1.0$, and $m_v = 0.2\ m^3m^{-3}$.

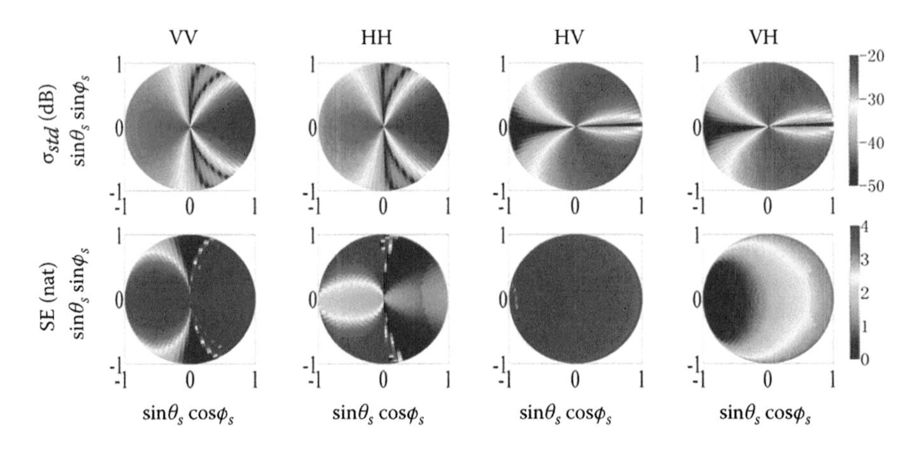

FIGURE 5.6. Hemispherical plots of the standard deviation σ_{std}, and the corresponding Shannon entropy (SE) with the normally disturbed m_v as the input: VV (first column); HH (second column); HV (third column); VH (fourth column), at $\theta_i = 50°$, with $kl = 5.0$, and $k\sigma = 1.0$.

incident angles for a smooth surface, and it shifts toward a greater incident angle for rougher cases. This finding indicates that, for roughness sensing, the backscattering observation at greater incident angles is preferable for a smooth surface, while smaller incident angles for rough surfaces, to retain the maximum information. The distribution of m_v is notably different from that of $k\sigma$. The SE curve with varying $k\sigma$ is monotone, and the scattering signal saturates more quickly for a more wet surface. The above results are based on VV polarization. Similar behavior with HH polarization is observed.

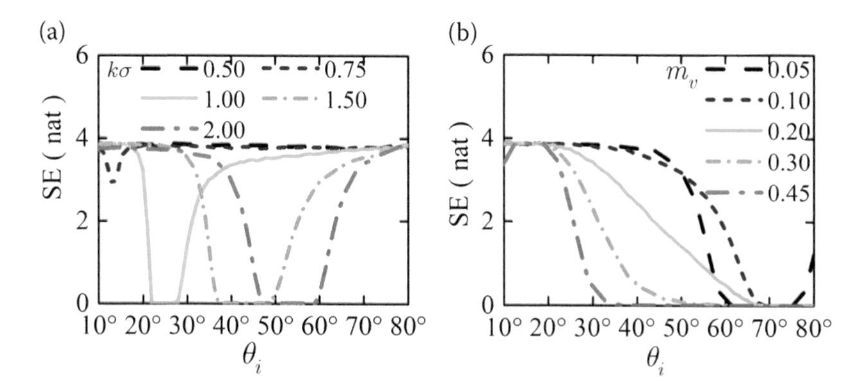

FIGURE 5.7 SE as a function of incident angles for varying (**a**) normalized rms height $k\sigma$ (kl = 5.0, m_v = 0.2 $m^3 \cdot m^{-3}$) and (**b**) moisture content m_v ($k\sigma$ = 1.0, kl = 5.0).

5.4.2 Sensitivity to Polarization

Two sets of simulation data with dual-polarization are combined (i.e., the polarization ratio and the polarization difference), as given in Figure 5.8. The figure shows that dual-polarization is expected to provide some improvements in estimating the moisture content compared to single-polarized measurements by reducing the scattering coefficient saturation for rough and wet surfaces, which is a problematic issue in surface parameter retrieving.

5.4.3 Sensitivity to Multi-Angle

For multi-angle observations, we investigate the SE response of two angles (θ_i = 30°, 45°), and three angles (θ_i = 30°, 45°, 50°) in simulations, as shown in Figure 5.9. Note that only varying the roughness $k\sigma$ and moisture content m_v were considered because the scattering response to kl is comparatively less sensitive. The figure shows that a multi-angle observation can help to address the issue of the scattering signal saturation for wet and rough surfaces, which is similar to the using dual-polarization. It is interesting to note that for multi-angle, which is the combination of backscattering coefficients at two incident angles ($\sigma_{VV(30°)}/\sigma_{VV(45°)}$), the SE is sensitive to $k\sigma$ but not to m_v, while the combination of three angles $(\sigma_{VV(30°)})^2/(\sigma_{VV(45°)}\sigma_{VV(50°)})$ shows sensitivity to both $k\sigma$ and m_v. This finding implies that one can use the ratio between the values of the backscatter data at two angles to estimate the roughness while assigning the soil moisture at a constant value. The soil moisture can then be estimated based on the ratio of the three backscatter values using the estimated roughness. Note that there is an obvious SE dip for a multi-angle observation due to the almost equal scattering coefficients of those angles.

To indicate quantitatively the effect of the angular combination, we compared the sensitivity of bistatic scattering observation, including single input–single output (SISO) and single input–multiple output (SIMO), with a comparison of the mono-static observation in Figure 5.10, which shows the effect of angular

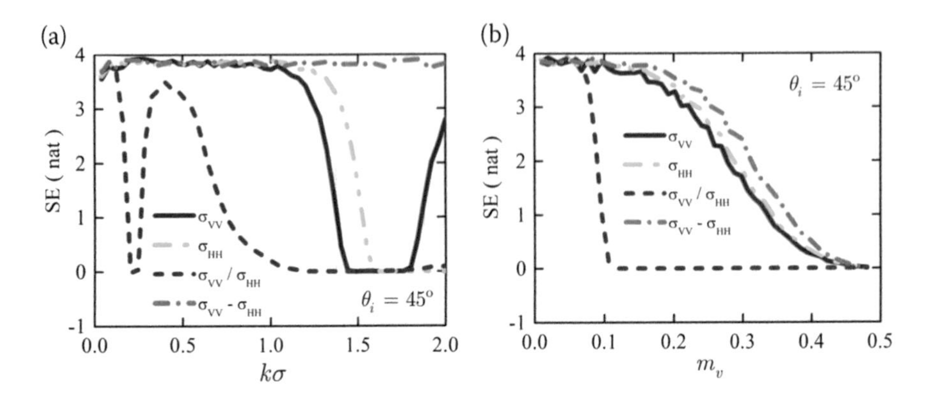

FIGURE 5.8. SE of single- and dual-polarization for varying (**a**) normalized rms height $k\sigma$ ($kl = 5.0$, $m_v = 0.2\ m^3{\cdot}m^{-3}$) and (**b**) moisture content m_v ($k\sigma = 1.0$, $kl = 5.0$) at $\theta_i = 45°$.

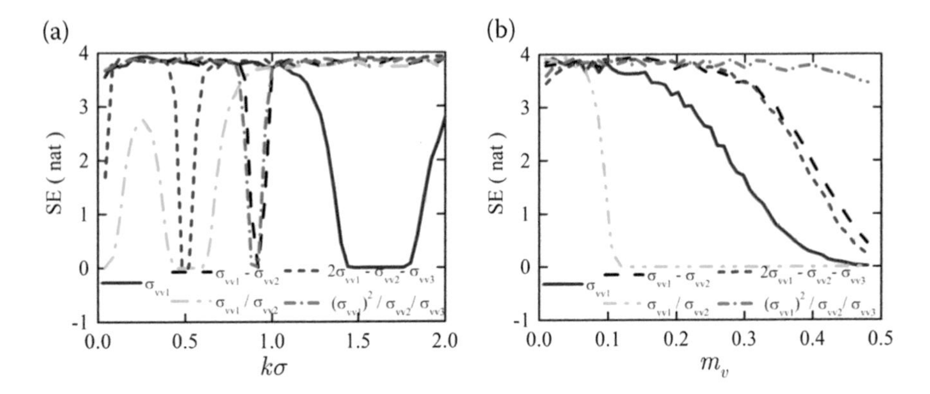

FIGURE 5.9. SE of the single and multi-angle when varying (**a**) normalized rms height $k\sigma$ ($kl = 5.0$, $m_v = 0.2\ m^3{\cdot}m^{-3}$) and (**b**) moisture content $m_v{\cdot}$ $k\sigma = 1.0$, $kl = 5.0$. σ_{VVi} is the VV-polarized scattering coefficient with the incident angle $\vartheta_i = 30°, 45°, 50°$.

combination of bistatic observation on sensitivity to $k\sigma$ and m_υ, with a comparison of mono-static observation for VV polarization. The parameters are given in Table 5.1. The results show that the angular combination can greatly improve the information of interest parameters of the rough surface.

5.5 DEPENDENCES ON SURFACE PARAMETERS

Three key parameters, i.e., soil moisture (m_v), root mean square height (σ), and correlation length (l), which exert significant influence on radar signals are selected, and their distributions and ranges in the AIEM model simulation are listed in Table 5.2. It is known that the ranges of the parameters impact the sensitivity indices and importance ranking of the parameters, which depends on the available information and

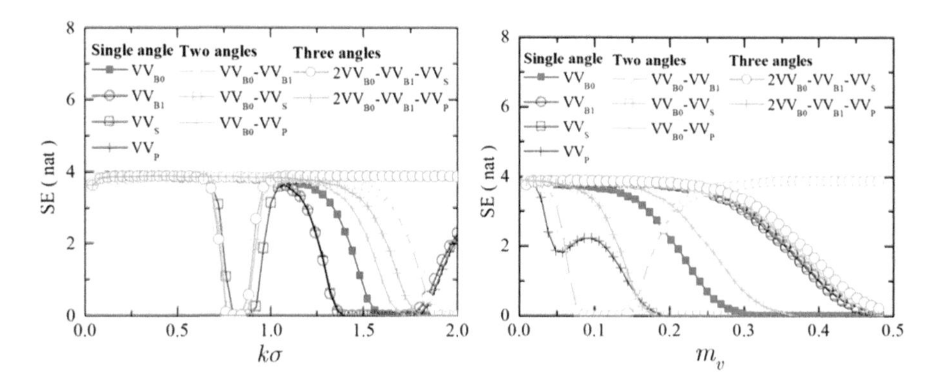

FIGURE 5.10. SE as a function of (**a**) $k\sigma$, with $m_\upsilon = 1$ and $m_\upsilon = 0.2\,m^3{\cdot}m^{-3}$ and (**b**) m_υ, with $k\sigma = 1$ and $m_v = 0.2\,m^3m^{-3}$.VV_n, n = B_0, B_1, S, P, B_0 and B_1 are backscattering observations with (ϑ_s, ϕ_s) equal to (50°, 180°) and (40°, 180°), respectively, S shows the observation in the specular scattering plane (50°, 0°), and P in perpendicular plane (50°, 90°) at $\theta_I = 50°$.

TABLE 5.1

Angular Combination of Bistatic Observations

Type		Receiving Direction	Receiving Angle	
mono-static	SISO	Backward	(50°, 180°)	B0
	SISO	Backward	(40°, 180°)	B1
	SISO	Perpendicular	(50°, 90°)	P
	SISO	Specular	(50°, 0°)	S
	SIMO	Backward	(50°, 180°), (40°, 180°)	B0, B1
Bistatic	SIMO	Backward, Perpendicular	(50°, 180°), (50°, 90°)	B0, P
	SIMO	Backward, Specular	(50°, 180°), (50°, 0°)	B0, S
	SIMO	Backward, Perpendicular	(50°, 180°), (40°, 180°), (50°, 90°)	B0, B1, P
	SIMO	Backward, Specular	(50°, 180°), (40°, 180°), (50°, 0°)	B0, B1, S

Note:
The incident angles are (50°, 0°). SISO means single input and single output observation; SIMO single input and multi-output.

the goal of the study. In some regional-dependent studies (e.g., [12], [24]), the ranges of the parameters are determined from local measurements. We may want to cover all the cases that can represent natural surface conditions (e.g., from a very smooth or dry surface to a rough or wet surface). It is a common practice to analyze the parameters covering a wide range of conditions [25].

We conduct SA tests on the whole upper half-space with $0° \leq \theta_s \leq 60°$ and $0° \leq \phi_s \leq 180°$ for the incident angles of 40°. It is known that bistatic scattering behavior is controlled by the surface correlation function together with σ and l, for characterizing

TABLE 5.2
Input Parameters and Their Distributions and Ranges in the AIEM Scattering Model

Parameter (unit)	Definition	Distribution	Range
m_v (m^3 m^{-3})	soil moisture	uniform	0.01~0.5
σ (cm)	RMS height	uniform	0.1~3
l (cm)	correlation length	uniform	2.5~35

surface roughness. Three commonly used correlation functions including exponential, Gaussian, and 1.5-power, are selected. Since the cross-polarized scattering coefficient is more sensitive to the volume scattering, it is often used to detect the growth of vegetation rather than soil moisture [26]–[27]. We calculate the SIs (TSI and MSI) and their difference (parameter interaction effects) for soil moisture, RMS height, and correlation length with respect to the bistatic scattering coefficient at both HH and VV polarizations respectively. Since the distributions of MSI and TSI on the whole upper half-pace are similar, we simply present the results of MSI for the three parameters.

5.5.1 Sensitivity to Soil Moisture

Figure 5.11 displays the distribution of MSI of soil moisture that is associated with VV-polarized and HH-polarized bistatic scattering coefficients on the whole upper half-space as a function of scattering angle θ_s and azimuth angle ϕ_s for incident angles of 40° with different correlation functions respectively. For VV polarization, as shown in Figure 5.11(a–c), the maximum sensitivity to soil moisture is in the forward direction regardless of incident angles and correlation functions. It was found that the bistatic scattering signals at or near an orthogonal configuration are most sensitive to soil moisture, and a small incident angle at VV polarization was suggested for SAOCOM to infer soil moisture [28]. However, the orthogonal or quasi-orthogonal directions may pose a problem because the co-polarized radar signals in these regions are small and maybe close to the receiver noise floor (i.e., the lowest measurable scattering coefficient due to system noise). Thus, from a practical point of view, observations near the cross-plane are not suggested [25]. Moreover, we can see there is always an obvious high-sensitivity point for MSI of soil moisture and a corresponding low-sensitivity point for RMS height near the cross-plane, regardless of the correlation function, when the scattering angle is equal to the incident angle. This phenomenon may be caused by the combined effects of both geometric parameters and ground surface conditions on radar signals, and may also be related to the "dip" feature in the azimuthal plane. Since the co-polarized radar signals in these regions are small, it is challenging to discriminate the return signals at such a narrow-angle. Meanwhile, from Figure 5.11(a–c) we can observe that correlation function does not have much impact on the bistatic sensitivity pattern of soil moisture, which leads to similar distributions of MSI of soil moisture under different correlation functions. As shown in Figure 5.11(d–f), we can see that the bistatic sensitivity pattern of soil moisture at HH polarization is generally different

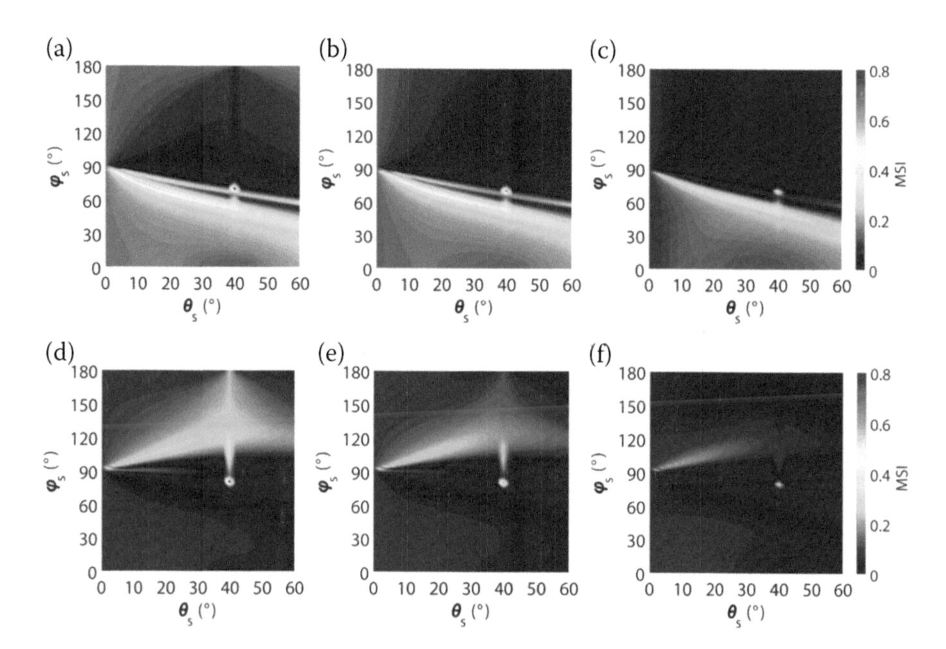

FIGURE 5.11 MSI of soil moisture derived from VV-polarized (top row: (**a–c**)) and HH-polarized (bottom row: (**d–f**)) scattering coefficients as a function of scattering angle θ_s and azimuth angle ϕ_s at L-band for incident angles, θ_i, of 40° with different correlation functions from left to right, i.e., (**a, d**) exponential, (**b, e**) 1.5-power, and (**c, f**) Gaussian.

from that at VV polarization, though with a few similarities. An intermediate incident angle (e.g., 40°) is recommended if HH polarization is used to estimate soil moisture. However, we can clearly see that the MSI values of soil moisture at HH polarization are generally small compared with those at VV polarization on the whole upper half-space, which is even more evident for a Gaussian-correlated surface. This finding is confirmed in more detail based on a global SA analysis, which demonstrates that bistatic VV polarization is preferable to HH polarization for soil moisture estimation. More numerical results and discussions can be found in [29].

In what follows, we examine the interactions between m_v, σ, and l and other parameters as functions of scattering angle θ_s and azimuth angle ϕ_s, with incident angles θ_i of 40° with exponential, 1.5-power and Gaussian correlation functions. We first present the results of VV polarization for soil moisture in Figure 5.12(a–c) and HH polarization in Figure 5.12(d–f). We can see that for all three parameters, the interaction effects are generally small and close to zero for most parts of the whole upper half-space. That is, in these scattering geometries, the first-order (MSI) and total-order (TSI) sensitivity indices are almost equal. Figure 5.12(a–c) displays the interactions of soil moisture with other parameters (e.g., soil moisture with RMS height, soil moisture with correlation length, and soil moisture with RMS height and correlation length). We see that the maximum interaction effects are very close to the cross-plane. As shown in Figure 5.12(d–f), for HH polarization, the difference between TSI and MSI for all the three parameters is very small on the whole upper

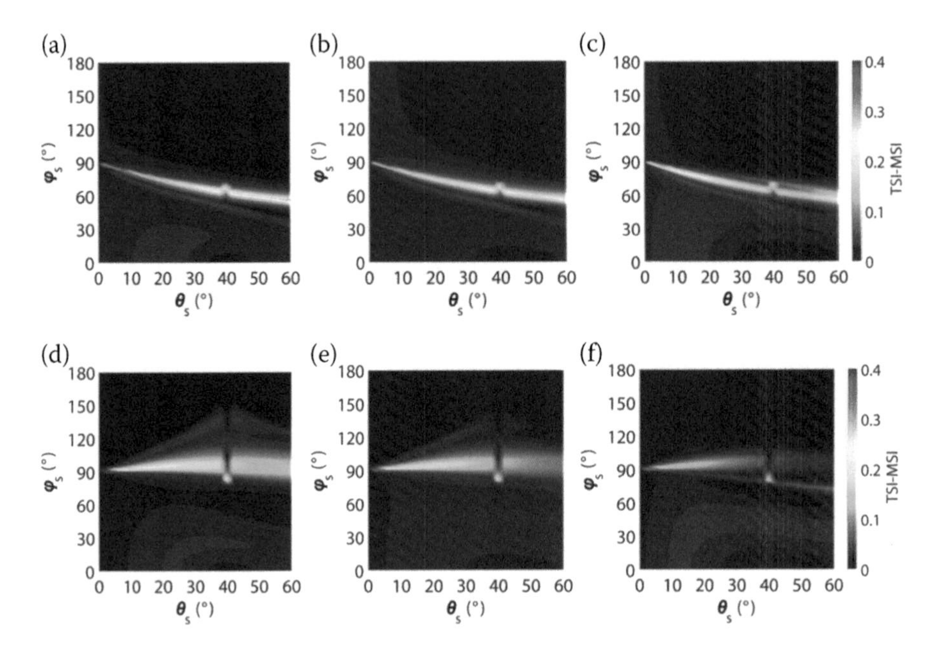

FIGURE 5.12 TSI-MSI (the difference between TSI and MSI) of soil moisture derived from VV-polarized (top row: (**a–c**)) and HH-polarized (bottom row: (**d–f**)) scattering coefficients as a function of ϑ_s and ϕ_s at *L*-band for incident angles ϑ_i of 40° with correlation functions from left to right, i.e., (**a, d**) exponential, (**b, e**) 1.5-power, and (**c, f**) Gaussian.

half-space at L-band [29]. It was found that the interactions are smallest at L-band and increase as the frequency increases. Since it is almost impossible to decompose these interactions from radar signals that introduce uncertainties into soil moisture retrieval, observations at L-band for soil moisture estimation are suggested.

5.5.2 Sensitivity to RMS Height

Figure 5.13 displays the distribution of MSI of RMSH that are associated with VV-polarized and HH-polarized bistatic scattering coefficients on the whole upper half-space as a function of scattering angle θ_s and azimuth angle ϕ_s for incident angles θ_i of 40° with different correlation functions respectively. For VV polarization, as shown in Figure 5.13(a–c), we can see that the RMSH exerts a significant influence on radar signals in a bistatic mode. Note that the sensitive zones of RMSH are largest for exponentially correlated surfaces and smallest for Gaussian-correlated surfaces. Moreover, we can see there is always an obvious low-sensitivity point for RMSH near the cross-plane, regardless of the correlation function, when the scattering angle is equal to the incident angle. This phenomenon may be caused by the combined effects of both geometric parameters and ground surface conditions on radar signals, and may also be related to the "dip" feature in the azimuthal plane. Since the co-polarized radar signals in these regions are small, it is challenging to discriminate the return signals at such a narrow-angle. Fortunately, these sensitive

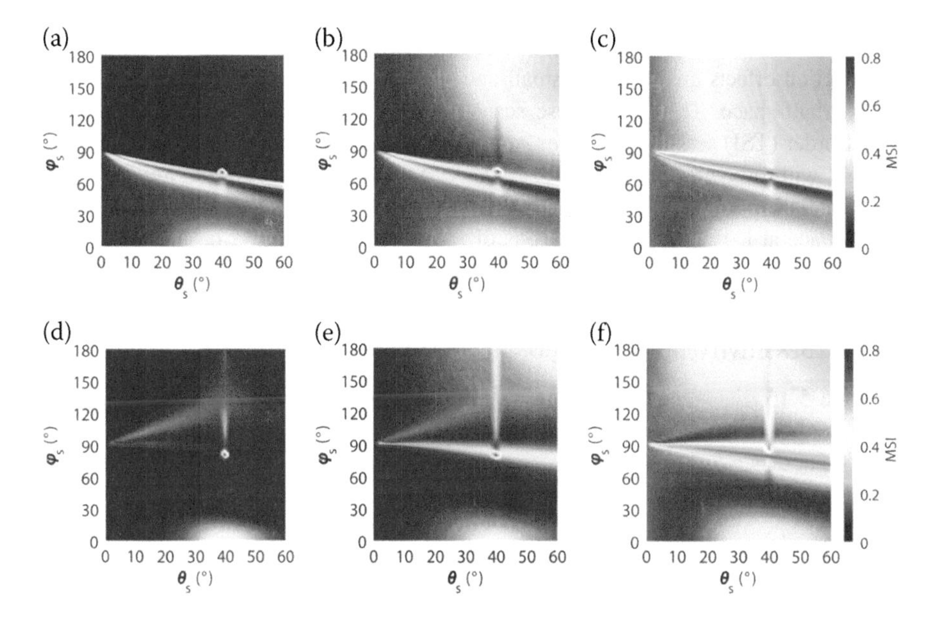

FIGURE 5.13 MSI of RMS height derived from VV-polarized (top row: (**a–c**)) and HH-polarized (bottom row: (**d–f**)) scattering coefficients as a function of scattering angle ϑ_s and azimuth angle ϕ_s at *L*-band for incident angles ϑ_i of 40° with different correlation functions from left to right, i.e., (**a, d**) exponential, (**b, e**) 1.5-power, and (**c, f**) Gaussian.

zones become closer to the forward direction at small azimuth scattering angles and large scattering angles. In these regions, the scattering signals are measurable and insensitive to RMS height regardless of the correlation function. In contrast to soil moisture, the MSI distributions of RMS height heavily depend on the correlation function. The bistatic scattering coefficients show the highest sensitivity to RMS height under the exponential correlation function. Compared with an exponential correlation function, a Gaussian correlation function is sensitive to RMS height. This phenomenon is clearly revealed by the EFAST algorithm. The sensitivity of RMS height always decreases near the specular region (i.e., $\theta_s = \theta_i$) and is independent of the incidence angles and correlation functions. For HH polarization, as shown in Figure 5.13(d–f), although the bistatic sensitivity patterns of RMS height at HH and VV polarizations are somewhat different, similar phenomena can be observed. The sensitivity of RMS height is reduced near the specular region, regardless of the incident angles and correlation functions. Also, the bistatic scattering shows the highest sensitivity to RMS height for exponential correlation function, for both HH and VV polarizations. This is because the exponentially correlated surface inherently contains rich micro-roughness scales and is prone to sensitivity to RMS height.

In what follows, we examine the difference between TSI and MSI (TSI-MSI) of RMSH derived from VV-polarized and HH-polarized scattering coefficients as a function of scattering angle θ_s and azimuth angle ϕ_s at L-band for incident angles θ_i of 40° with exponential, 1.5-power and Gaussian correlation functions, as shown in

Figure 5.14. For VV polarization, as shown in Figure 5.14(a–c), we can see that the interaction effects are generally small and close to zero for most parts of the whole upper half-space. That is, in these scattering geometries, the first-order (MSI) and total-order (TSI) sensitivity indices are almost equal. The interaction effects of RMS height with other parameters for HH polarization are examined in Figure 5.14(d–f). In general, the difference between TSI and MSI is very small on the whole upper half-space at L-band. The interactions of RMS height with other parameters are the highest at the exponentially correlated surface [29].

5.5.3 Sensitivity to Correlation Length

Figure 5.15 displays the MSI of correlation length that is associated with VV-polarized and HH-polarized bistatic scattering coefficients on the whole upper half-space as a function of scattering angle θ_s and azimuth angle ϕ_s for incident angles θ_i of 40° with exponential, 1.5-power and Gaussian correlation functions, respectively. For VV polarization, as shown in Figure 5.15(a–c), the sensitive zones of soil moisture are the largest for exponentially correlated surfaces and smallest for Gaussian-correlated surfaces. This is attributed to the significantly increased sensitivity of correlation length when the Gaussian correlation function is applied, implying that the sensitivity index of correlation

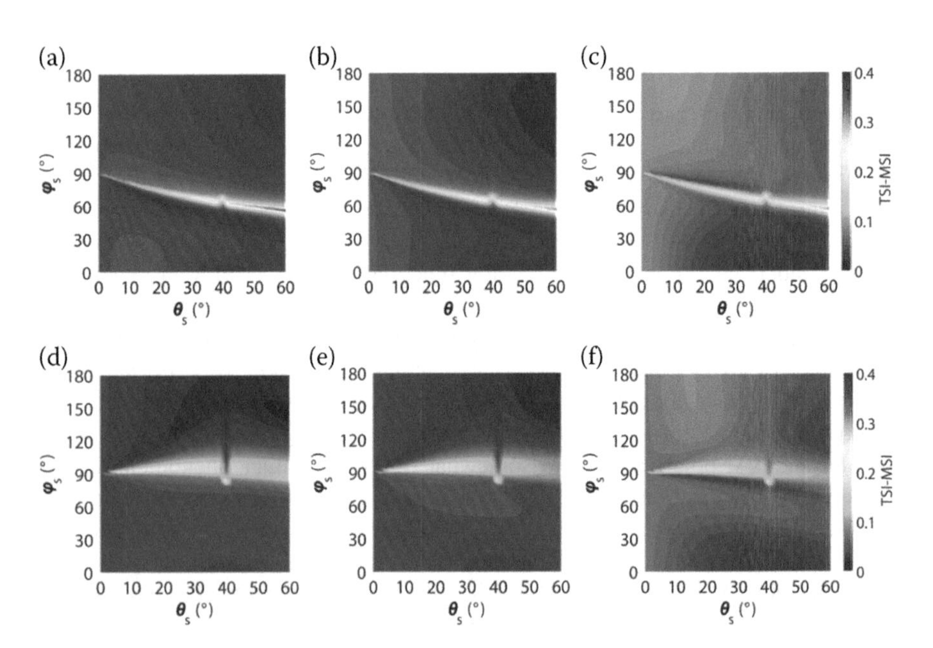

FIGURE 5.14 TSI-MSI (the difference between TSI and MSI) of RMSH derived from VV-polarized (top row: (**a–c**)) and HH-polarized (bottom row: (**d–f**)) scattering coefficients as a function of scattering angle θ_s and azimuth angle ϕ_s at L-band for incident angles θ_i of 40° with different correlation functions from left to right, i.e., (**a, d**) exponential, (**b, e**) 1.5-power, and (**c, f**) Gaussian.

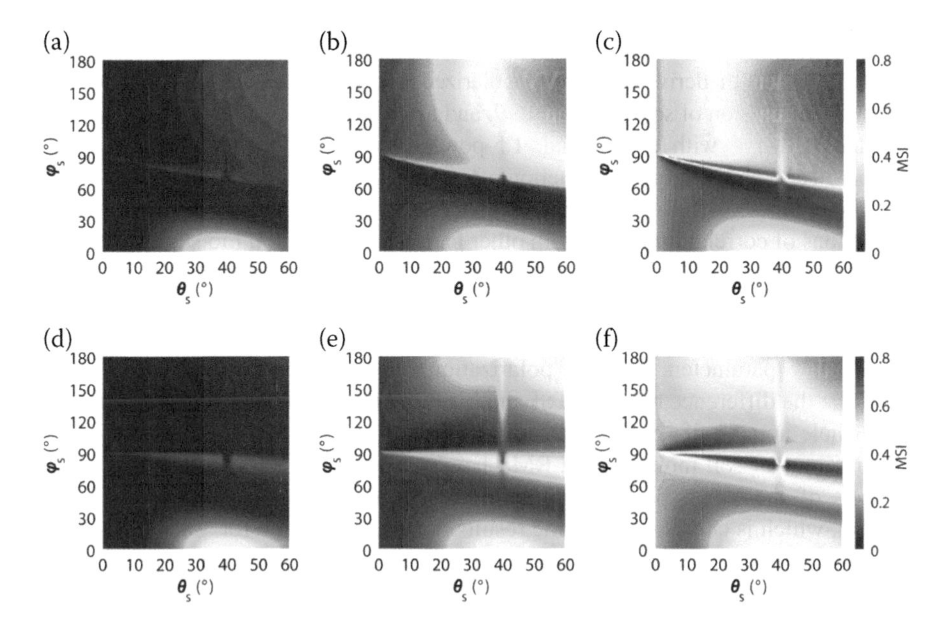

FIGURE 5.15 MSI of the correlation length derived from VV-polarized (top row: (**a–c**)) and HH-polarized (bottom row: (**d–f**)) scattering coefficients as a function of scattering angle θ_s and azimuth angle ϕ_s at L-band for incident angles θ_i of 40° with different correlation functions from left to right, i.e., (**a, d**) exponential, (**b, e**) 1.5-power, and (**c, f**) Gaussian.

length is much higher with the Gaussian correlation function than exponential correlation function for backscattering [29]. In contrast to soil moisture, the MSI distributions of the correlation length strongly depend on the correlation function. We can see that the sensitivities of correlation length are generally complementary, that is, the bistatic scattering coefficients show the lowest sensitivity to correlation length under exponential correlation function. The sensitivity of radar signals to correlation length increases significantly for the Gaussian correlation function. Compared with an exponential surface, a Gaussian surface is much more sensitive to the correlation length. Moreover, the sensitivity of correlation length near the specular region (i.e., $\theta_I = \theta_s$) always increases. This indicates that the specular region or quasi-specular region may be not preferred for soil moisture estimation due to the influence of the correlation length. In Figure 5.15(d–f), we investigate the distributions of MSI of the correlation length for HH-polarized scattering coefficients on the whole upper half-space as functions of scattering angle θ_s and azimuth angle ϕ_s for three incident angles and correlation functions. The sensitivity of correlation length is generally enhanced, regardless of the incident angles and correlation functions. In addition, the bistatic scattering shows the highest sensitivity to correlation length under the Gaussian correlation function for both HH and VV polarizations. This is because the Gaussian-correlated surface is more sensitive to correlation length, as already shown.

In Figure 5.16, we examine the difference between TSI and MSI (TSI-MSI) of correlation length derived from VV-polarized and HH-polarized scattering coefficients as a function of scattering angle θ_s and azimuth angle ϕ_s at L-band for incident angles θ_i of 40° with exponential, 1.5-power and Gaussian correlation functions, respectively. Note that the interaction effects are generally small and close to zero for most parts of the whole upper half-space. Figure 5.16(a–c) displays the interactions of correlation length with other parameters (e.g., soil moisture with RMS height, soil moisture with correlation length, and soil moisture with RMS height and correlation length). We see that the maximum interaction effects are very close to the cross-plane. Next, we investigate the interaction effects of correlation length with other parameters for HH polarization, as shown in Figure 5.16(d–f). In general, the difference between TSI and MSI for correlation length is very small on the whole upper half-space at *L*-band. In addition, we see that the distributions of interactions of soil moisture and RMS height are very similar, and the interaction effects of these two parameters are generally higher than those of correlation length, which is consistent with the results that were obtained for VV polarization. The interactions of correlation length with other parameters are the highest at the Gaussian-correlated surface.

At this point, we may draw a short summary regarding the parameter sensitivity conducted for the rough surface. Extensive analysis [29] show the SIs

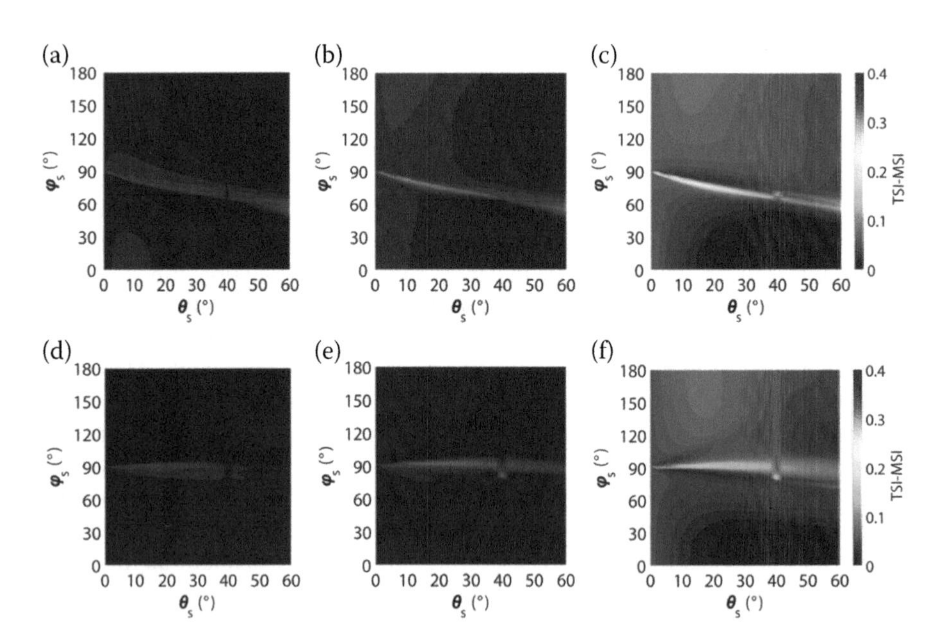

FIGURE 5.16 TSI-MSI (the difference between TSI and MSI) of the correlation length derived from VV-polarized (top row: (**a–c**)) and HH-polarized (bottom row: (**d–f**)) scattering coefficients as a function of scattering angle ϑ_s and azimuth angle ϕ_s at L-band for incident angles ϑ_i of 40° with different correlation functions from left to right, i.e., (**a, d**) exponential, (**b, e**) 1.5-power, and (**c, f**) Gaussian.

(TSI, MSI, and their difference) of soil moisture and surface roughness vary significantly for different bistatic configurations. In bistatic scattering, VV polarization is generally more sensitive to soil moisture than HH polarization, and the sensitivity becomes more remarkable as the incident angle increases. Both VV- and HH-polarized bistatic scattering show the highest sensitivity to soil moisture near the orthogonal plane at a small incident angle (e.g., 20°), but with very low returned signals and undesirable interaction effects. For VV polarization, the sensitive zone of soil moisture increases significantly as the incident angle increases. Moreover, these sensitive zone shifts toward the forward direction with small azimuth scattering angles and large scattering angles (i.e., farther away from the cross-plane), particularly at large incident angles (e.g., 60°). The scattering signals are measurable and preserve rich information on soil moisture at these geometries, making them favorable configurations for soil moisture retrieval. For HH polarization, as the incident angle increases, the high-sensitivity zone of soil moisture gradually becomes closer to the backward direction with large azimuth scattering angles and large scattering angles. However, an intermediate incident angle (e.g., 40°) is recommended for retrieving soil moisture when using HH-polarized bistatic scattering because low sensitivity is exhibited at small (e.g., 20°) and large incident angles (e.g., 60°) and in the presence of interaction effects.

It has been pointed out [30] that great care must be taken to estimate the sensitive parameters, while to assign those less sensitive parameters to a constant value for calibrating and for simplifying the scattering model. A study in [30] suggests that the parameter ranges affect both the SIs and their relative importance. A wider range of a given parameter corresponds to a larger SIs of that parameter. Parameters with the same range of variance but span different intervals exhibit different degrees of sensitivity. Smaller soil moisture and roughness generally exhibit higher sensitivity.

A final call is that specific receiver sensitivity, dynamic range, antenna footprint tracking, and synchronization, among others, in the context of the complex bistatic radar system. To this end, efforts to validate the optimal bistatic radar configurations by conducting indoor and field experiments as well as using real satellite bistatic measurements (e.g., Cyclone Global Navigation Satellite System) deserves further study.

REFERENCES

1. Ulaby, F. T., and Long, D. G., *Microwave Radar and Radiometric Remote Sensing*, University of Michigan Press: Ann Arbor, MI, 2014.
2. Crosson, W. L., Limaye, A. S., and Laymon, C. A., Parameter sensitivity of soil moisture retrievals from airborne L-band radiometer measurements in SMEX02. *IEEE Trans. on Geosci. Remote Sens.*, 43, 1517–1528, 2005.
3. Du, Y., Ulaby, F. T., and Dobson, M. C., Sensitivity to soil moisture by active and passive microwave sensors. *IEEE Trans. on Geosci. Remote Sens.*, 38, 105–114, 2000.
4. Zribi, M., and Dechambre, M., A new empirical model to retrieve soil moisture and roughness from C-band radar data. *Remote Sens. Environ.*, 84, 42–52, 2003.

5. Verhoest, N. E., Lievens, H., Wagner, W., Álvarez-Mozos, J., Moran, M. S., and Mattia, F., On the soil roughness parameterization problem in soil moisture retrieval of bare surfaces from synthetic aperture radar. *Sensors*, 8, 4213–4248, 2008.
6. Lievens, H., Verhoest, N. E., De Keyser, E., Vernieuwe, H., Matgen, P., Álvarez-Mozos, J., and De Baets, B., Effective roughness modeling as a tool for soil moisture retrieval from C-and L-band SAR. *Hydrolol. Earth Sys.*, 15, 151–162, 2011.
7. Zribi, M., Gorrab, A., and Baghdadi, N., A new soil roughness parameter for the modelling of radar backscattering over bare soil, *Remote Sens. Environ.*, 152, 62–73, 2014.
8. Pianosi, F., Sarrazin, F., and Wagener, T., A Matlab toolbox for global sensitivity analysis, *Environ. Modeling Softw.*, 70, 80–85, 2015.
9. Saltelli, A., Ratto, M., and Andres, T., *Global Sensitivity Analysis: The Primer*, Chichester, Wiley, London, UK, 2008.
10. Saltelli, A., and Annoni, P., How to avoid a perfunctory sensitivity analysis, *Environ. Modelling and Software*, 25(12), 1508–1517, 2010.
11. Saltelli, A., Tarantola, S., and Chan, K. S., A quantitative model-independent method for global sensitivity analysis of model output, *Technometrics*, 41(1), 39–56, 1999.
12. Wang, J., Li, X., Lu, L., and Fang, F., Parameter sensitivity analysis of crop growth models based on the extended Fourier amplitude sensitivity test method. *Environ. Model. and Softw.*, 48, 171–182, 2013.
13. Li, D., Jin, R., Zhou, J., and Kang, J., Analysis and reduction of the uncertainties in soil moisture estimation with the L-MEB model using EFAST and ensemble retrieval. *IEEE Geosci. Remote Sens. Lett.*, 12, 1337–1341, 2015.
14. Lentini, N. E., and Hackett, E. E., Global sensitivity of parabolic equation radar wave propagation simulation to sea state and atmospheric refractivity structure. *Radio Sci.*, 50, 1027–1049, 2015.
15. Neelam, M., and Mohanty, B. P., Global sensitivity analysis of the radiative transfer model. *Water Resour. Res.*, 51, 2428–2443, 2015.
16. Petropoulos, G., and Srivastava, P. K., (Eds.), *Sensitivity Analysis in Earth Observation Modelling*, Elsevier, Oxford, UK, 2016.
17. Liu, H., Chen, W., and Sudjianto, A. Relative Entropy Based Method for Probabilistic Sensitivity Analysis in Engineering Design, *Journal of Mechanical Design*, 128(2), 326. doi:10.1115/1.2159025, 2006.
18. Liu, Y., and Chen, K. S., An information entropy-based sensitivity analysis of radar sensing of rough surface. *Remote Sensing*, 10(2), 286, 2018.
19. Sobol, I. M., Sensitivity estimates for nonlinear mathematical models, *Math. Model. Comput. Experim.*, 1(4), 407–414, 1993.
20. Cook, C. E., and Bernfeld, M., *Radar Signals: An Introduction to Theory and Application*, Artech House, Norwood, MA, 1993.
21. Lee, J. S., and Pottier, E., *Polarimetric Radar Imaging: From Basics to Applications*. CRC Press, Boca Raton, FL, 2009.
22. Cover, T. M., and Thomas, A., *Elements of Information Theory*, 2nd Edition, Wiley, New York, 2006.
23. Silverman, B. W., *Density Estimation for Statistics and Data Analysis*, Chapman and Hall/CRC Press, Boca Raton, FL, 1998.
24. Neelam, M., and Mohanty, B. P., Global sensitivity analysis of the radiative transfer model, *Water Resour. Res.*, 51(4), 2428–2443, 2015.
25. Zeng, J. Y., Chen, K. S., Bi, H. Y., Chen, Q., and Yang, X. F., Radar response of off-specular bistatic scattering to soil moisture and surface roughness at L-band, *IEEE Geosci. Remote Sens. Lett.*, 13(2), 1945–1949, 2016.
26. Kim, Y., and van Zyl, J. J., A time-series approach to estimate soil moisture using polarimetric radar data, *IEEE Trans. Geosci. Remote Sens.*, 47(8), 2519–2527, 2009.

27. Arii, M., van Zyl, J. J., and Kim, Y., A general characterization for polarimetric scattering from vegetation canopies, *IEEE Trans. Geosci. Remote Sensing*, 48(9), 3349–3357, 2010.
28. Pierdicca, N., Guerriero, L., Comite, D., Brogioni, M., and Paloscia, S., Bistatic radar with large baseline for bio-geophysical parameter retrieval, *Proc. IEEE IGARSS*, 125–128, 2017.
29. Zeng, J. Y., Chen, K. S., Theoretical Study of Global Sensitivity Analysis of L-Band Radar Bistatic Scattering for Soil Moisture Retrieval, *IEEE Geosci. Remote Sens. Letters*, 15(11), 1710–1714, 2018.
30. Ma, C., Li, X., and Wang, S., A global sensitivity analysis of soil parameters associated with backscattering using the advanced integral equation model, *IEEE Trans. Geosci. Remote Sens.*, 53(10), 5613–5623, 2015.

6 Geophysical Parameters Estimation

6.1 BAYESIAN ESTIMATION

Parameter estimation has a long history of research and is an ultimate goal in the remote sensing of the Earth's surface. Multiple parameter estimations from radar measurements becomes a difficult task [1,2]. Several reasons make this so. Because radar echoes are corrupted by noise, this will profoundly, among other factors, affect the accuracy of any parameter estimation. Models that accurately describe the characteristics of radar–surface interactions are usually nonexistent. Well-calibrated in situ measurements, on the other hand, are too small in number to be reliably used. Parameter retrieval from radar returns which are embedded in a noise background are probabilistic in nature. Conventionally, inversion methods, such as those discussed earlier, have been developed to extract a single parameter at a time. However, radar return signals from a natural soil surface are simultaneously affected by surface characteristics such as soil roughness, correlation length, or dielectric constant. It is generally difficult to take hold of the dependence of the measured scattered signal on these parameters within a system limitation. Data enhancement may help to decouple one parameter from another to certain degrees. As a coherent system, radar measurement is always subject to both multiplicative noise and additive noise. We discuss the additive noise first. For radar measurements (observations) $\mathbf{y}$ that is noisy and may be expressed as

$$\mathbf{y} = \mathbf{y}_t + \mathbf{y}_n \tag{6.1}$$

where $\mathbf{y}_t$ is "truth" value to be estimated and $\mathbf{y}_n$ is noisy and random component.

From the radar equation for a distributed target, Equation 3.38, we see that the received power P_r is the sum of expected signal power P_s and unwanted component, the noise power P_n from system noise: $\overline{P_r} = P_s + P_n$. A good measure to evaluate the impact of noise power on the estimated signal power is the measurement precision [3]:

$$\kappa_p = \sqrt{\frac{\mathrm{var}(\hat{P}_s)}{P_s}} = \sqrt{\frac{1}{N_r}\left(1 + \frac{1}{SNR}\right)^2 + \frac{1}{N_n}\left(1 + \frac{1}{SNR}\right)^2} \tag{6.2}$$

where N_r, N_n are, respectively, the independent samples of the averaged received power and the noise power, and the signal-to-noise ratio, $SNR = P_s/P_n$. For most of the system operation, it is set to $N_r = N_n = N$. The independent samples can be obtained by either spatial sampling or frequency sampling, or both such that $N = N_k N_f$, where N_k is the number of spatial samples and N_f is the number of frequency samples. To warrant the spatial samples independence, it is required that $d_s > \ell$, where d_s is the spacing between two measurements corresponding to two different spots within the same distributed targets, and ℓ is horizontal correlation length of the surface or volume. The number of frequency-independent samples indeed is the ratio of total system modulation bandwidth to the decorrelation bandwidth. To avoid confusion, in what follows, we shall call N as look number L, though for non-imaging radar measurements such as from scatterometer and altimeter, look number is not as commonly used as in imaging radar.

At this point, it is worth mentioning that estimating σ° from P_s via P_r, (see Equation 3.38), involves several processes, each process, to a different extent, introduces error. For example, solving the integral Equation 3.38 induces error which may be affected by the antenna pattern, the illumination area size, etc. In short, it is affected by the radiometric and geometric properties of the system and of the scene being observed. Note that the integration over an illuminated area overlapping by transmitting and receiving antenna patterns also contributes estimation error. For backscattering and a narrow-beam system, error can be smaller and sometimes ignorable as in many practical applications. Care must be exercised, however, for bistatic measurements at which time-phase-beam synchronizations are critical.

In addition to the additive noise, the inherent multiplicative noise is embedded in the scattering coefficient. For a spatially homogeneous scene with reflectivity variance of σ^2 within a resolution cell or pixel, the returned radar amplitude follows the Rayleigh distribution, while the returned power follows exponential distribution [1]. The stronger the speckle noise σ^2, the wider the distribution, making the estimation of returned power or amplitude more difficult and less accurate. To reduce the speckle strength, more independent samples (look number) of measurement can be made to average down the variance, called multi-looking process, in particular in radar imagery data. If L looks are independent and have equal power, then the distribution of power (intensity) square-law detector is [1]

$$p_I(I|\sigma^2) = \left(\frac{L}{\sigma^2}\right)^L \frac{1}{\Gamma(L)} \exp\left(-\frac{LI}{\sigma^2}\right) I^{L-1}, \; I \geq 0, \tag{6.3}$$

which is the Gamma distribution with degree of freedom $2L$. The speckle strength is given by the ratio the standard deviation to mean:

$$\gamma_s = \frac{\sqrt{\langle I^2 \rangle - \langle I \rangle^2}}{\langle I \rangle} = \frac{1}{\sqrt{L}}. \tag{6.4}$$

For single look data, the mean equals to standard deviation. Similarly, the L-look amplitude has a distribution of the form (linear detector):

$$p_A(A|\sigma^2) = 2\left(\frac{L}{\sigma^2}\right)^L \frac{A^{(2L-1)}}{\Gamma(L)} \exp\left(-\frac{LA^2}{\sigma^2}\right), \quad A \geq 0. \tag{6.5}$$

The ratio of standard deviation to mean for the amplitude distribution is:

$$\gamma_a = \sqrt{\frac{L\Gamma^2(L)}{\Gamma^2(L+1/2)} - 1} \tag{6.6}$$

For a Gamma distributed speckle, by Bayesian estimation, the L-look amplitude distribution is a K-distribution [4,5]

$$p_A(A) = \int p_A(A|\sigma^2)p(\sigma^2)d\sigma^2 = \frac{4(L\alpha)^{(\alpha+L)/2}}{\Gamma(L)\Gamma(\alpha)} A^{(\alpha+L)-1} K_{\alpha-L}(2A\sqrt{\alpha L}), \; A \geq 0, \tag{6.7}$$

where $K_n(\cdot)$ is the modified Bessel function of the second kind, ordered n; α is an order parameter. Figure 6.2 plots the distributions with α values. It is seen that K-distribution has longer tails than Rayleigh distribution. The K-distribution is consistent with a large number of coherent scattering experiments, over wavelengths from optical to sonar, and types of scatterer from atmospheric turbulence to natural radar clutter (both sea and land clutters) [1]. When $\alpha \rightarrow \infty$, the K-distribution approaches to Rayleigh distribution, i.e., tends to be more spatially homogeneous and contains purer speckle.

Notice that the mth moment of the amplitude in Equation 6.7

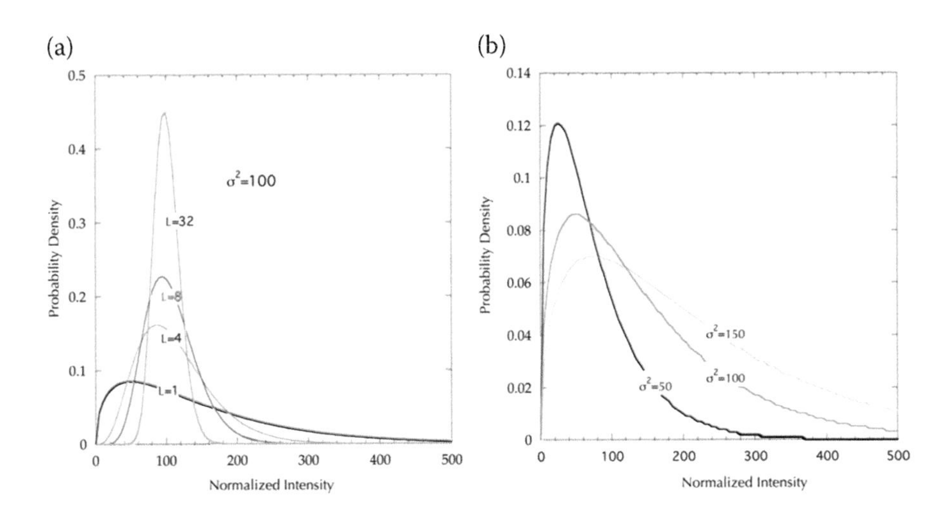

FIGURE 6.1 Amplitude distributions for (a) different look with $\sigma^2 = 100$; (b) different speckle with $L = 1$.

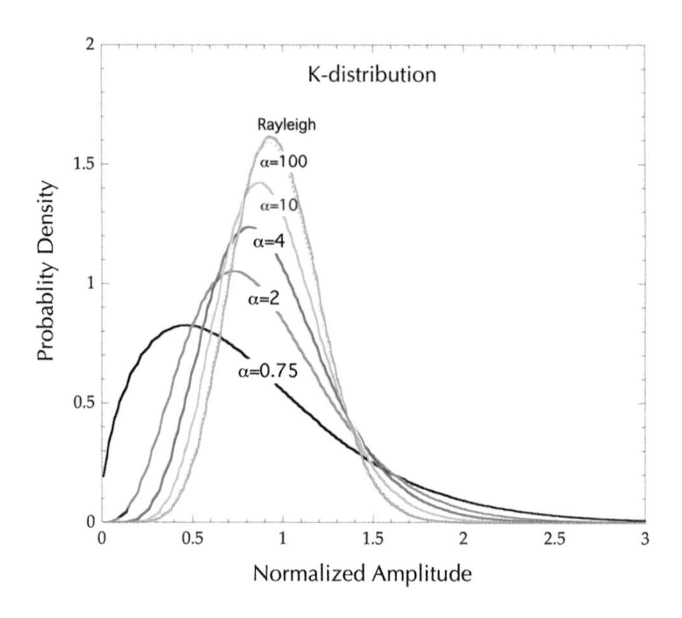

FIGURE 6.2 K-distributions with α values of 0.75, 2, 4, 10, and 100. Also included is a Rayleigh distribution.

$$\langle A^m \rangle = \int A^m p_A(A) dA = \frac{\Gamma(L + m/2)\Gamma(\alpha + m)}{L^{m/2}\Gamma(L)\alpha^{m/2}\Gamma(\alpha)}. \tag{6.8}$$

Using the preceding formula, we can estimate the order parameter α from observed data and test the goodness of fit, which may be done by Kolmogorov–Smirnov (K–S) test [4,5], for example. We also note that the RCS fluctuations are assumed on a much greater spatial scale than speckle so that multi-looking averages speckle without affecting RCS fluctuations.

Now back to Equation 6.3, the value of the speckle strength, σ^2, is nothing more than the scattering strength from a rough surface. For remote sensing of soil or sea surface, σ^2 is the desired quantity to estimate through Equation 6.3 or Equation 6.4. To better infer the surface parameters (roughness, permittivity, etc.), it is a prerequisite to have a good estimate of the scattering coefficients vector from radar measurements. Before proceeds, we note that as an analog to Equation 6.2, the performance measure for the estimated σ^o is [6]:

$$\kappa_\sigma = \sqrt{\frac{\text{var}(\hat{\sigma}^o)}{\sigma^o}} \tag{6.9}$$

Following the conventional notation in statistical signal processing, let the unknown parameters of interest be θ, $\theta \in \Theta$. Then the estimator is any transformation

$$\hat{\theta}(\mathbf{A}) = \hat{\theta}\{A_1, A_2, \ldots A_m\} \tag{6.10}$$

where $\mathbf{A} = \{A_1, A_2, \ldots A_m\}$ is formed by m independent samples. As the radar received signal (amplitude) A is a random variable, so is the estimate $\hat{\theta}$; that is, no measurements yield the same values of scattering coefficients. Two criteria are commonly used to obtain the estimate $\hat{\theta}$ depending on the availability of the probability density function of $\hat{\theta}$, $p(\hat{\theta})$.

$$p(\hat{\theta}(\mathbf{A})|\mathbf{A}) \geq p(\theta(\mathbf{A})|\mathbf{A}) \tag{6.11}$$

If the *a priori* $p(\hat{\theta})$ is known, then posteriori $p(\theta(\mathbf{A})|\mathbf{A})$ is possibly attainable by Bayes' rule:

$$p(\theta(\mathbf{A})|\mathbf{A}) = \frac{p(\theta)p(\mathbf{A}|\theta)}{p(\mathbf{A})} \tag{6.12}$$

The optimal estimate $\hat{\theta}$ is possible obtained by maximizing the right-hand side of Equation 6.12, and is called maximum *a posteriori* (MAP) estimator:

$$\hat{\theta}_{MAP} = \arg\max_{\theta} p(\theta|\mathrm{A}) \tag{6.13}$$

Instead of maximizing the probability of *a posteriori* $p(\theta(\mathbf{A})|\mathbf{A})$, we can take the mean value of it, and is called the minimum mean square error (MMSE) estimator:

$$\hat{\theta}_{MMSE} = \int_{\Theta} \theta p(\theta|\mathbf{A})d\theta \tag{6.14}$$

which yields the estimate with minimum deviation from the true value.

Both MAP and MMSE estimators belong to the Bayesian estimator. For a specific scene to be measured, in general the *a priori* $p(\hat{\theta})$ is unknown or difficult to obtain. In such a case, we may and perhaps have to ignore the *a priori* $p(\hat{\theta})$. Then we can obtain the estimate simply by maximizing $p(\mathbf{A}|\theta(\mathbf{A}))$, the so-called maximum likelihood estimator:

$$\hat{\theta}_{ML} = \arg\max_{\theta} p(\mathbf{A}|\theta) \tag{6.15}$$

For a polarimetric radar, e.g., SAR or scatterometer system, the parameters of interest are the scattering matrix, which can be also multi-frequency, multi-angle, and multi-temporal, etc., as already mentioned above.

6.2 CRAMER–RAO BOUND

In all approaches, the mean square error induced by the estimator is

$$MSE = \langle \|\hat{\theta} - \theta\|^2 \rangle = \mathrm{cov}[\hat{\theta}] + \mathbf{b}(\hat{\theta}) \cdot \mathbf{b}^{\mathrm{H}}(\hat{\theta}) \tag{6.16}$$

where $\langle\,\rangle$ is ensemble average over number of measurements; the bias $\mathbf{b}$ is $\mathbf{b} = \langle\hat{\theta}\rangle - \langle\theta\rangle$; cov is covariance; H denote Hermitian transpose operator.

At this point, it is understood that the estimate can be either unbiased or biased. The MSE serves as a performance measure of a parameter's estimator. The Cramér–Rao bound (CRB) sets the lower value of the covariance for an unbiased estimator with parametric pdf that can be asymptotically attained by the ML estimator for a large number of observations.

Let $\hat{\theta}$ be the unbiased estimator from the set $\theta = \theta(\mathbf{A})$; the CRB sets the lower bound of the error covariance:

$$\mathrm{cov}(\hat{\theta}) \geq C_{CRB} = \mathbf{J}^{-1}(\theta(\mathbf{A})) \tag{6.17}$$

where $\mathbf{J}$ is Fisher information matrix and assume positive definite:

$$\mathbf{J}(\theta) = -\left\langle \frac{\partial}{\partial\theta}\left(\frac{\partial}{\partial\theta}\ln p(\mathbf{A}|\theta)\right)^{\mathrm{T}} \right\rangle \tag{6.18}$$

The diagonal terms of $\mathrm{cov}(\hat{\theta})$ is the mean square error of the unbiased estimator for θ_p with lower bound:

$$\langle(\hat{\theta}_p - \theta_p)^2\rangle \geq J_{pp}^{-1} \tag{6.19}$$

It is seen that for a biased estimator, the minimum square error is larger than CRB.

6.3 LEAST SQUARE ESTIMATION

There are applications where it is necessary to estimate the values of a set of parameters based on the generation model of the deterministic part of the data but without a knowledge of the involved pdfs. In this case one can estimate the parameter values by minimizing the error between the data and the model, according to some metric. A common metric is the sum of the square of the error, and the estimation method is the least-squares (LS) technique. The LS is used in a large variety of applications due to its simplicity in the absence of any statistical model, and for this reason it can be considered an optimization technique for data-fitting rather than an estimation approach. Furthermore, LS has no proof of optimality, except for some special cases where LS coincides with MLE as for additive Gaussian noise, and it is the first pass approach in complex problems.

In radar remote sensing of parameters vector x, it is through a geophysical model function GMF, g;

$$\mathbf{y} = g(\mathbf{x}) \tag{6.20}$$

where $\mathbf{y}$, $\mathbf{y} \in Y$, is radar measurement vector, containing the radar scattering coefficients at n different channels of measurements by combinations of polarization, angular, frequency, temporal, or spatial diversities, including angular and frequency correlations. Here the radar scattering coefficients are not limited by backscattering only, as was done traditionally.

In general, g is a complex and highly nonlinear function that relates the input vector (observation space) and the output vector (parameter space) includes an electromagnetic scattering model, dielectric model (e.g., soil and ocean surface), and wind-wave model (ocean), where the electromagnetic scattering model describes the relationship between the radar echo and the target parameters (both geometrical and electrical), while the dielectric model or the wind-wave model converts the electrical parameters to the physical parameters of interest (e.g., soil moisture content, ocean winds). Reminding from the uncertainties of the roughness parameters, and RCS measurements in previous chapters, what we discussed about (6.1) is an ideal case. In practice, uncertainties must, more or less, be attached to $\mathbf{y}$, g, and $\mathbf{x}$. Physically, the noise sources come from different aspects. For roughness parameters, please refer to Equations 2.40 and 2.41 in Chapter 2.

A data model, in a proper way, is to enable a mapping of measurement space to parameter space, which we are of interest. Here, following, we detail the procedures. To be more explicitly and realistic, Equation 6.20 is re-expressed by

$$\mathbf{y} = \mathbf{W}\mathbf{x} + \mathbf{u} \tag{6.21}$$

where $\mathbf{x}$ is the surface parameters vector; matrix $\mathbf{W}$ relates the surface parameters vector $\mathbf{x}$ to radar scattering measurements $\mathbf{y}$, and $\mathbf{u}$ represents the measurement error vector induced by system and calibration errors, and speckle noise. Note that the radar response is formed by, in general, the scattering matrix. Knowing that small values for κ_p or κ_σ will result in a more accurate estimate and that both $\mathbf{x}$ and $\mathbf{y}$ are random variables, we now deal with the mapping between x and y. That is, we need to find a good estimator for $\mathbf{x}$ from $\mathbf{y}$. Building the relationship between $\mathbf{x}$ and $\mathbf{y}$, both experimental and theoretical methods may be applied, but both will introduce the issues of resolution, distortion, fuzziness and noise. This, equivalently, requires a function, $\mathrm{f} = \mathrm{g}^{-1}$. In practice, we are never able to get a noise-free inverse function f. An approximate $\hat{\mathrm{f}}$, in statistical or deterministic form, is usually applied.

From previous discussions, we may summarize the estimation error sources in terms of the remote sensing of surfaces as (1) the surface parameter uncertainty and (2) the measurement uncertainty, as described in Equation 6.21. Fur one often faces the situation where one measurement set must be mapped to more than one parameter set; i.e., an ill-posed problem. As shown in Figure 6.3 the parameter sets (object domain) correspond to the same measurement result $\mathbf{y}$ (data domain). If we want to determine one representative value $\hat{x}$, in the minimum squared error sense, a sample average of the known parameter sets will be used as the retrieval result. As in mapping problem, there always come with issues of uniqueness, existence, and stability. Regularization theory and neural networks with deep learning are often applied to resolve these problematic issues.

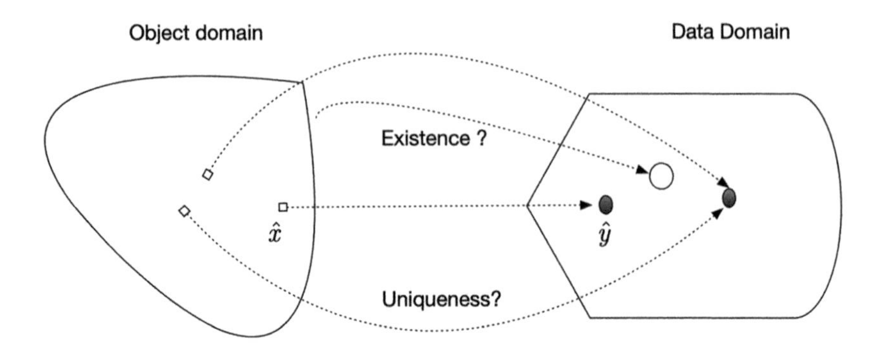

FIGURE 6.3 Mapping of parameters from measurements.

6.4 KALMAN FILTER-BASED ESTIMATION

It has been illustrated from previous chapters, the inverse mapping from radar measurements to surface parameters will be highly nonlinear and an analytic function does not usually exist. Indeed, the retrieval of surface parameters can be viewed as a problem of mappings from the measured signal domain onto the range of surface characteristics that quantify the observation [7–9]. To solve the problems of multidimensional retrieval, the application of a neural network to microwave remote sensing has been carried out by many investigators, because of its ability to adapt to geophysical multidimensionality and its noise robustness in realistic remote sensing. It becomes apparent that the parameter estimation problem invokes two issues that must be simultaneously solved. One is the need for an inversion scheme that provides the mapping, between the parameters to be inferred and the measurements, in a least-squares error sense. Because parameter estimation must come from a finite set of radar observations and measurements, an ill-posed must be solved. Neural networks seem to be well suited for these purposes. Another is the need for GMF models that relate the radar echoes and the geophysical parameters. Thus, the GMF includes two parts: the scattering model relating to the dependence of radar returns to surface roughness and permittivity; the dielectric model relating the permittivity to the moisture content under various conditions.

The Kalman filter (KF) is the optimal MMSE estimate of the state for time-varying linear systems and Gaussian processes in which the evolution of the state is described by a linear dynamic system. In a statistical sense, $\mathbf{x}$ constitutes a random variable due to spatially and temporally varying properties, such that

$$\mathbf{x} = \mathbf{x}_t + \mathbf{x}_n \tag{6.22}$$

where $\mathbf{x}_t$ is true variable, and $\mathbf{x}_n$ is noise term.

In practice, the "truth" is never obtainable; it is always vague. Statistically, $\mathbf{x}_t$ and $\mathbf{x}_n$ may be assumed to be, as they usually are, uncorrelated, such that $\mathbf{x}$ is an unbiased estimate of $\mathbf{x}_t$, i.e.,

$$E(\mathbf{x}) = E(\mathbf{x}_t), \forall \ \mathbf{x}_n \sim N(0, \sigma^2_{\mathbf{x}_n}) \tag{6.23}$$

where E denotes statistical mean, $\sigma^2_{\mathbf{x}_n}$ is variance of $\mathbf{x}_n$.

Note that the radar response is formed by, in general, the scattering matrix. It has been argued that [10] non-quadratic regularization is practically effective in minimizing the clutter while enhancing the target features via:

$$\hat{\mathbf{x}} = \arg\min\{\|\mathbf{y} - \mathbf{W}\mathbf{x}\|_2^2 + \gamma^2\|\mathbf{x}\|_p^p\} \tag{6.24}$$

where $\|\|_p$ denotes ℓ_p – norm $(p \le 1)$, and γ^2 is a scalar parameter and $\{\|\mathbf{y} - \mathbf{W}\mathbf{x}\|_2^2 + \gamma^2\|\mathbf{x}\|_p^p\}$ is recognized as the cost or objective function.

Direct solving of Equation 6.24 perhaps is possible but demands intensive computation resources. From preceding chapters, we also see that the scattering behavior, both pattern and strength is complicatedly determined, in a stochastic sense, by three surface parameters. Hence, in search of the cost function minima in Equation 6.24, we may seek a neural network approach. Perhaps one disadvantage of the neural network is that it constitutes a black box for most users. Extensive studies show that it is a powerful tool for handling complex problems involving bulky volume data in high dimensional feature space. The inverting parameters vector from measurements, a neural network offers an effective and efficient approach and will be detailed in what follows.

Modified from a multilayer perceptron (MLP), a dynamic learning neural network (DLNN) was proposed [11] and is adopted. Figure 6.4 schematically depicts the configuration of a dynamic learning neural network (DLNN). It features: (1) every node at the input layer and all hidden layers are fully connected to the output layer; (2) the activation function is removed from each output node; the output of the modified network can be characterized as the weighted sum of the polynomial basis vectors. Such modifications form a condensed model of the MLP in which the output is a weighted sum of compositions of polynomials. Hence, with measurement error matrix $\mathbf{u}$, as in Equation 6.1, $\mathbf{W}$ is the network weight matrix.

The network training or learning scheme, based on the Kalman filter technique [12] that lends itself to a highly dynamic and adaptive merit during the learning stage, is described below. To begin, the basic concept of Kalman filtering is briefly described, with notation given in Figure 6.5.

The measurement equation takes the form for one-step n:

$$\mathbf{y}_n = \mathbf{W}\mathbf{x}_n + \mathbf{u}_n \tag{6.25}$$

where the subscript n denotes the measurement at a discrete nth time step. The process equation relates to the transition states of the surface parameters vector x:

$$\mathbf{x}_{n+1} = \Phi_n\mathbf{x}_n + \mathbf{B}_n\mathbf{v}_n \tag{6.26}$$

where Φ_n is a transition matrix, $\mathbf{v}_n$ is the process error vector, and $\mathbf{B}_n$ is the error matrix which, in our case, is a diagonal matrix.

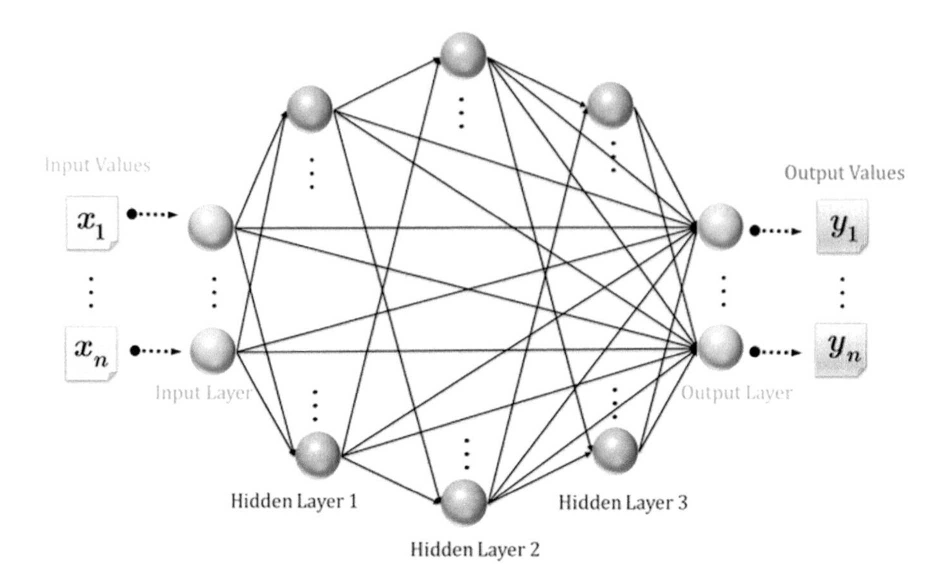

FIGURE 6.4 Network structure of a dynamic learning neural network (DLNN). It features every node at input layer and all hidden layers are fully connected to the output layer; the activation function is removed from each output node; the output of the modified network can be characterized as the weighted sum of the polynomial basis vectors.

The measurement error and process error may be assumed to be statistically independent, i.e.,

$$\mathrm{E}[\mathbf{u}_m \mathbf{v}_n^{\mathrm{t}}] = 0, \ \ \forall m, n. \tag{6.27}$$

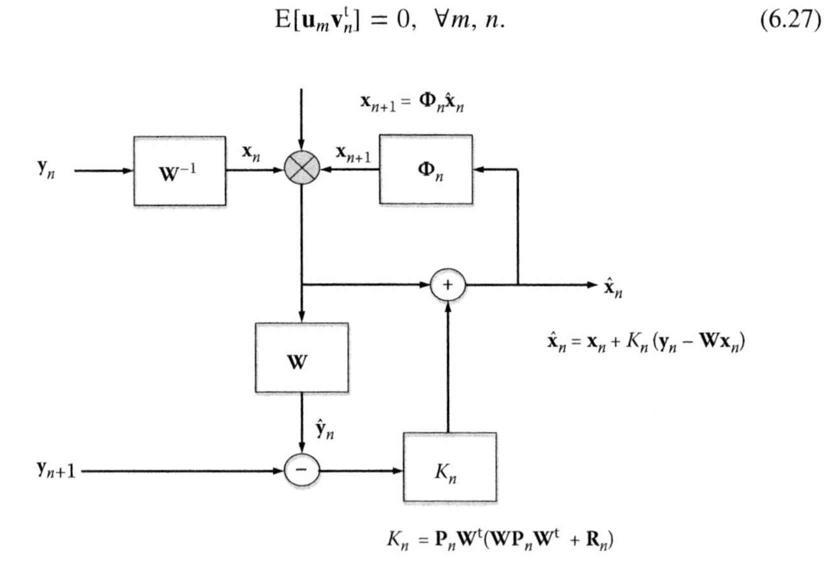

FIGURE 6.5 Schematic description of Kalman filtering.

$$\hat{\mathbf{x}}_n = \tilde{\mathbf{x}}_n + K_n(\mathbf{y}_n - \mathbf{W}\tilde{\mathbf{x}}_n) \tag{6.28a}$$

$$\tilde{\mathbf{x}}_{n+1} = \Phi_n \hat{\mathbf{x}}_n \tag{6.28b}$$

where $\tilde{\mathbf{x}}_n$ is the one-step predicted estimate, $\hat{\mathbf{x}}_n$ is the filter estimate of the desired $\mathbf{x}_n$ and is the computed Kalman gain. For numerical stability, its computation takes the following steps:

$$K_n = \tilde{\mathbf{P}}_n\mathbf{W}^t(\mathbf{W}\tilde{\mathbf{P}}_n\mathbf{W}^t + \mathbf{R}_n) \tag{6.29a}$$

$$\hat{\mathbf{P}}_n = \tilde{\mathbf{P}}_n - K_n(\tilde{\mathbf{P}}_n\mathbf{W}^t)^t \tag{6.29b}$$

$$\tilde{\mathbf{P}}_{n+1} = \Phi_n\hat{\mathbf{P}}_n\Phi_n^t + \mathbf{B}_n\mathbf{Q}_n\mathbf{B}_n^t \tag{6.29c}$$

where $\tilde{\mathbf{P}}_n$, $\hat{\mathbf{P}}_n$ are the one-step predicted and filter estimate error covariance matrices, respectively. The initial state can be set as $\tilde{\mathbf{P}}_1 = \mathrm{E}[\mathbf{x}_1\mathbf{x}_1^t]$.

For the error vectors, it is physically reasonable to assume that for both measurement and process error vectors:

$$\mathrm{E}[\mathbf{u}_m\mathbf{u}_n^t] = \begin{cases} \mathbf{R}_n, & m = n \\ 0, & m \neq n \end{cases} \tag{6.30a}$$

$$\mathrm{E}[\mathbf{v}_m\mathbf{v}_n^t] = \begin{cases} \mathbf{Q}_n, & m = n \\ 0, & m \neq n \end{cases} \tag{6.30b}$$

where $\mathbf{R}_n$, $\mathbf{Q}_n$ are error covariance matrices, respectively. The process error and the measurement error in radar observation may be reasonably assumed to be statistically independent and can be modeled as zero-mean, white noise process.

From the modified MLP structure, each updated estimate of the neural network weight is computed from the previous estimate and the new input data. The weights connected to each output node can be updated independently such that the vector problem can, therefore, be decomposed into L scalar problems as

$$\mathbf{y}_\kappa = \mathbf{w}_\kappa\mathbf{x} \quad (\kappa = 1, 2, ..., L) \tag{6.31}$$

Applying to the Kalman filtering technique, the network structure can be modeled by

$$y_\kappa^i = \mathbf{w}_\kappa^i\mathbf{x} + v_\kappa^i \tag{6.32}$$

$$\mathbf{w}_\kappa^{i+1} = \mathbf{w}_\kappa^i\mathbf{A}^i + \boldsymbol{u}_\kappa^i\mathbf{B}^i \tag{6.33}$$

where the superscript i denotes the ith training pattern with total of N, $\mathbf{A}^i$ is a M by M state transition matrix, $\mathbf{B}^i$ is a M by M diagonal matrix, $\boldsymbol{u}_\kappa^i$ represents a 1 by M process error vector, with M the dimension of concatenated activations in modified MLP, and v_κ^i is a scalar measurement error.

The update of network weights is according to the following recursions:

$$\hat{\mathbf{w}}_{\kappa}^{i} = \tilde{\mathbf{w}}_{\kappa}^{i} + \mathbf{g}_{\kappa}^{i}[d_{\kappa}^{i} - \tilde{\mathbf{w}}_{\kappa}^{i}\mathbf{x}] \quad (i = 1, 2, ..., N) \tag{6.34}$$

$$\tilde{\mathbf{w}}_{\kappa}^{i+1} = \hat{\mathbf{w}}_{\kappa}^{i}\mathbf{A}^{i} \tag{6.35}$$

where d_{κ}^{i} is the desired output, $\tilde{\mathbf{w}}_{\kappa}^{i}$ is the one-step predicted estimate, and $\hat{\mathbf{w}}_{\kappa}^{i}$ is the filter estimate of $\mathbf{w}_{\kappa}^{i}$, respectively, and $\mathbf{g}_{\kappa}^{i}$ is the computed Kalman gain, which is viewed as an adaptive learning rate and its computation is according to Equations 6.31–6.35.

From the previous chapter, the sensitivity analysis in response to radar scattering is essential. Knowing the parameter bounds is important to the network performance. A neural-based estimation seems powerful and yet efficient as long as training data is effective. The MLP together with learning by Kalman filter is expected to resolve the highly nonlinear, and complex decision boundary problems, as will be demonstrated in the following chapter.

REFERENCES

1. Lee, J. S., and Pottier, E., *Polarimetric Radar Imaging—From Basics to Applications*, CRC Press, Boca Raton, FL, 2009.
2. Chen, K. S., Wu, T. D., and Shi, J. C., A model-based inversion of rough soil surface parameters from radar measurements, *Journal of Electromagnetic Waves and Applications*, 15(2), 173–200, 2001.
3. Ulaby, F. T., Moore, R. K., and Fung, A. K., *Microwave Remote Sensing*, Addison-Wesley, Reading, MA, 1982.
4. Lehmann, E. L., and Romano, J. P., *Testing Statistical Hypotheses*, 3rd Edition, Springer, New York, 2005.
5. Spagnolini, U., *Statistical Signal Processing in Engineering*, John Wiley & Sons, New York, 2018.
6. Long, D. G., and Spencer, M. W., Radar backscatter measurement accuracy for a spaceborne pencil-beam wind scatterometer with transmit modulation, *IEEE Transactions on Geoscience and Remote Sensing*, 35(17), 102–114, 1997.
7. Chen, K. S., Kao, W. L., and Tzeng, Y. C., Retrieval of surface parameters using dynamic learning neural network, *International Journal of Remote Sensing*, 16, 801–809, 1995.
8. Thiria, S., Mejia, C., and Badran, F., A neural network approach for modeling non-linear transfer functions: Application for wind retrieval from space borne scatter meter data, *Journal of Geophysical Research: Oceans*, 98, 22827–22841, 1993.
9. Chen, K. S., Tzeng, Y. C., and Chen, P. C., Retrieval of ocean winds from satellite scatter meter by a neural network. *IEEE Transactions on Geoscience and Remote Sensing*, 37, 247–256, 1999.
10. Çetin, M., and Karl, W. C., Feature-enhanced synthetic aperture radar image formation based on nonquadratic regularization. *IEEE Transactions on Image Processing*, 10, 623–631, 2001.
11. Tzeng, Y. C., Chen, K. S., Kao, W. L., and Fung, A. K., A dynamic learning neural network for remote sensing applications. *IEEE Transactions on Geoscience and Remote Sensing*, 32, 1096–1102, 1994.
12. Scharf, L. L., *Statistical Signal Processing—Detection, Estimation, and Time-Series Analysis*, Addison Wesley Publishing Co., Reading, MA, 1991.

7 Selected Model Applications to Remote Sensing

An analytical scattering model presented in Chapter 4 has been validated by comparison with numerical simulations and measurements from ground-based, airborne, and spaceborne sensors. The sensitivity test on the surface parameters response to radar return was presented in Chapter 5. In the previous chapter, issues regarding the parameter estimation from radar measurements have been discussed. The mapping from parameter domain to data (or image) domain, and vice versa, is a highly nonlinear problem. Various techniques for information retrieval from remotely sensed data have been proposed in many recent studies—they are too numerous to cite all of them here. Some of them are based on an empirical relationship between the measured return signals and the ground truth. Because they developed from a limited number of observations, these models are generally valid only for the conditions under which those measured data were taken. For example, some of these models also appear that no dependence on the roughness parameter, correlation length. In this chapter, we presented a few model application examples: soil moisture retrieval and directional finding of the incident source.

If one perceives radar remote sensing—a stochastic electromagnetic wave scattering problem—more closely and deeply, the following the 4Vs characteristics may be recognized: volume, variety, velocity, and veracity. As pointed out by Hey et al. [1], the data science in the big data era combines and synergizes the observation, model prediction, and numerical simulation such that data transforms into information, and knowledge can be assured. Within a context of radar imaging, as we see from the 4Vs, more advanced data analytics apparently should be developed to explore richer information offered by fully bistatic radars with fully polarized remote sensing data, which is the main objective of this paper. Hence, it can be realized that a better solution of inverting rough surface parameters is that, by knowing the scattering patterns, one may be able to detect the presence of undesired random roughness of a reflective surface (e.g., an antenna reflector), and thus accordingly, and perhaps effectively, devise a means to correct or compensate phase errors. In many aspects, parameters retrieval from radar measurements is a highly nonlinear problem, yet an important objective in radar remote sensing of surfaces. The advance of neural network and deep learning [2–4] has provided an effective and yet efficient approach to retrieving these surface parameters.

7.1 SURFACE PARAMETER RESPONSE TO RADAR OBSERVATIONS—A QUICK LOOK

To get started, we present the AIEM backscattering predictions with experimental measurements from ground-based, airborne, and spaceborne sensors. Sample data from POLARSCAT Data [5], EMSL data [6], and SMOSREX06 Data [7] are used to illustrate the radar response to the surface parameters, though more details have been presented in previous chapters. These data were summarized and discussed in [8].

7.1.1 Comparison with POLARSCAT Data

Remember that in Chapter 4, we illustrated a model prediction comparing with these data sets, but in a manner of correlation plot. The correlation plots of backscattering coefficients between the AIEM and the measured data for HH, VV, and HV polarizations are shown in Figures 4.12 and 4.13. Figure 4.12 shows the data for three frequency bands, and Figure 4.13 shows the data by separating the wet and dry conditions. Here, we would like to expand the comparison into more detail [8].

7.1.1.1 For Surface 1 (S1)

A total of three different soil surfaces with exponential autocorrelation function measured at three different frequencies (1.5, 4.75, and 9.5 GHz) ranging from 20° to 50° were used. The comparisons of AIEM predictions with the POLARSCAT measurements under different soil surfaces. The backscattering coefficients simulated by AIEM for an exponentially correlated surface with $\sigma = 0.4$ cm, $\ell = 8.4$ cm at L/C/X-band under dry and wet soil conditions, are shown in Figures 7.1 and 7.2, respectively. It is seen visually that the angular trends of AIEM predictions generally coincide with the POLARSCAT data at all three frequencies. As expected, the backscattering coefficient decreases with increasing incident angle and increases with decreasing frequency from a smooth surface. Besides, it can be observed that the AIEM predictions agree better with the POLARSCAT data at smaller incident angles (e.g., 20° and 30°) while slightly deviate at lager incident angle (e.g., 40° and 50°), especially for HH polarization.

7.1.1.2 For Surface 2 (S2)

We compare the model predictions with the measurement data for the second soil surface (i.e., S2), and the results are shown in Figures 7.3 and 7.4. In general, the AIEM predictions are in good agreement with the POLARSCAT measurements over the most of the angular range under consideration, but deviates for HH polarization for incident angle larger than 40° under the dry soil conditions. It can be observed that the experimental measurements do not decrease monotonically with the increase of incident angle particularly for the L-band under dry soil conditions. This perhaps was attributed from measurement noises. As observed from the results of surface 1, the AIEM performs much better under wet soil conditions than under dry conditions for surface 2. This may be explained by the gradually increase of volume scattering in the dry soil surfaces that is somehow difficult to characterize [9].

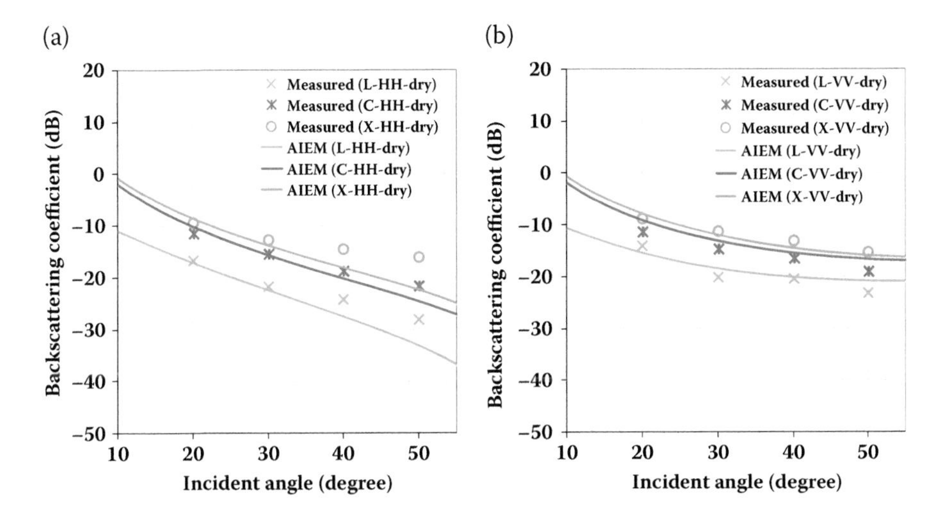

FIGURE 7.1 Comparison of backscattering coefficient between AIEM and POLARSCAT data for an exponential correlated surface with $\sigma = 0.4$ cm, $\ell = 8.4$ cm at 1.5 GHz ($\varepsilon_r = 7.99\text{-}j2.02$), 4.75 GHz ($\varepsilon_r = 8.77\text{-}j1.04$) and 9.5 GHz ($\varepsilon_r = 5.70\text{-}j1.32$) for (a) HH polarization and (b) VV polarization.

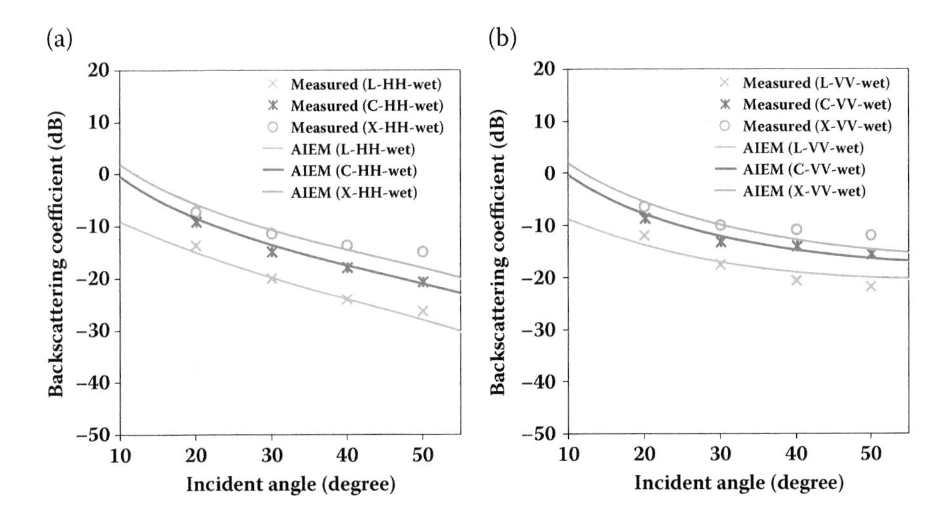

FIGURE 7.2 Comparison of backscattering coefficient between AIEM and POLARSCAT data for an exponential correlated surface with $\sigma = 0.4$ cm, $\ell = 8.4$ cm at 1.5 GHz ($\varepsilon_r = 15.57\text{-}j3.71$), 4.75 GHz ($\varepsilon_r = 15.42\text{-}j2.15$) and 9.5 GHz ($\varepsilon_r = 12.31\text{-}j3.55$) for (a) HH polarization and (b) VV polarization.

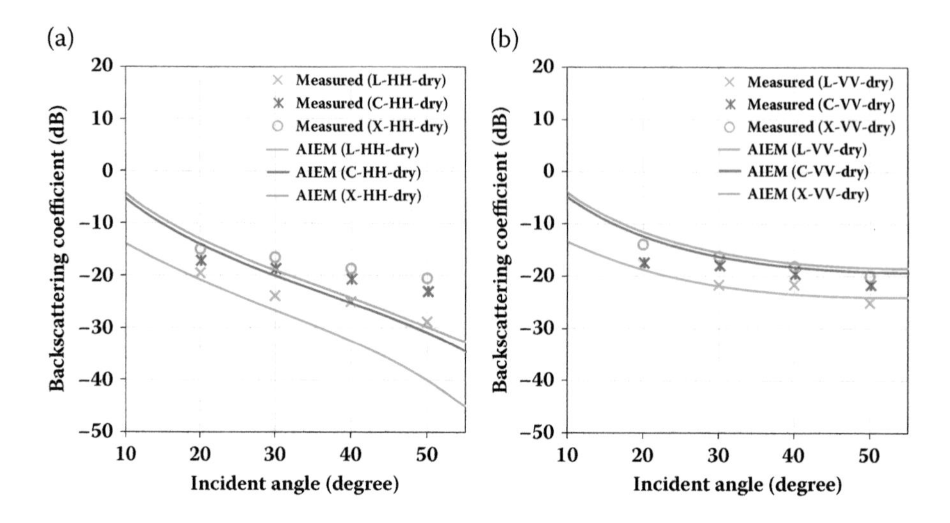

FIGURE 7.3 Comparison of backscattering coefficient between AIEM and POLARSCAT data for an exponential correlated surface with $\sigma = 0.32$ cm, $\ell = 9.9$ cm at 1.5 GHz ($\varepsilon_r = 5.85\text{-}j1.46$), 4.75 GHz ($\varepsilon_r = 6.66\text{-}j0.68$) and 9.5 GHz ($\varepsilon_r = 4.26\text{-}j0.76$) for (a) HH polarization and (b) VV polarization.

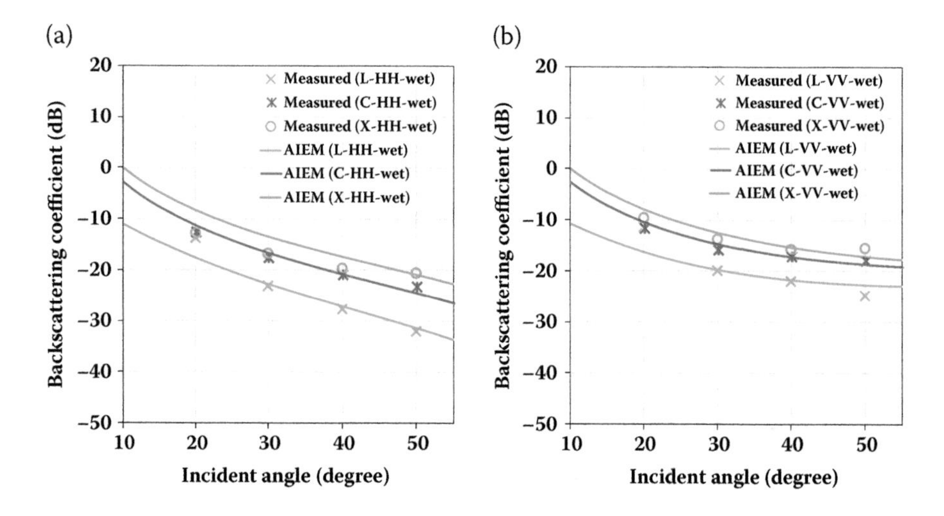

FIGURE 7.4 Comparison of backscattering coefficient between AIEM and POLARSCAT data for an exponential correlated surface with $\sigma = 0.32$ cm, $\ell = 9.9$ cm at 1.5 GHz ($\varepsilon_r = 14.43\text{-}j3.47$), 4.75 GHz ($\varepsilon_r = 14.47\text{-}j1.99$) and 9.5 GHz ($\varepsilon_r = 12.64\text{-}j3.69$) for (a) HH polarization and (b) VV polarization.

7.1.1.3 For Surface 3 (S3)

In what follows, we continue to assess the accuracy of AIEM backscattering simulations with the POLARSCAT data, shown in Figures 7.5 and 7.6. In this

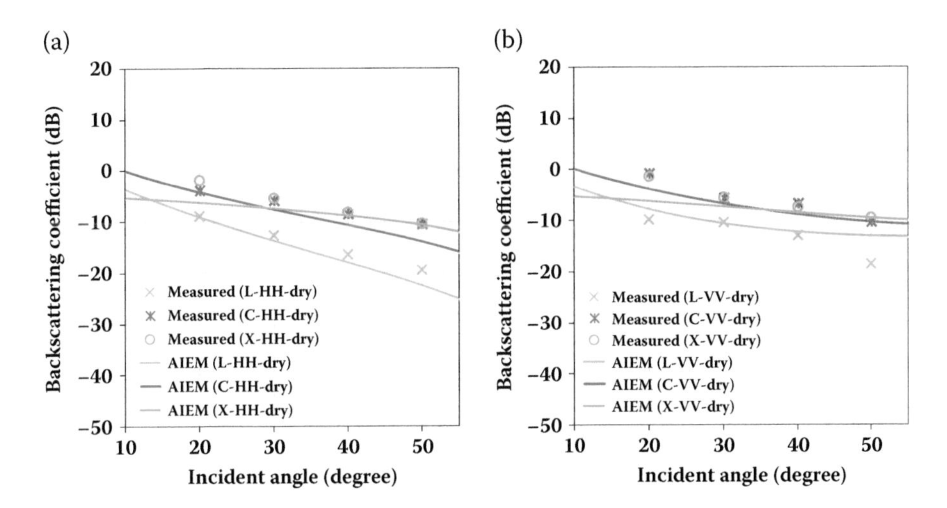

FIGURE 7.5 Comparison of backscattering coefficient between AIEM and POLARSCAT data for an exponential correlated surface with $\sigma = 1.12$ cm, $\ell = 8.4$ cm at 1.5 GHz ($\varepsilon_r = 7.70$-j1.95), 4.75 GHz ($\varepsilon_r = 8.50$-j1.00) and 9.5 GHz ($\varepsilon_r = 6.07$-j1.46) for (a) HH polarization and (b) VV polarization.

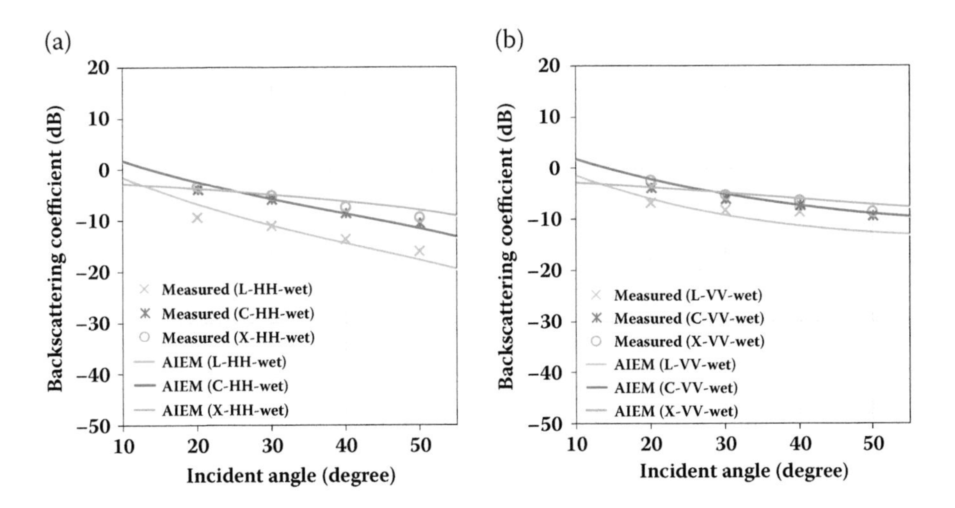

FIGURE 7.6 Comparison of backscattering coefficient between AIEM and POLARSCAT data for an exponential correlated surface with $\sigma = 1.12$ cm, $\ell = 8.4$ cm at 1.5 GHz ($\varepsilon_r = 15.34$-j3.66), 4.75 GHz ($\varepsilon_r = 15.23$-j2.12) and 9.5 GHz ($\varepsilon_r = 13.14$-j3.85) for (a) HH polarization and (b) VV polarization.

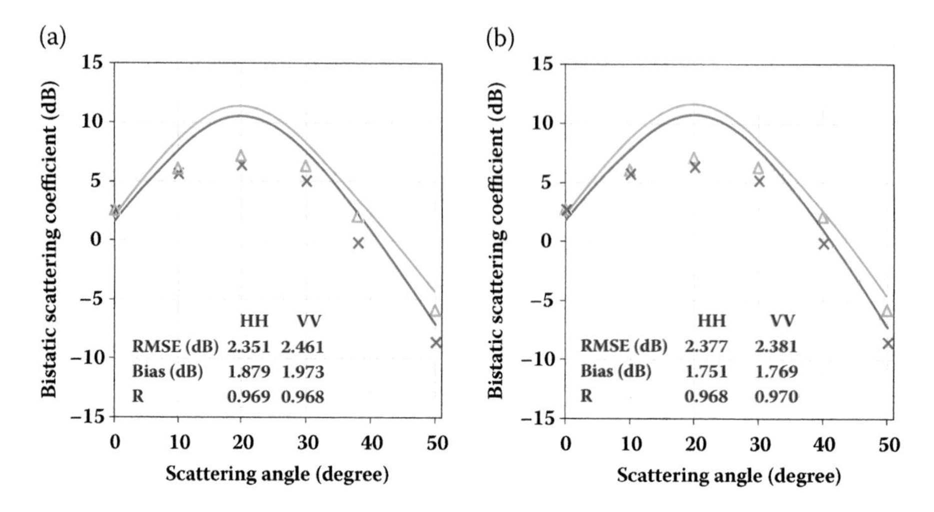

FIGURE 7.7 Comparison of bistatic scattering coefficient between AIEM and EMSL measurements for a Gaussian correlated surface with ε_r = 5.5-j2.1, σ = 0.4 cm, ℓ = 6 cm, and incident angle of 20° for (a) 11 GHz and (b) 13 GHz.

case, the RMS height is 1.12 cm, and the correlation length is 8.4 cm, which corresponds to the roughest surface among the three experiments. Due to the increased roughness, the scattering curve over all angles becomes flat. Again, the AIEM predictions well capture the angular trends of the POLARSCAT data at all frequencies for both HH and VV polarizations. It is also observed that in this case, the measurements at C-band and X-band are nearly the same regardless of the incident angle, whereas there is an obvious separation between the HH and VV polarized backscattering coefficient predicted by the AIEM over the incident angles. Insofar, we only illustrate the co-polarization. For discussion on cross-polarized backscattering, please refer to Chapter 4, and reference [10] for details.

7.1.2 Comparison with EMSL Data

The bistatic scattering measurement data used here was acquired by the EMSL [6]. The EMSL data were performed on a Gaussian correlated surface with $\sigma = 0.4$ cm, $\ell = 6$ cm, $\varepsilon_r = 5.5 - j2.1$ and incident angle of 20° and scattering angle between 0° and 50° at two frequencies (i.e., 11 and 13 GHz). The comparison between the AIEM predictions and the EMSL measurements is shown in Figure 7.7. It can be observed that the bistatic scattering behavior of AIEM is generally in agreement with the EMSL data. The flatness of the angular trend between (10° and 30°), a forward scattering region, in measurements seems unexplainable, because in these scattering angles, a stronger scattering signal is physically expected. Nevertheless, the AIEM well predicts the bistatic scattering measurements for both HH and VV polarizations with a favorable RMSE value within 2.5 dB and a correlation value greater than 0.96.

7.1.3 Comparison with SMOSREX06 Data

For a rough surface, the noncoherent reflectivity is obtained by integrating the bistatic scattering coefficient over the upper hemisphere. Because of the reciprocity, the emissivity is the same as absorptivity, the amount of power absorbed by the dielectric in a scattering problem. Hence, it would be useful to examine the model prediction of the surface emissivity. Also, in this regard, the active and passive microwave remote sensing is interplayed.

The comparisons of the emissivity predicted by AIEM and the SMOSREX06 measurements are illustrated in Figure 7.8. Here, the performances of AIEM are observed in terms of angular and polarization dependences. It is seen that, in general, the AIEM predictions are fairly close to the measured emissivity at small to intermediate look angles, from 20° to 40°, for both H and V polarizations. However, when the look angle increases to 50° and 60°, the gap between these two data becomes larger, especially for H polarization. It is found that both the H and V polarized emissivities simulated by AIEM underestimate the SMOSREX06 measurements. Increasing the look angle, the dry bias at H polarization also increases, which is not observed at V polarization, however. Indeed, the RMSE is not quite satisfactory, particularly at large look angles (e.g., 50° and 60°) for H polarization. The AIEM predictions correlate well with the measured emissivity with the correlation coefficient close to 0.9 for both H and V polarizations. In Figure 7.8, we see a fairly well linear relationship between the AIEM predictions and measurements. Since the observation period of the measurements is very long, collecting a total of 8125 samples and the comparison, as presented here, is made without adjusting the given input parameters to the AIEM model, it is convinced that the AIEM is able to correctly predict the temporal dynamic of the observed surface emissivity over a wide range of look angle and surface parameters. More on the error measures are referred to [8].

7.2 SURFACE PARAMETER RETRIEVAL

7.2.1 Data Input–Output and Training Samples

In this demonstration, two neural network configurations were devised, with three input parameters for backscattering and four input parameters for bistatic scattering (see Table 7.1). The outputs of the network were normalized surface roughness (RMS height $k\sigma$, correlation length $k\ell$), and soil moisture m_v. Three roughness spectra—Gaussian, exponential, and 1.5-power—were all included in the simulation. The inputs are given in Table 7.1.

Backscattering: In this case, HH-, VV-, and HV (=VH)-polarized backscattering coefficients were simultaneously fed into the dynamic learning neural network. The incident angles were set at 10~60°.

Bistatic Scattering: In this setup, inputs are chosen from HH-, HV-, VH-, and VV-polarized bistatic scattering coefficients at different incident angles, scattering angles, and scattering azimuthal angle were fed into the network. Both the incident angles and scattering angle were selected at 10~60°. The scattering azimuthal angle

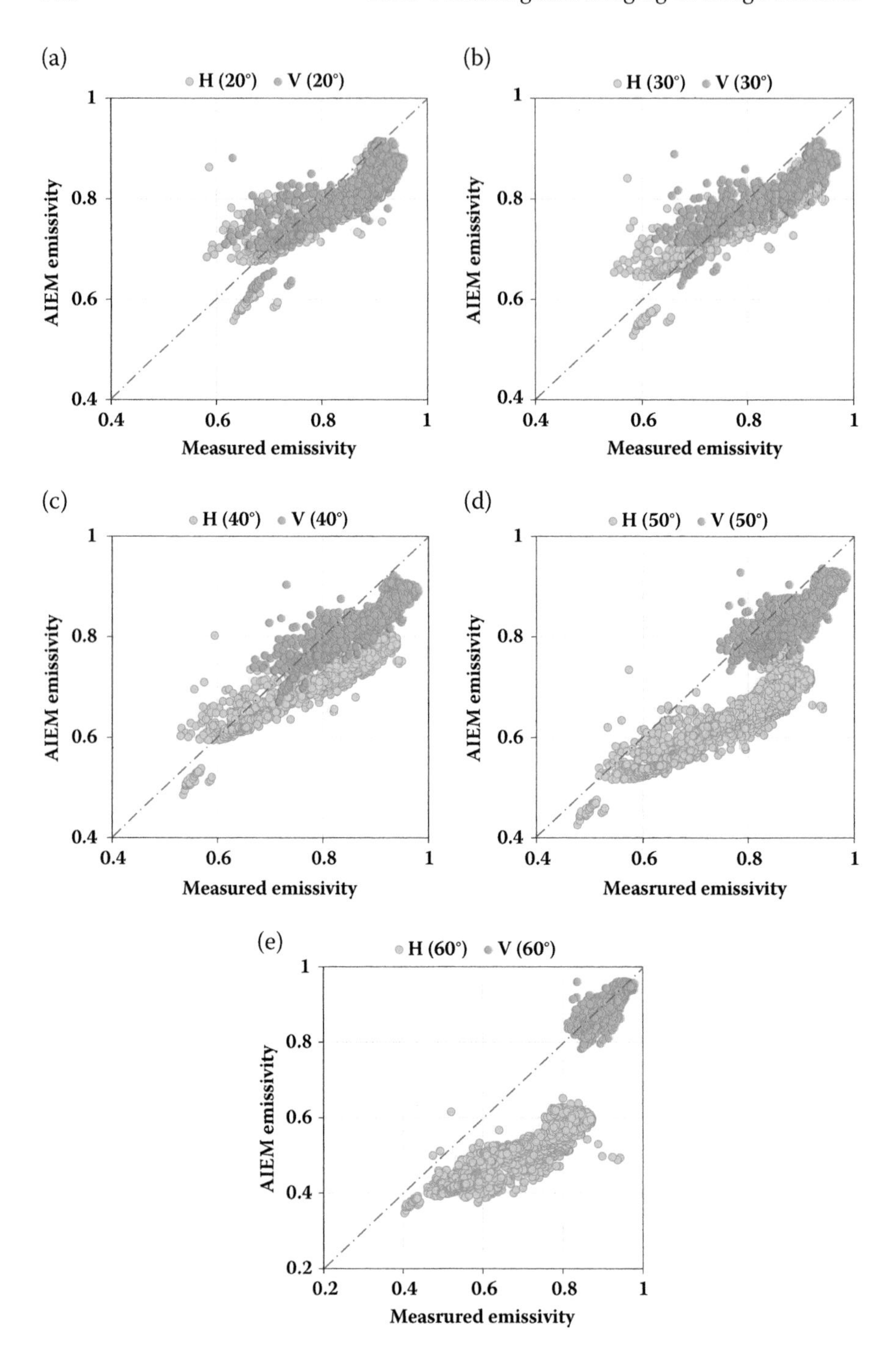

FIGURE 7.8 Comparison of H and V polarized emissivity between AIEM and SMOSREX06 measurements for an exponential correlated surface at 1.41 GHz with look angle of (a) 20°, (b) 30°, (c) 40°, (d) 50°, and (f) 60°.

TABLE 7.1
Input–Output of a Neural Network

Backscattering (Three-Input)		Bistatic Scattering (Three-Input & Four-Input)	
Input	**Output**	**Input**	**Output**
HH, VV, and HV polarized backscattering coefficients of $\mathbf{b} = [\sigma_{hh}^{o}, \sigma_{hv}^{o}, \sigma_{vv}^{o}]^{t}$	Normalized surface roughness and soil moisture $\mathbf{x} = [k\sigma, k\ell, m_v]^{t}$	HH, HV, VH, and VV polarized bistatic coefficients $\mathbf{b} = [\sigma_{hh}^{o}, \sigma_{hv}^{o}, \sigma_{vv}^{o}]^{t}$ $\mathbf{b} = [\sigma_{hh}^{o}, \sigma_{hv}^{o}, \sigma_{vh}^{o}\sigma_{vv}^{o}]^{t}$	Normalized surface roughness and soil moisture $\mathbf{x} = [k\sigma, k\ell, m_v]^{t}$

TABLE 7.2
Neural Network Configurations for Three-Input and Four-Input Cases

Configuration	Three-Input	Four-Input
Nodes of input layer	3	4
Nodes of output layer	3	3
Hidden layer	2	2
Nodes of each hidden layer	100	100
Threshold	0.1%	0.1%

was set for the forward region at 0~90° and for the backward region at 90~180°. For bistatic scattering, we devised three inputs (as in backscattering) for comparison of inversion performance under the same number of inputs.

It follows that the network configuration, after trial and error, was given with a predetermined threshold of 0.1% (see Table 7.2). A total of 10,450 data sets were selected randomly as training data from the 15,014 data sets for backscattering, with 4564 data sets used as testing data. As a rule of thumb, 70% of the total data sets were used for training (i.e., 127,240 data sets are chosen randomly as training data from the 181,770 data sets for bistatic scattering), and 54,530 data sets were used as testing data, from which the backward region and forward region cases each accounted for half of the data. The data sets are randomly selected from the database simulated by the AIEM model with the ranges of the surface and radar parameters listed in Table 7.3. The step size of the discretized scattering azimuthal angle was set to 10°, and the discretized incident and scattering angles were both set to 1°. The network training was accomplished by mapping input–output pairs that were randomly selected from the database simulated by the AIEM model with the range of surface and radar parameters listed in Table 7.3. As a rule of thumb, 70% of the data was selected, randomly, as the training set, with the rest as the testing set.

TABLE 7.3
Radar Parameters for Generating Training Samples

Parameters	Backscattering	Bistatic Scattering
$k\sigma$	0.1~0.8	0.1~0.8
$k\ell$	1~7	1~7
m_v	0~0.5	0~0.5
θ_i	10°~60°	10°~60°
θ_s	$=\vartheta_i$	10°~60°
ϕ_s	180°	0°~90° (forward)
		90°~180° (backward)
$k\sigma/k\ell$	0.1~0.4	0.1~0.4
S	Gaussian, Exponential, 1.5-Power	

7.2.2 Retrieval Results Using Backscattering Coefficients

After completing the training, the DLNN entered the process stage. By randomly selecting 30% testing sample, the surface parameter retrieval was performed via the DLNN. This was indeed a highly nonlinear mapping of feature sets (by training) onto the surface domain (by process). The retrieval performance between the network-inverted result and the model-observed data, using backscattering data only (i.e., three-input), can be seen in Figure 7.9. For three surface parameters of interest, the normalized root-mean-squared errors (nRMSE) were 0.074, 0.075, and 0.070 for RMS height, correlation length, and soil moisture, respectively, and correlation coefficients were larger than 0.95, which was quite satisfactory; among the three parameters, the inversion of $k\ell$ was poorer. The reason for this is that, as discussed previously, the estimation of correlation length always poses higher errors due to a higher uncertainty of the functional form of the correlation function. Several samples of inversion results between the measured data (POLARSCAT) and the network inversion are presented in Table 7.4. As we can see, the retrieval performance of these three parameters can achieve satisfactory accuracy.

7.2.3 Retrieval Using Bistatic Scattering Coefficients

When it comes to inversion from bistatic scattering data, we can examine the performance of using forward, backward, and full (backward plus forward), respectively. This might be practically useful since, in setting up the bistatic observation, the transmitter and receiver may be in specular or off-specular geometries. The correlation between the network-inverted result and model-observed data is shown in Figure 7.10. The inversion results were relatively close to the 1:1 line, indicating a quite-favorable correlation between the inversion and the truth data. It is worth noting that retrieval from bistatic scattering data performed better than from backscattering,

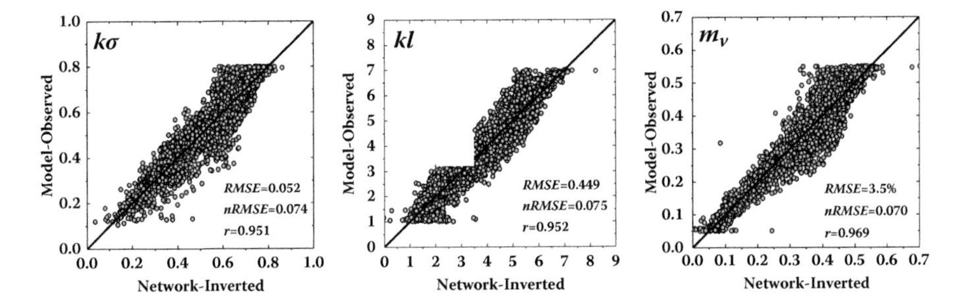

FIGURE 7.9 Retrieval of surface roughness and soil moisture from backscattering data by three inputs–three outputs DLNN configuration. Both root-mean-squared errors (RMSE) and normalized root-mean-squared errors (nRMSE) for each case are shown.

TABLE 7.4
Comparison of Inverted and Truth Data (POLARSCAT)

Truth (POLARSCAT Data)			Inverted (Network-Inverted)		
$k\sigma$	$k\ell$	m_v	$k\sigma$	$k\ell$	m_v
0.949	2.768	0.142	0.907804	2.75213	0.141224
3.004	8.765	0.172	7.1636	12.1984	0.232222
0.352	2.617	0.266	0.346407	2.61012	0.265967
0.126	2.62	0.126	0.126175	2.62442	0.125997
0.352	2.617	0.14	0.371045	2.6396	0.14006
0.101	3.098	0.09	0.103238	3.15896	0.08993
1.114	8.287	0.14	1.20734	8.85712	0.137385
0.796	16.594	0.253	1.66042	5.38294	0.364186
3.004	8.765	0.172	3.03834	8.83309	0.172412
0.949	2.768	0.172	0.948958	2.76862	0.172035
6.009	17.529	0.172	2.69795	5.08291	0.16386
6.009	17.529	0.142	5.91543	17.6258	0.142792
0.352	2.617	0.266	0.351636	2.61618	0.266002
2.228	16.574	0.14	2.22803	16.5737	0.140001
0.796	16.594	0.253	0.797609	16.6051	0.252946
2.228	16.574	0.14	2.24075	16.5635	0.139998
0.398	8.297	0.253	0.396886	8.30057	0.252988
3.004	8.765	0.172	3.06484	8.54212	0.172106

Error	$k\sigma$	$k\ell$	m_v
RMSE	1.2701	4.0330	2.987%
nRMSE	0.2149	0.2705	0.2134
r	0.7706	0.7761	0.9150

especially for soil moisture inversion. Interestingly, the inversion results performed better at the forward region than in the backward region. This can also be seen quantitatively in Table 7.5, from which we can read the nRMSE being 0.060, 0.088, and 0.044 in the backward region and 0.045, 0.071, and 0.018 in the forward region for RMS height, correlation length, and soil moisture, respectively. The overall nRMSE and correlation coefficient (r) for backscattering and bistatic scatter are also given in Table 7.5. Typically, among the three surface parameters being inverted, the soil moisture tended to experience the least error, regardless of retrieval from backscattering data or bistatic scattering data. Better inversion performance seems to have been gained by the use of scattering data from the forward region. Based on this point and compared to the backscattering, it is not surprising that if we combine the scattering data from the forward and backward regions, the best retrieval accuracy in terms of nRMSE and the correlation coefficient can be obtained. This is also evident from Table 7.5, where the smaller nRMSE and higher correlation coefficient can be read as "Full" when compared to those for backscattering. It is apparently a conviction that the dynamic learning neural network as presented is able to tackle a bulky volume of data, either the training stage or operational stage, and thus achieve superior retrieval accuracy.

There is a much higher degree of freedom to test the inversion from bistatic scattering compared to backscattering, as well as to choose the inputs, depending on the physical feasibility of the measurements. We can have measurements at the backward, forward, or combined forward and backward regions to input to the DLNN. In terms of polarization, we can have four polarizations in bistatic—two

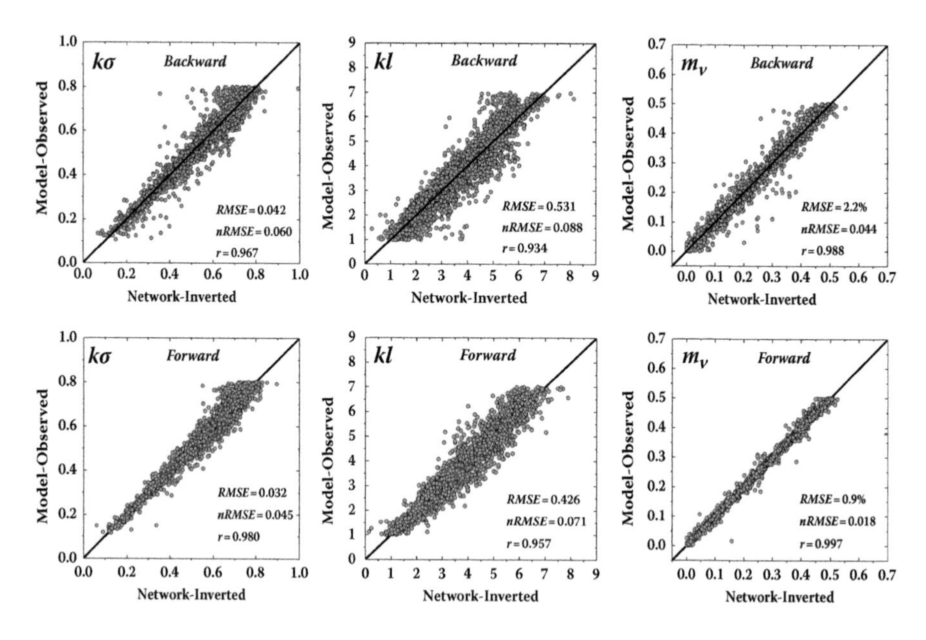

FIGURE 7.10 Retrieval performance of surface roughness and soil moisture from bistatic scattering data (four-input). Top row: backward region. Bottom row: forward region.

TABLE 7.5
DLNN Inversion Performance Using Backscattering and Bistatic Scattering

Backscattering & Bistatic Scatter		**Error**	$k\sigma$	$k\ell$	m_v
Backscattering		nRMSE	0.074	0.075	0.070
(Three-input)		r	0.951	0.952	0.969
Bistatic scattering	Backward	nRMSE	0.075	0.087	0.039
(Three-input)		r	0.951	0.936	0.990
	Forward	nRMSE	0.040	0.070	0.044
		r	0.985	0.959	0.988
	Full (Backward + Forward)	nRMSE	0.058	0.097	0.042
		r	0.970	0.949	0.991
Bistatic scattering	Backward	nRMSE	0.060	0.088	0.044
(Four-input)		r	0.967	0.934	0.988
	Forward	nRMSE	0.045	0.071	0.018
		r	0.980	0.957	0.997
	Full (Backward + Forward)	nRMSE	0.053	0.080	0.034
		r	0.974	0.945	0.992

co-polarizations, two cross-polarizations, two-polarizations, or one cross-polarization. Figure 7.11 shows the retrieval performance of surface roughness and soil moisture from bistatic scattering data (three-input). The simulation test confirms the inversion performance of the neural network in terms of training and operation. From Table 7.5, we can see that the retrieval accuracy is higher using forward bistatic data than backward. The impact of polarization for bistatic cases seems not as significant as that for backscattering cases. More importantly, for the bistatic case, three inputs do not necessarily produce higher retrieval accuracy than four inputs do. It is the number of features that determine the training effectiveness and thus the retrieval accuracy. It becomes clear at this point that under the same number of inputs, inversion accuracy is higher from bistatic scattering than from backscattering. This is likely due to the fact that the bistatic scattering measurements can better separate the coupling effect of roughness and dielectric constant. For this, the sensitivity analysis given in Chapter 5 is useful for feature selection to train the neural network.

At this point, it is of relevance to demonstrate the retrieval of soil surface parameters from bistatic radar measurements [6,11], including roughness and moisture content, using the model-trained DLNN. A bistatic measurement facility (BMF) [11] was designed and constructed consisting of a 10 and 35 GHz polarimetric radar system for the purpose of characterizing the 3D bistatic scattering response of rough dielectric surfaces. According to, [11] a measurement with 10 GHz from a Gaussian random rough surface, with a normalized RMS height of $k\sigma = 0.2$ and a normalized correlation length of $kl = 1.0$, was water-soaked bricks in which the real part of the dielectric constant was 62. The incident angle and scattering angle were both set as

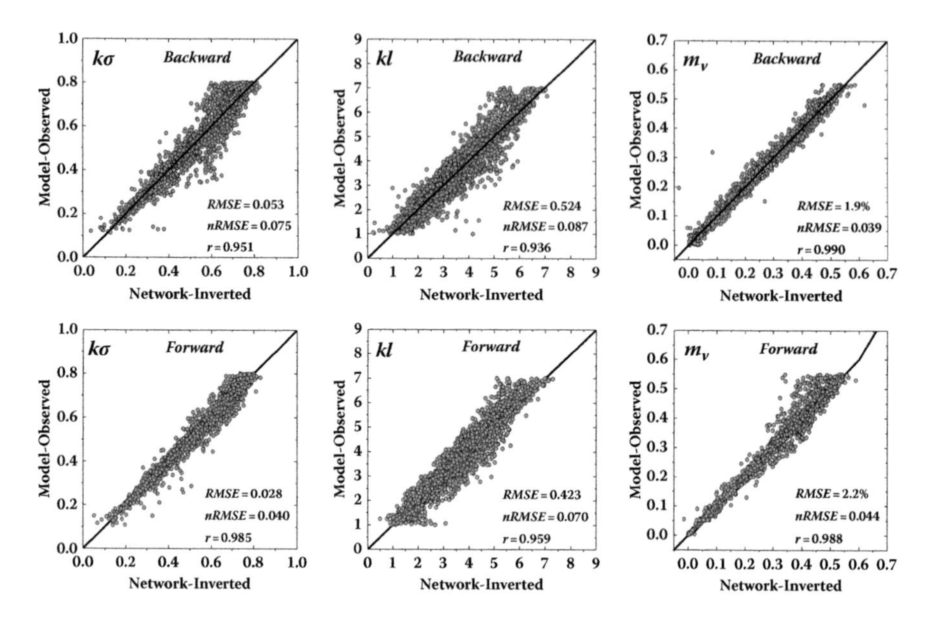

FIGURE 7.11 Retrieval performance of surface roughness and soil moisture from bistatic scattering data (three-input). Top row: backward region. Bottom row: forward region.

45°, and the scattering azimuth angle was changed from 180° to −180°. A study in [11] gave the measurements of a rough soil surface with 35 GHz using an indoor bistatic measurement facility (IBMF). At 35 GHz, the surface was assumed to be an exponentially correlated function with the normalized RMS height and correlation length being 3.28 and 29.6, respectively. The effective soil permittivity was 3.5-*j*0.05. The measured bistatic scattering coefficient as a function of the scattering angle (0~70°) in the backward direction (ϕ_s = 180°) and forward direction (ϕ_s = 0°), and the incident angle was 20°. Another data set is taken from [6,12], where the bistatic rough dielectric surface scattering was measured at the European Microwave Signature Laboratory (EMSL) at (−10~−50°) of incidence and scattering angles with 1.5–18.5 GHz of frequency. The surface was characterized by a Gaussian correlation function with surface roughness of σ = 2.5 cm, l = 6 cm.

To assess the inversion performance and be more quantitative, we list the numeric results in Table 7.6, showing the measured $k\sigma$, $k\ell$, and m_v, along with the inverted data. Of the eight total test sets, the inverted results are in good agreement with the measured data. Relatively, the difference between the measured and inverted data is well within 20% error. The RMSE, nRMSE, and correlation coefficient (r) for each individual parameter, $k\sigma$, $k\ell$, and m_v, are also given. Of the three parameters, $k\ell$ had the largest RMSE. This is consistent with our previous discussion on the measurement uncertainty of the correlation length. Nevertheless, numeric values in Table 7.6 confirm the good performance of surface parameter inversion by model-trained DLNN with inputs from bistatic radar scattering measurements.

TABLE 7.6
Comparison of Surface Parameters between Measured and Inverted by Model-Trained DLNN

Truth (Measured)			Inverted (Network-Inverted)			Difference (Absolute)		
$k\sigma$	$k\ell$	m_v	$k\sigma$	$k\ell$	m_v	$k\sigma$	$k\ell$	m_v
1.05	2.51	0.16	1.24	2.48	0.17	0.19	0.03	0.59%
2.09	5.02	0.12	1.99	5.27	0.12	0.09	0.25	0.28%
2.62	6.28	0.11	2.83	8.74	0.11	0.21	2.46	0.41%
4.19	10.05	0.12	4.25	9.99	0.13	0.06	0.06	1.43%
5.23	12.56	0.10	7.49	12.56	0.09	2.25	0.004	0.11%
8.37	20.09	0.09	7.64	22.46	0.09	0.73	2.37	0.48%
0.20	1.00	0.87	0.19	1.10	0.94	0.006	0.10	6.57%
3.28	29.60	0.07	3.10	32.98	0.06	0.18	3.38	0.34%
	$k\sigma$			$k\ell$			m_v	
RMSE	0.85		RMSE	1.70		RMSE	2.41%	
nRMSE	0.10		nRMSE	0.09		nRMSE	0.03	
r	0.95		r	0.99		r	0.99	

7.2.4 Comparison with Image-Based Surface Parameter Estimation from Polarimetric SAR Image Data

The trained DLNN is now applied to the measured data acquired at L-band (1.3 GHz) from E-SAR over the floodplain of River Elbe located in North-Eastern Germany [13–15] as shown in Figure 7.12. At L-band, the spatial resolution of the single look complex data is in azimuth about 0.75 m and in the range of about 1.5 m. The data were acquired in April and August of 1997 along with two 15 km long and 3.2 km wide strips. Ground data has been collected in August 1997 over agriculture test fields with different roughness conditions. Soil moisture measurements have been performed on five different locations at each test field. The fields were viewed with incidence angle θ_i ranging from 48° to 50°. Four fields were selected due to the vegetation-covered and the choices of them were constrained by the image-based model [13] (see Table 7.7).

The DLNN results show that the varied inversion results using different correlated surfaces in the AIEM model. All the retrieval results (mean value for each test area) are listed in Table 7.8. First, we can see the deviation of roughness ($k\sigma$) between the inversion results and ground truth values. The largest deviation occurs for the case of field A 5/16. Gaussian correlated surface matches best for the cases. It is interesting to note that 1.5 power correlated surfaces fall in between Gaussian and exponential correlated surfaces that represent two extremes of roughness spectra in terms of their bandwidth. For horizontal roughness scale, correlation length, there is no ground truth available, the comparison is excluded. Nevertheless,

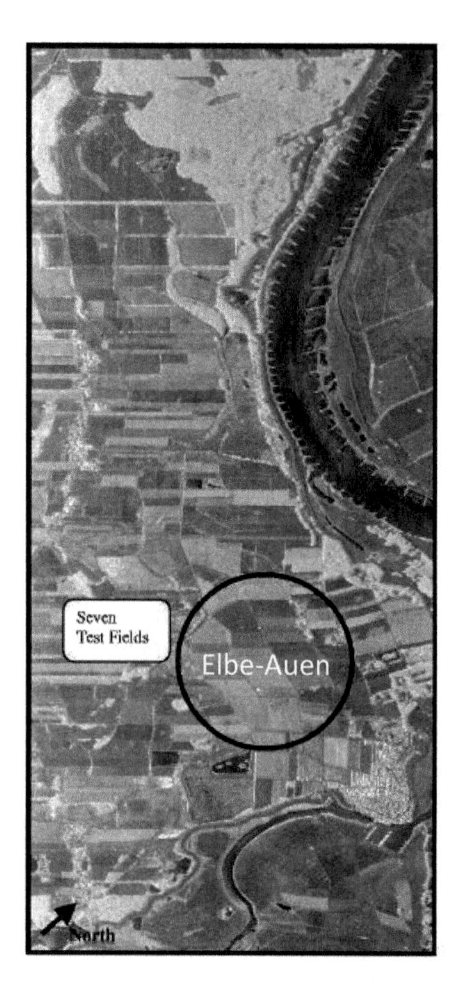

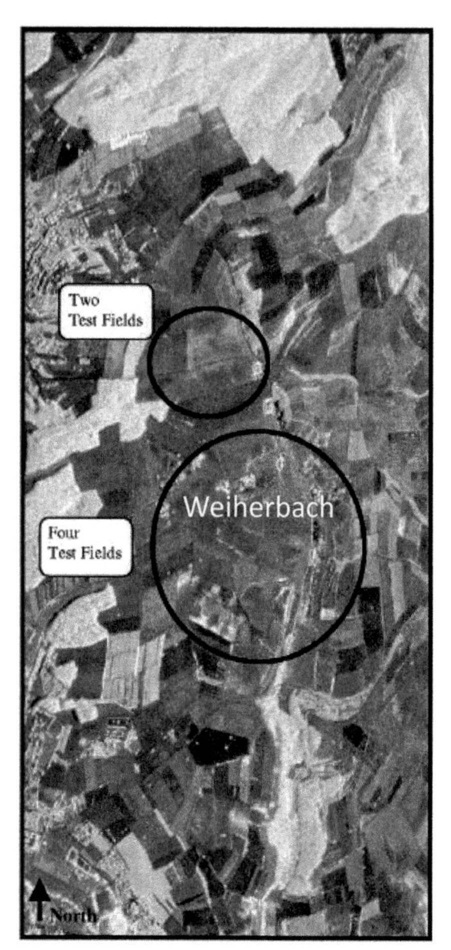

FIGURE 7.12 Total power image acquired by L-band E-SAR over the floodplain of River Elbe located in North-Eastern Germany [13].

TABLE 7.7
Ground Measurements for the Elbe-Auen Test Site

Field ID	θ_i	σ (cm)	$k\sigma$	ε'(0–4 cm)	ε'(4–8 cm)
A 5/10	49.20	1.66	0.45	10.79	9.28
A 5/13	50.03	2.1	0.57	5.34	9.84
A 5/14	49.99	2.77	0.75	4.51	10.82
A 5/16	48.56	3.5	0.95	5.86	12.19

TABLE 7.8
The DLNN Inversion Results Using Testing the Different Correlation Functions at Different Test Sites

A5/10	$k\sigma$	$k\ell$	ε'
Gaussian	0.51106	2.9856	8.8747
Exponential	0.36842	3.5310	10.552
1.5 Power	0.39530	3.0111	8.6624
A5/13	$k\sigma k\ell\varepsilon'$		
Gaussian	0.61287	4.12425	7.6096
Exponential	0.30934	3.9752	5.5954
1.5 Power	0.36935	4.0491	7.8188
A5/14	$k\sigma k\ell\varepsilon'$		
Gaussian	0.53227	3.7903	7.7906
Exponential	0.31289	4.0010	5.4194
1.5 Power	0.35864	4.1336	10.283
A5/16	$k\sigma k\ell\varepsilon'$		
Gaussian	0.61799	3.6400	7.9737
Exponential	0.37857	3.8496	7.6764
1.5 Power	0.43115	3.7744	10.025

the inversion outputs are listed in Table 7.8 for reference. Next, we check the retrieved dielectric constants which may be related to moisture content. It is observed that the inversion results agree well with the ground truth. To indicate this point more clearly, we plot the inverted dielectric constants by model-based DLNN and image-based [13] along with ground truth values (0–4 and 4–8 cm), as shown in Figure 7.13. The image-based results and the model-based DLNN results reasonably fall within the range of two different depths [16]. A short note is that the neural network inversion can explain more closely the observed data and hence give a good inversion result.

7.3 DIRECTION ESTIMATION OF INCIDENT SOURCE: A DATA ANALYTIC EXAMPLE

Here, we estimate the source of the incident field in terms of incident angle and azimuthal angle from quad-pol radar measurements under the framework of a convolutional neural network (CNN) [3,4]. The CNN is adopted to implement the inversion of an AIEM model that relates the bistatic scattering to surface wind speed and direction. In such a case, the CNN is constructed as an inverse model. For inputs, co- and cross-polarized bistatic scattering coefficients (σ^o_{hh}, σ^o_{hv}, σ^o_{vh}, σ^o_{vv}) and scattering direction (scattering angle θ_s and scattering azimuth angle ϕ_s) are selected.

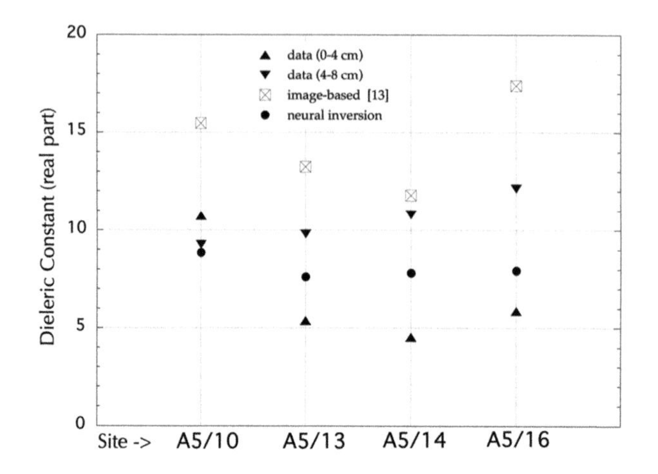

FIGURE 7.13 Retrieval versus measured dielectric constants at depths (0–4 cm) and (4–8 cm). Also included are results from image-based method [13] from four test sites: A5/10, A5/13, A5/14, A5/16.

The outputs of the network are incident angle θ_i and incident azimuthal angel ϕ_i. The network training is accomplished by mapping input-output pairs that are randomly selected from the database. Note that the model predictions of the scattering represent the mean value. To generate the training samples, to be realistic, the speckle must be taken into account by using the K-distribution as discussed in the previous chapter. The radar parameters for generating training samples are given in Table 7.9.

Figure 7.14 illustrates the network structure of a convolutional neural network (CNN). The retrieval result of incident direction from microwave observations with a frequency of 10 GHz arise shown in Figure 7.15, and frequency of 13.99 GHz

TABLE 7.9
Radar Parameters for Generating Training Samples

Paremeters	Range	Step Size
Incident Angle θ_i	20°~70°	5°
Incident Azimuth Angle ϕ_i	10°~170°	5°
Scattering Angle θ_s	10°~70°	5°
Scattering Azimuth Angle ϕ_s	10°~170°	5°
Wind Direction φ	0°~180°	5°
Frequency f	10 GHz, 13.99 GHz	
Wind Speed U	10 GHz: 3.2 m/s, 9.3 m/s, 14.5 m/s; 13.99 GHz: 5 m/s, 10 m/s, 20 m/s, 30 m/s	

in Figure 7.16. The root-mean-squared error (RMSE) is about 1°~2° for incident angle and 3°~4° for incident azimuthal angle, indicating that the directional estimation of the incident source is possible to achieve a satisfactory accuracy by means of CNN. Indeed, several other Backpropagation-based neural networks had been tested, such as MLP-BP, dynamic learning neural networks (DLNN), but none of them is able to give a good estimate, and most of the time, they even failed come up with outputs.

It brings to our attention from demonstrations of the parameter retrieval that the task of rough-surface parameter (geometric and dielectric) retrieval from

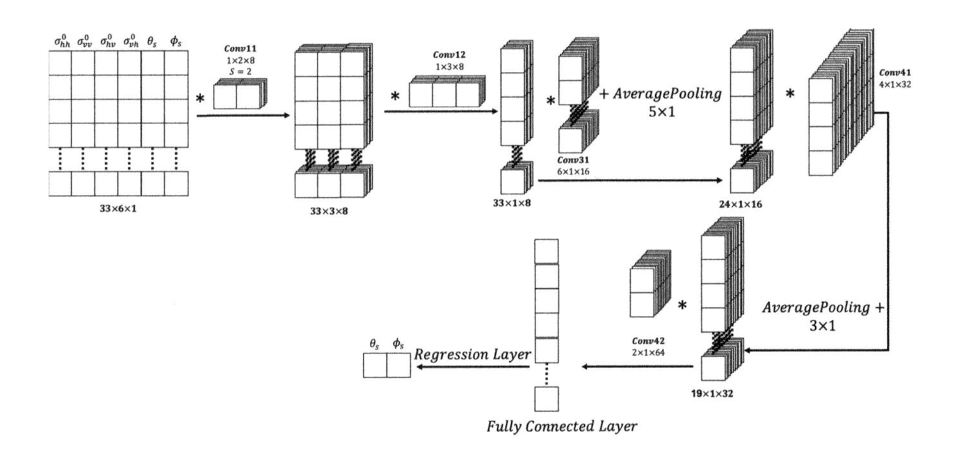

FIGURE 7.14 Network structure of a convolutional neural network (CNN).

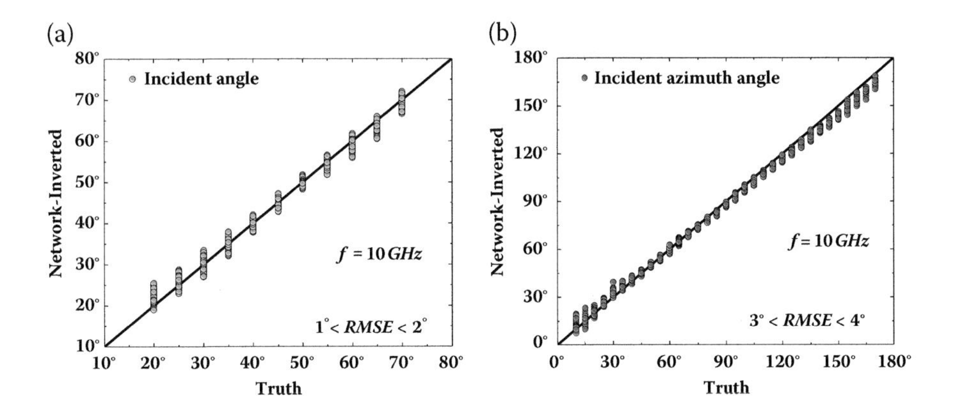

FIGURE 7.15 Correlation of wind direction between truth and reconstructed value by CNN with two frequency of $f = 10\ GHz$: (a) inverted incident angle, (b) inverted azimuth angle.

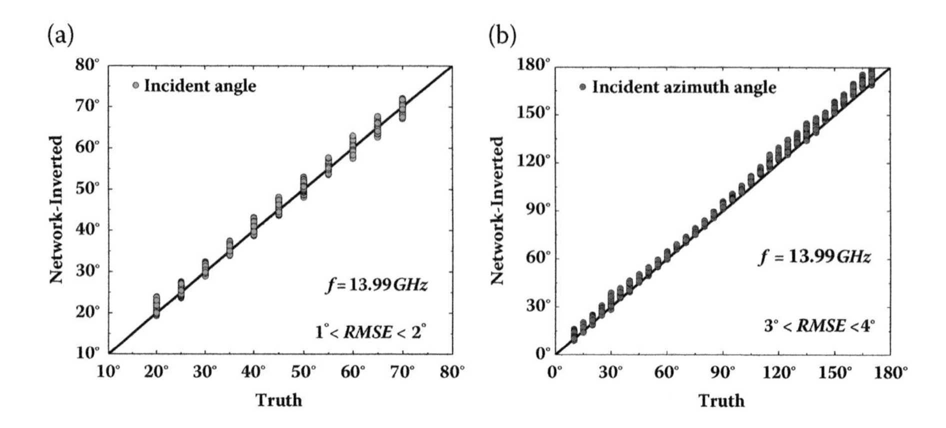

FIGURE 7.16 Same as Figure 7.15 except at frequency of 13.99 GHz.

bistatic radar scattering data is a highly nonlinear problem. It has been shown that the use of the neural network enhances the separability for highly nonlinear boundary problems. The neural network was able to tackle the complex inversion problem with very fast learning. Experimental testing showed good performance of surface parameter inversion by the model-trained DLNN or CNN with inputs from bistatic radar scattering measurements. In the future study, through the model simulation and the proposed inversion scheme, the optimal bistatic observation configuration (in the sense of most efficient multiple inputs–multiple outputs (MIMO) to come up with high retrieval accuracy) should be investigated. It should be emphasized that, to our best understanding, real bistatic radar measurements are scant, or only taken in very limited and yet specific radar configurations (e.g., specular direction). We have demonstrated how useful bistatic data is in retrieving multiple surface parameters of interest, as compared to the backscattering data which occupies almost all of the data available, at least currently. It is therefore our objective to promote the focus more on the bistatic scattering measurements, which are shown to be powerful in inferring the surface parameters.

REFERENCES

1. Hey, T., Tansley, S., Toll, K., *The Fourth Paradigm: Data-Intensive Scientific Discovery*, 1st Edition, Microsoft Research, Mountain View, CA, 2009.
2. Haykin, S. O., *Neural Networks and Learning Machines*, Prentice Hall, Upper Saddle River, NJ, 2008.
3. Marsland, S., *Machine Learning: An Algorithm Perspective*, 2nd Edition, Chapman & Hall/CRC, Boca Raton, FL, 2015.
4. Aggarwal, C. C., *Neural Networks and Deep Learning: A Textbook*, Springer, New York, 2018.
5. Oh, Y., Sarabandi, K., and Ulaby, F. T., An empirical model and an inversion technique for radar scattering from bare soil surfaces, *IEEE Transactions on Geoscience and Remote Sensing*, 30(2), 370–381, 1992.

6. Macelloni, G., Nesti, G., Pampaloni, P., Sigismondi, S., Tarchi, D., and Lolli, S., Experimental validation of surface scattering and emission models. *IEEE Transactions on Geoscience and Remote Sensing,* 38(1), 459–469, 2000.
7. Mialon, A., Wigneron, J. P., De Rosnay, P., Escorihuela, M. J., and Kerr, Y. H., Evaluating the L-MEB model from long-term microwave measurements over a rough field, SMOSREX 2006, *IEEE Transactions on Geoscience Remote Sensing*, 50(5),1458–1467, 2012.
8. Zeng, J. Y., Chen, K. S., Bi, H. Y., Zhao, T. J., and Yang, X. F., A comprehensive analysis of rough soil surface scattering and emission predicted by AIEM with comparison to numerical simulations and experimental measurements. *IEEE Transactions on Geoscience and Remote Sensing*, 55(3), 1696–1708, 2017.
9. Fung, A. K., and Chen, K. S., *Microwave Scattering and Emission Models for Users*, Artech House, Norwood, MA, 2010.
10. Yang, Y., Chen, K. S., Tsang, L., and Liu, Y., Depolarized backscattering of rough surface by AIEM model. *IEEE Journal of Selected Topics in Applied Earth Observations and Remote Sensing*, 10(11), 4740–4752, 2017.
11. Nashashibi, A. Y., and Ulaby, F. T. MMW polarimetric radar bistatic scattering from a random surface. *IEEE Transactions on Geoscience and Remote Sensing*, 45(6), 1743–1755, 2007.
12. Brogioi, M., Pettinato, S., Macelloni, G., Paloscia, S., Pampaloni, P., Pierdicca, N., and Ticconi, F., Sensitivity of bistatic scattering to soil moisture and surface roughness of bare soils. *International Journal of Remote Sensing*, 31, 4227–4255, 2010.
13. Hajnsek, I., Pottier, E., and Cloude, S. R., Inversion of surface parameters form polarimetric SAR, *IEEE Transactions on Geoscience and Remote Sensing*, 41(4), 727–744, 2003.
14. Hajnsek, I., Jagdhuber, T., Schon, H., and Papathanassiou, K. P., Potential of estimating soil moisture under vegetation cover by means of PolSAR. *IEEE Transactions on Geoscience and Remote Sensing*, 47(2), 442–454, 2009.
15. Jagdhuber, T., Hajnsek, I., Bronstert, A., and Papathanassiou, K. P., Soil moisture estimation under low vegetation cover using a multi-angular polarimetric decomposition, *IEEE Transactions on Geoscience and Remote Sensing*, 51(4), 2201–2214, 2012.
16. Lee, H. W., Chen, K. S., Lee J. S., Shi, J. C., Wu, T. D., and Hajnsek, I., A comparisons of model based and image-based surface parameters estimation from polarimetric SAR. In *Proceedings of the 2005 IEEE International Geoscience and Remote Sensing Symposium*, 2005. doi:10.1109/igarss.2005.1526460.

8 Radar Imaging Techniques

8.1 STOCHASTIC WAVE EQUATIONS

Radar as a coherent imager is physically based on the wave equations that govern the behavior of wave-targets interactions through the process of radiation, transmission, propagation, and reflection/refraction. For a randomly rough surface being imaged, the governing waver equations become stochastic [1], as we already considered in Chapter 5. Here, we briefly introduce the stochastic wave equations at a minimum but sufficient level to serve as fundamentals of radar imaging.

Consider the radar imaging of a nonmagnetic random medium as shown in Figure 8.1. The locations of the transmitter T_x and receiver R_x are in bistatic mode. For more the general random medium, it is specified by the constitute relations as given in 3.2 of Chapter 3.

Similarly to the problem we deal with the Huygen's principle, the total field $\mathbf{E}$ received at Rx is the sum of the transmitted field $\mathbf{E}^t$ at Tx and scattered field $\mathbf{E}^s$ from the random medium under the presence of the incident field, as already introduced in Equation 1.18:

$$\mathbf{E}(\vec{r}) = \mathbf{E}^i(\vec{r}) + k_0^2 \int \bar{\bar{\mathbf{G}}}(\vec{r}, \vec{r}')\cdot[\bar{\bar{\varepsilon}}_r(\vec{r}') - 1]\mathbf{E}(\vec{r}')d\vec{r}' = \mathbf{E}^i(\vec{r}) + \mathbf{E}^s(\vec{r}) \quad (8.1)$$

where $\bar{\bar{\mathbf{G}}}$ is tensor Green's function; k_0 is the free space wavenumber. $\bar{\bar{\varepsilon}}_r(\vec{r}) - 1 = 0$ is objection function, and may be any other forms. The random medium is described by permittivity $\bar{\bar{\varepsilon}}(\vec{r}) = \langle\bar{\bar{\varepsilon}}(\vec{r})\rangle + \Delta\bar{\bar{\varepsilon}}(\vec{r})$, where $\langle\bar{\bar{\varepsilon}}(\vec{r})\rangle$ is the mean permittivity and $\Delta\bar{\bar{\varepsilon}}(\vec{r})$ is the fluctuating permittivity, with $\bar{\bar{\varepsilon}}_r(\vec{r}) = \bar{\bar{\varepsilon}}(\vec{r})/\varepsilon_0$. In the absence of the random medium, the objection function $\bar{\bar{\varepsilon}}_r(\vec{r}) - 1 = 0$, the total field equals the transmitted field. Noted that depending on the location of the receiver, the transmitted field may propagate through the medium. But it would not affect the problem solving, which is by way of applying the boundary conditions.

The scattered field appearing in Equation 8.1 is given rise by scattered by random scatters, scattering from random rough surface, or fluctuating propagation in random continua inside the random medium intercepted by the transmitting antenna and receiving antenna. Because of the random properties of the permittivity, the total field is composed of a mean-field and a fluctuating field [1–3].

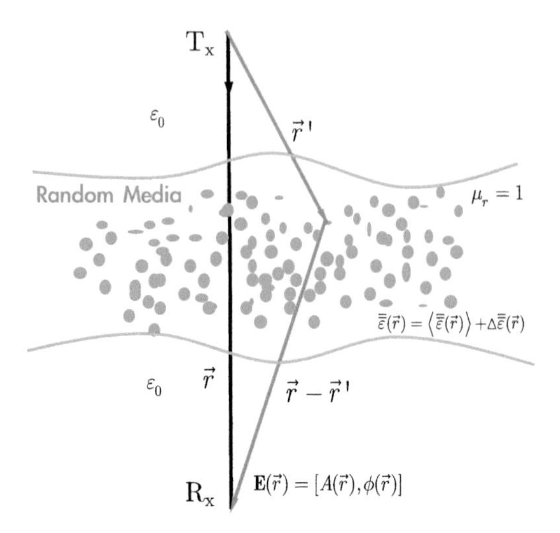

FIGURE 8.1 Radar imaging of a random medium.

In radar, it is commonly named as a coherent and inherent field, correspondingly, similar to our treatment of scattering field in Chapters 4 and 5. Theoretically, decomposing of the total field into a coherent field and incoherent field is mathematically straightforward. In practice, it is not so as the antenna patterns always exercise influence, making it is difficult to differentiate the coherent returns from incoherent returns in total returns. In transmitting the signal toward the medium, an unmodulated waveform is rarely used. Broadly, there are two basic types of waveform: pulse and continuous wave. As such, in microwave remote sensing, linear frequency modulated (LFM) pulse (chirp), frequency modulated continuous wave (FMCW), and step frequency modulated (SFM), are among commonly used waveforms. Proper choice of waveform determines the radar capability of resolving range and azimuth ambiguities and spatial resolutions, and of course depends on the application scenario. It also involves the radar sensitivity, measurement precision, and dynamic ranges [4].

Equation 8.1 constitutes a stochastic vector wave equation to determine the received field. The problem risen here is how to solve Equation 8.1. This is indeed a computational electromagnetic problem. Many excellent books are available to offer solutions for various kinds of boundary values problems [5–9]. A thorough treatment of such problems is not possible and is beyond the scope of this book. Readers can always consult the source of these books to seek a preferred approach. Indeed, with the advances of both software programing (e.g., massively parallel) and hardware development (e.g., graphic computation unit), the computational electromagnetic problem is still in its accelerating progress toward offering higher computationally efficient and accurate solutions as never before. For random medium imaging, an analytical solution to Equation 8.1 is rarely feasible. Two types of numerical solutions are usually devised: full-wave solution and approximate solution. In a full-wave solution, the problem can be treated in time domain or frequency domain, or hybrid. The method of moment (MoM) in the frequency domain and Finite difference time domain (FDTD) is well-known. The full wave approach implies that the solution contains all bounces of scattering, between radar-targets, and among targets. Imaging formation based on full wave solution therefore poses much lesser information loss in terms of preserving the complete scattering process.

Major advances have been made in numerical solutions of Maxwell equations in three-dimensional simulations (NMM3D) [10,11] for rough surface scattering. As an application example, with the fast and yet accurate numerical simulation method at hand, it is not difficult to study the effects of the surface inhomogeneity. By viewing the Green's function of layered inhomogeneous media, it is noted that the reflect term that accounts for the inhomogeneous effects is needed in addition to a direct term. Mathematically, this is simple and straightforward. Numerically, it is not really so, however. After some mathematical manipulations, it is possible to split the reflect term and cast them into the original algorithm with minimum modifications. we describe the governing equations and the calculation of incident fields and scattered fields of a dielectric rough surface with a finite extent. In numerical solutions of Equation 8.1, only a surface of finite area can be considered. An area of L_x by L_y is discretized. A finite surface will cause edge diffraction of the incident wave. However, considering the practical case of an antenna radiation pattern that usually falls off, or making so, to the edge of the surface, the diffraction may be ignorable. We chose to apply MoM to solve Maxwell equations and the surface electric fields and magnetic fields are calculated. These surface waves are the "final" surface fields on the surface and include the multiple scattering of the monochromatic tapered wave within the surface area of L_x by L_y. The electric fields of the scattered waves are calculated by using Huygen's principle and are obtained by integration of the "final" surface fields, weighted by the Green's functions, over the area of L_x by L_y. The simulation of the scattering matrix is the calculation of scattered fields for two incident polarizations, vertical and horizontal polarizations. The two incident waves v and h are of the same frequency and incident on the same realization of a rough surface. The scattering matrix elements include all the multiple scattering within the area L_x by L_y. The scattering matrix elements are normalized by the square root of the incident power as they are calculated from the scattered electric field.

Consider a plane wave, $\vec{E}^i(\vec{r})$ and $\vec{H}^i(\vec{r})$, with a time dependence of $e^{-j\omega t}$, impinging upon a two-dimensional dielectric rough surface with a random height of $z = \xi(x, y)$. The incident fields can be expressed in terms of the spectrum of the incident wave [10,11]

$$\vec{E}^i(\vec{r}) = \int_{-\infty}^{+\infty} dk_x \int_{-\infty}^{+\infty} dk_y \exp(ik_x x + ik_y y - ik_z z)\tilde{E}(k_x, k_y)\hat{e}(-k_z) \quad (8.2)$$

$$\vec{H}^i(\vec{r}) = -\frac{1}{\eta}\int_{-\infty}^{+\infty} dk_x \int_{-\infty}^{+\infty} dk_y \exp(ik_x x + ik_y y - ik_z z)\tilde{E}(k_x, k_y)\hat{h}(-k_z) \quad (8.3)$$

For horizontally polarized (TE) wave incidence

$$\hat{e}(-k_z) = \frac{1}{k_\rho}(\hat{x}k_y - \hat{y}k_x) \tag{8.4}$$

$$\hat{h}(-k_z) = \frac{k_z}{kk_\rho}(\hat{x}k_x + \hat{y}k_y) + \frac{k_\rho}{k}\hat{z} \tag{8.5}$$

and for vertically polarized (TM) wave incidence

$$\hat{h}(-k_z) = -\frac{1}{k_\rho}(\hat{x}k_y - \hat{y}k_x) \tag{8.6}$$

$$\hat{e}(-k_z) = \frac{k_z}{kk_\rho}(\hat{x}k_x + \hat{y}k_y) + \frac{k_\rho}{k}\hat{z} \tag{8.7}$$

with $k_z = \sqrt{k^2 - k_\rho^2}$ and $k_\rho = \sqrt{k_x^2 + k_y^2}$. The incident wave vector is $\hat{k}_i = \hat{x}\sin\theta_i\cos\phi_i + \hat{y}\sin\theta_i\sin\phi_i - \hat{z}\cos\theta_i$, and $\hat{e}$, $\hat{h}$ denote the polarization vectors. In the above k and η are the wavenumber and wave impedance of free space, respectively. In practical cases, the incident field is tapered so that the illuminated rough surface can be confined to the surface size $L_x \times L_y$. The spectrum of the incident wave, $\tilde{E}(k_x, k_y)$, is given as

$$\tilde{E}(k_x, k_y) = \frac{1}{4\pi^2}\int_{-\infty}^{\infty} dx \int_{-\infty}^{\infty} dy \exp(-ik_x x - ik_y y)\exp[i(k_{ix}x + k_{iy}y)(1 + w)]\exp(-\Theta) \tag{8.8}$$

where $\Theta = \Theta_x + \Theta_y = (x^2 + y^2)/\beta_e^2$ and

$$\Theta_x = \frac{(\cos\theta_i\cos\phi_i x + \cos\theta_i\sin\phi_i y)^2}{\beta_e^2\cos^2\theta_i} \tag{8.9}$$

$$\Theta_y = \frac{(-\sin\varphi_i x + \cos\varphi_i y)^2}{\beta_e^2} \tag{8.10}$$

$$w = \frac{1}{k_i^2}\left(\frac{2\Theta_x - 1}{\beta_e^2\cos^2\theta_i} + \frac{2\Theta_y - 1}{\beta_e^2}\right) \tag{8.11}$$

The parameter β_e, a fraction of surface length L, controls the tapering of the incident wave such that the edge diffraction is negligible. Physically, β_e is similar to finite beamwidth of the antenna. The tapered wave is close to a plane wave if β_e is much larger than a wavelength. The tapered incident wave still obeys, as should be,

Maxwell equation because only the wave-vector spectrum of the incident wave is modified from that of a plane wave. With the incident waves defined above, now the surface fields satisfy equations:

$$\vec{E}_i + \int [j\omega\mu_0 \bar{\bar{\mathbf{G}}}_0 \cdot \hat{n}' \times \vec{H}(\vec{r}') + \nabla' \times \bar{\bar{\mathbf{G}}}_0 \cdot \hat{n}' \times \vec{E}(\vec{r}')] dS' = \begin{cases} \vec{E}(\vec{r}), \ z < \xi(x, y) \\ 0, \ z > \xi(x, y) \end{cases} \tag{8.12}$$

$$\int [-j\omega\mu_1 \bar{\bar{\mathbf{G}}}_t \cdot \hat{n}' \times \vec{H}_1(\vec{r}') - \nabla' \times \bar{\bar{\mathbf{G}}}_t \cdot \hat{n}' \times \vec{E}_1(\vec{r}')] dS' = \begin{cases} 0, \ z < \xi(x, y) \\ \vec{E}_1(\vec{r}), \ z < \xi(x, y) \end{cases} \tag{8.13}$$

$$\vec{H}_i + \int [-jk/\eta \bar{\bar{\mathbf{G}}}_0 \cdot \hat{n}' \times \vec{E}(\vec{r}') + \nabla' \times \bar{\bar{\mathbf{G}}}_0 \cdot \hat{n}' \times \vec{H}(\vec{r}')] dS'$$
$$= \begin{cases} \vec{H}(\vec{r}), \ z > \xi(x, y) \\ 0, \ z < \xi(x, y) \end{cases} \tag{8.14}$$

$$\int [-jk/\eta \bar{\bar{\mathbf{G}}}_t \cdot \hat{n}' \times \vec{E}_1(\vec{r}') + \nabla' \times \bar{\bar{\mathbf{G}}}_t \cdot \hat{n}' \times \vec{H}_1(\vec{r}')] dS' = \begin{cases} 0, \ z > \xi(x, y) \\ \vec{H}_1(\vec{r}), \ z < \xi(x, y) \end{cases} \tag{8.15}$$

where S' denotes the rough surface, $\vec{r}'$ a source point, and a field point on the rough surface. The unit normal vector $\hat{n}'$ refers to primed coordinate and points away from the second medium.

The dyadic Green's functions of free space, $\bar{\bar{\mathbf{G}}}_0$, is

$$\bar{\bar{\mathbf{G}}}_0(\vec{r}, \vec{r}') = \frac{[\bar{\bar{\mathbf{I}}} + \nabla\nabla/k^2] e^{jk|\vec{r} - \vec{r}'|}}{4\pi|\vec{r} - \vec{r}'|} \tag{8.16}$$

where $\bar{\bar{\mathbf{I}}}$ is unit dyadic. The dyadic Green's function of the transmitted medium, an inhomogeneous layer, $\bar{\bar{\mathbf{G}}}_t$, consists of two parts: a direct part and a reflected part:

$$\bar{\bar{\mathbf{G}}}_t(\vec{r}, \vec{r}') = \bar{\bar{\mathbf{G}}}_t^d(\vec{r}, \vec{r}') + \bar{\bar{\mathbf{G}}}_t^r(\vec{r}, \vec{r}') \tag{8.17}$$

The direct part $\bar{\bar{\mathbf{G}}}_t^d(\vec{r}, \vec{r}')$ is the same as the one for homogeneous medium

$$\bar{\bar{\mathbf{G}}}_t^d(\vec{r}, \vec{r}') = \frac{[\bar{\bar{I}} + \nabla\nabla/k_1{}^2] e^{jk_t|\vec{r} - \vec{r}'|}}{4\pi|\vec{r} - \vec{r}'|}, \tag{8.18}$$

which can be put into a vector form and the reflected part $\bar{\bar{\mathbf{G}}}_t^r(\vec{r}, \vec{r}')$ that accounts for the layered effects is given as [5,7]

$$\bar{\bar{\mathbf{G}}}_t^r(\vec{r}, \vec{r}') = \frac{-j}{8\pi^2} \iint dk_x dk_y e^{jk_x(x - x')+jk_y(y - y')} e^{jk_{tz}(z+z')}$$
$$[R_h \hat{e}(- k_{tz}) + R_v \hat{h}(k_{tz}) \hat{h}(-k_{tz})] \quad (8.19)$$

where

$$\hat{e}(k_{tz}) = \hat{e}(-k_{tz}) = \frac{1}{k_\rho}(\hat{x}k_y - \hat{y}k_x) \quad (8.20)$$

$$\hat{h}(k_{tz}) = \frac{-k_{tz}}{k_t k_\rho}(\hat{x}k_x + \hat{y}k_y) + \frac{k_\rho}{k_t}\hat{z} \quad (8.21)$$

$$\hat{h}(-k_{tz}) = \frac{k_{tz}}{k_t k_\rho}(\hat{x}k_x + \hat{y}k_y) + \frac{k_\rho}{k_t}\hat{z} \quad (8.22)$$

The reflection coefficients for horizontal and vertical polarizations, R_h and R_v, respectively, are readily obtained through the recurrence relation assuming that the general inhomogeneous layer is represented by N layers of piecewise constant regions. Please refer Chapter 3 for details. This formulation is applicable if the first layered medium interface is below the lowest point of the rough surface $\xi(x, y)$, i.e., $\xi(x, y) \geq -d_1$. To evaluate Equation 8.19, a double infinite integral is required to compute. The numerical approach proposed in [12] is applied in this paper. As a result, a total of nine elements of $\bar{\bar{\mathbf{G}}}_t^r$ and a total of eight elements of $\nabla \times \bar{\bar{\mathbf{G}}}_t^r$ are numerically evaluated. These elements represent the contributions from the inhomogeneous layers to the total scattering. The reformulation is given in the next section, while the complete components for numerical computation are given in Appendix 8B. When the lower medium is homogeneous, $\bar{\bar{\mathbf{G}}}_t^r(\vec{r}, \vec{r}')$ vanishes, and the problem reduces to those homogeneous rough surface scattering. We then show that how the reflected part of the dyadic Green's function is cast into formulation such that the modification of the numerical coding can be minimized within the framework of the physics-based two-grid method (PBTG) [12,13]. Equations 8.12–8.14 are written in the matrix equations using moment of method with pulse function as basis function and point matching method

$$\sum_{n=1}^{N} [Z_{mn}^{p1} I_n^{(1)} + Z_{mn}^{p2} I_n^{(2)} + Z_{mn}^{p3} I_n^{(3)} + Z_{mn}^{p4} I_n^{(4)} + Z_{mn}^{p5} I_n^{(5)} + Z_{mn}^{p6} I_n^{(6)}] = I_m^{(p)inc} \quad (8.23)$$

where $I_n^{(q)}$ are unknown surface fields needed to be solved, and $I_m^{(p)inc}$ are given by the incident fields, and the parameter N is the number of points we use to sample the rough surface. For $p = 1, 2, 3$ which correspond the surface integral equation when approaching the surface from free space and for $p = 4, 5, 6$ when approaching the surface from the lower medium. The quantities of $I_m^{(p)inc}$ are zero for $p = 4, 5, 6$.

where

$$I_n^{(1)} = F_x(\vec{r}) = S_{xy}(\vec{r}_n)[\hat{n} \times \vec{H}(\vec{r}_n)]\cdot\hat{x} \tag{8.24}$$

$$I_n^{(2)} = F_y(\vec{r}) = S_{xy}(\vec{r}_n)[\hat{n} \times \vec{H}(\vec{r}_n)]\cdot\hat{y} \tag{8.25}$$

$$I_n^{(3)} = I_n(\vec{r}) = S_{xy}(\vec{r}_n)\hat{n}\cdot\vec{E}(\vec{r}_n) \tag{8.26}$$

$$I_n^{(4)} = I_x(\vec{r}) = S_{xy}(\vec{r}_n)[\hat{n} \times \vec{E}(\vec{r}_n)]\cdot\hat{x} \tag{8.27}$$

$$I_n^{(5)} = I_y(\vec{r}) = S_{xy}(\vec{r}_n)[\hat{n} \times \vec{E}(\vec{r}_n)]\cdot\hat{y} \tag{8.28}$$

$$I_n^{(6)} = F_n(\vec{r}) = S_{xy}(\vec{r}_n)\hat{n}\cdot\vec{H}(\vec{r}_n) \tag{8.29}$$

are surface unknowns and the slope tilting factor is $S_{xy} = \sqrt{1 + \xi_x^2 + \xi_y^2}$, and $\xi_x = \frac{\partial \xi}{\partial x}$, $\xi_y = \frac{\partial \xi}{\partial y}$ are surface slopes along x and y directions, respectively.

The Z_{mn}^{pq} in Equation 8.23 is the impedance elements and are determined by the free-space Green's function and the lower medium Green's function. The parameter N is the number of points we use to digitize the rough surface.

To solve Equation 8.23, we obtain the surface fields. Traditionally, the matrix equation is solved by matrix inversion or Gaussian elimination methods, which requires $O(N^3)$ operations and $O(N^2)$ memory. To keep the structure of the PBTG algorithm as much as possible, the surface fields associated with the reflected part of the dyadic Green's function is decomposed into tangential and normal components for $\vec{E}$ and $\vec{H}$ fields. For inhomogeneous surfaces, there are additional tangential fields, $\hat{n} \times \vec{E}$, $\hat{n} \times \vec{H}$ associated with the $\bar{\bar{\mathbf{G}}}_t^r$, $\nabla \times \bar{\bar{\mathbf{G}}}_t^r$ and normal fields, $\hat{n}\cdot\vec{E}$, $\hat{n}\cdot\vec{H}$ associated with $\nabla\bar{\bar{\mathbf{G}}}_t^r$ need to be solved. In the following, we illustrate our derivation for the reflected part of the Green's function. The final results can be put into the structure of PBTG and thus the fast method can be applied.

Consider the integral equation of the form

$$\vec{E} = \vec{E}^i + \int \{\nabla \times \bar{\bar{\mathbf{G}}}_t^r\cdot(\hat{n}' \times \vec{E}) + j\omega\mu\bar{\bar{\mathbf{G}}}_t^r\cdot(\hat{n}' \times \vec{H})\}\, dS' \tag{8.30}$$

$$\vec{H} = \vec{H}^i + \int \{-j\omega\varepsilon\bar{\bar{\mathbf{G}}}_t^r\cdot(\hat{n}' \times \vec{E}) + \nabla \times \bar{\bar{\mathbf{G}}}_t^r\cdot(\hat{n}' \times \vec{H})\} dS' \tag{8.31}$$

Taking the tangential projection, we reach the following forms

$$\hat{n} \times \vec{E} = \hat{n}' \times \vec{E}^i + \hat{n} \times \int \{\nabla \times \bar{\bar{\mathbf{G}}}_t^r \cdot (\hat{n}' \times \vec{E}) + j\omega\mu \bar{\bar{\mathbf{G}}}_t^r \cdot (\hat{n}' \times \vec{H})\}\, dS' \tag{8.32}$$

$$\hat{n} \times \vec{H} = \hat{n} \times \vec{H}^i + \hat{n} \times \int \{ - j\omega\varepsilon \bar{\bar{\mathbf{G}}}_t^r \cdot (\hat{n}' \times \vec{E}) + \nabla \times \bar{\bar{\mathbf{G}}}_t^r \cdot (\hat{n}' \times \vec{H})\}\, dS' \tag{8.33}$$

Following the notations of PBTG, Equations 8.32 and 8.33 can be rewritten as

$$\vec{I}^r = (\vec{s} \times \vec{E}^i) + \vec{s} \times \iint (\nabla \times \bar{\bar{\mathbf{G}}}_t^r \cdot \vec{I}^{r'} + j\omega\mu \bar{\bar{\mathbf{G}}}_t^r \cdot \vec{F}^{r'})dx'dy' \tag{8.34}$$

$$\vec{F}^r = (\vec{s} \times \vec{H}^i) + \vec{s} \times \iint (-j\omega\varepsilon \bar{\bar{\mathbf{G}}}_t^r \cdot \vec{I}^{r'} + \nabla \times \bar{\bar{\mathbf{G}}}_t^r \cdot \vec{F}^{r'})dx'dy' \tag{8.35}$$

What we need is to write up explicit forms of $\vec{I}^r$ and $\vec{F}^r$. This is given in Appendix 8A. Now that I_x^r, I_y^r, I_n^r, $F_x{}^r$, F_y^r, $F_n{}^r$ are new terms resulting from $\bar{\bar{\mathbf{G}}}_t^r$ and are readily added on the original six scalar integral equations for the homogeneous medium. The rewritten I_x^r, I_y^r, I_n^r, F_x^r, F_y^r, F_n^r makes the inclusion of the inhomogeneous effects without difficulties by simply casting them into the numerical computation framework. The inclusions of these terms in the generation of impedance matrix slightly increase the computation time. Numerically calculations involving $\bar{\bar{\mathbf{G}}}_t^r$ and $\nabla \times \bar{\bar{\mathbf{G}}}_t^r$ are illustrated below.

As the formulations made in the above section, the inclusion of the reflected part of the dyadic Green's function that accounts for the inhomogeneity of the lower medium under the framework PBTG method is straightforward. In formulating the impedance matrix and thus in the calculation of the matrix elements, additional efforts must be exercised to compute the matrix elements with double infinite integral involving $\bar{\bar{\mathbf{G}}}_t^r$ and $\nabla \times \bar{\bar{\mathbf{G}}}_t^r$. In this aspect, we adopted the method proposed by Tsang et al. [12]. The method evaluates the matrix elements by numerically integrating the Sommerfeld integrals along the Sommerfeld with higher-order asymptotic extraction. Written in spectral form, $\bar{\bar{\mathbf{G}}}_t^r$ is given by

$$\begin{aligned}
\bar{\bar{\mathbf{G}}}_t^r(\bar{r}, \bar{r}') &= \frac{j}{8\pi^2} \iint dk_x dk_y \frac{1}{k_{tz}} [R_h \hat{e}(k_{tz})\hat{e}(-k_{tz}) + R_v \hat{h}(k_{tz}) \\
&\quad \hat{h}(-k_{tz})] e^{ik_x(x - x') + ik_y(y - y') + ik_{tz}(z + z')} \\
&= \frac{i}{8\pi^2} \iint dk_x dk_y \frac{1}{k_{tz}} [R_h e^{-2ik_{tz}d_1} \hat{e}(k_{tz})\hat{e}(-k_{tz}) + R_v e^{-2ik_{tz}d_1} \hat{h}(k_{tz}) \\
&\quad \hat{h}(-k_{tz})] e^{ik_x(x - x') + ik_y(y - y') + ik_{tz}(z + z' + 2d_1)}
\end{aligned} \tag{8.36}$$

and the curl of Equation 8.36 is

$$\nabla \times \bar{\bar{\mathbf{G}}}_t^r(\vec{r}, \vec{r}') = \frac{j}{8\pi^2} \iint dk_x dk_y \frac{1}{k_{tz}} [R_h \hat{h}(k_{tz}) \hat{e}(-k_{tz}) + R_v \hat{e}(k_{tz})$$
$$\hat{h}(-k_{tz})] e^{ik_x(x - x') + ik_y(y - y') + ik_{tz}(z + z')}$$
$$= \frac{j}{8\pi^2} \iint dk_x dk_y \frac{1}{k_{tz}} [R_h e^{-2ik_{tz}d_1} \hat{h}(k_{tz}) \hat{e}(-k_{tz})$$
$$+ R_v e^{-2ik_{1z}d_1} \hat{e}(k_{tz}) \hat{h}(-k_{tz})] e^{ik_x(x - x') + ik_y(y - y') + ik_{tz}(z + z' + 2d_1)} \tag{8.37}$$

More explicit forms of Equations 8.36 and 8.37 are given in Appendix 8B. Following the numerical procedures proposed in [12], they are calculated in high numerical stability and accuracy. Finally, the reflection coefficients R_v, R_h are given for completeness [2].

$$R_p = \frac{e^{i2k_z d_0}}{R_{p01}} + \frac{[1 - (1/R_{p01})^2] e^{i2(k_{tz} + k_z) d_0}}{(1/R_{p01}) e^{i2k_{1z} d_0}} + \frac{e^{i2k_{1z} d_1}}{R_{p12}} + \frac{[1 - (1/R_{p12})^2] e^{i2(k_z + k_{1z}) d_1}}{(1/R_{p12}) e^{i2k_{2z} d_1}} + \ldots..$$
$$+ \frac{e^{i2k_{(l-1)z} d_{l-1}}}{R_{p(l-1)l}} + \frac{[1 - (1/R_{p(l-1)l})^2] e^{i2(k_{lz} + k_{(l-1)z}) d_{l-1}}}{(1/R_{p(l-1)l}) e^{i2k_{lz} d_{l-1}}} + R_{lt} e^{i2k_{lz} d_l} \tag{8.38}$$

where $p = h$ or $p = v$ polarization, $l = 1, 2, \ldots N$ and d_l represents region depth in region l.

We have insofar presented a full-wave solution of Equation (8.1) by the technique of Galerkin method, in particular the PBTG-based MoM for an inhomogeneous rough surface, as an illustrated example. Another approach to solving Equation (8.1) is an approximate solution. Since Equation (8.1) is Fredholm's integral equation of the second kind [14], an iterative approach is suitable, and perhaps, is the most appropriate to seek an approximate solution T.

The solution to a general Fredholm integral equation of the second kind is called an integral equation Neumann series. By iteration, we can account for the order solutions to which we are satisfied with specific problems at hand. In view of the iterative approach to solving Equation (8.1), if we start with the initial guess for the unknown field inside the integral using the incident field, we have the following total field as

$$\mathbf{E}(\vec{r}) \approx \mathbf{E}^i(\vec{r}) + k_0^2 \int \bar{\bar{\mathbf{G}}}(\vec{r}, \vec{r}') \cdot [\bar{\bar{\varepsilon}}_r(\vec{r}') - 1] \mathbf{E}^i(\vec{r}') d\vec{r}' \tag{8.39}$$

Equation (8.39) states for the known transmitted field, the total field, and hence the scattered field is determined immediately by carrying out the integration over the space covering the random medium within the reach of antenna footprint. Such a solution is the first-order solution to account for the single scattering process and is known as the first-order Born approximation [15]. To make the first-order Born approximation applicable, the scattered field must be much weaker than the incident field:

$$|\mathbf{E}^s(\vec{r})| \ll |\mathbf{E}^i(\vec{r})| \tag{8.40}$$

For our object function of interest in Equation (8.1), condition of Equation (8.40) imposes that

$$|\varepsilon_r(\vec{r}) - 1| \ll 1 \tag{8.41}$$

To be more strictly, we not only require the dielectric contrast against the background that supports the object must be small, but the electrical size of the object also matters, where the size means the maximum extent subtended by the antenna footprint, as illustrated in Figure 3.11. Hence, if the object is electrically larger, the range of dielectric contrast of Equation 8.41 becomes even smaller.

If we keep on the iterative process until reaching convergence, we obtain the Neumann series. So the Born approximation the first term of Neumann series.

Another commonly used approximation is Rytov approximation [15], which was supposed to improve the predictions by the geometric optics and Born approximation. For the first-order solution, Rytov approximation reduces to geometric optics solution if the diffraction scattering is ignored, and is found to reduce to Born approximation if the scattered field amplitude and phase fluctuations are both small. In this context, it has been argued that Rytov approximation and Born approximation have the same domain of validity [16,17]. The advantages of one over the other approximation still count on the degree of media inhomogeneity and the wavefields. The basic Rytov approximation is to replace the unknow field inside the integral of Equation 8.1 using an exponential form:

$$\mathbf{E} = \mathbf{E}^i e^{\psi} \tag{8.42}$$

Then expanding ψ into power series in the fluctuated part of the permittivity in random media. Noted that in Born approximation, the series expansion does not converge when the phase fluctuation is greater than unity, both first-order Born approximation and Rytov approximation remain still linear integral equations to the object function $\varepsilon_r(\vec{r}) - 1$. Hence, the accuracy of the Rytov approximation is more sensitive to the phase variations due to the dielectric contrast. However, unlike Born approximation, the Rytov approximation releases the limitation of the electrical size of the object. For dealing with wave scattering from the rough surface or random media, Born approximation is usually preferred. On the other hand, if we concern with the wave propagation in the media, namely, dealing with the transmitted filed, then Rytov approximation gives a more accurate solution.

Another popular approximate solution to the wave equation is the parabolic equation (PE) approximation [18,19]. Instead of treating it exhaustively, we simply brief it. For those interested readers, an excellent book can be referred [18]. The basic assumptions to make PE work fine include that most of the energy propagating is confined to the so-called paraxial direction, and the dielectric contrast against the surrounding background is small and smooth. Whether we take the full-wave solution or approximate solution, once the surface fields are solved, the scattered fields at the far-zone can be computed according to the Stratton and Chu formula given in Equations 4.9–4.10 of Chapter 4.

8.2 TIME-REVERSAL IMAGING

As briefly introduced in Chapter 1, it is of great practical significance to image the targets obscured by complex random media using the time-reversal (TR) technique [15,20–24], which is essentially a spatiotemporal matched filter to focus the target image adaptively. However, a space-space multi-static data matrix TR is somewhat difficult to achieve both high resolution under highly noisy interference. A modified time-reversal imaging method, formed by space-frequency multi-static data matrix, or space-frequency time-reversal multiple signal classification (TR-MUSIC) may be preferred. In Chapter 1, we have illustrated that the TR-MUSIC can offer a high capability of imaging targets in the presence of a random medium.

The space-frequency TR-MUSIC imaging utilizes the full backscattered data, including the contributions of all multiple sub-matrices, and is found to be statistically stable. Using the backscattered data collected by an antenna array, a space-frequency multi-static data matrix (SF-MDM) is possibly configured. Then the singular value decomposition is applied to the matrix to obtain the noisy subspace vector, which is then employed to image the target. Based on the statistical modeling of random media, the space-space TR-MUSIC and space-frequency TR-MUSIC imaging of the target obscured by random media are compared and analyzed. Numerical simulations show that the imaging performance of the space-frequency TR-MUSIC is better than that of the traditional space-space TR-MUSIC in both free space and random media.

Referring to Figure 1.6, the time-reversal array, consisting of N antenna units, is considered an N-input and an N-output of the linear time-invariant system. Let $h_{lm}(t)$ be the impulse response between the antenna array element m and the antenna array element l, including all propagation effects of the random medium and system noise between the array elements. Assume the input signal is $e_j(t)$, $1 \le j \le N$, and the output signal is given by

$$r_l(t) = \sum_{j=1}^{N} h_{lj}(t) \otimes_t e_j(t), \; 1 \le l \le N \tag{8.43}$$

where $\otimes_t$ denotes the convolution over time. The Fourier transform of $r_l(t)$ is

$$R_l(w) = \sum_{j=1}^{N} H_{lj}(w) E_j(w), \; 1 \le l \le N \tag{8.44}$$

In matrix notation, Equation (8.44) is expressed by

$$\mathbf{R}(w) = \mathbf{H}(w)\mathbf{E}(w) \tag{8.45}$$

where $\mathbf{R}(w)$ and $\mathbf{E}(w)$ are transmitting and receiving signals vector in the frequency domain, respectively, $\mathbf{H}(w)$ is the transmission matrix, whose dimension is the number of antennas in time-reversal array. By duality in Green function, the transmission matrix $\mathbf{H}(w)$ is symmetric, namely, for all matrix elements l, m, $H_{lm}(w) = H_{ml}(w)$.

Assuming the nth element of TR array transmits a Gaussian beam of the form as given in Equation 1.37 or the time-domain of the form:

$$s(t) = \exp\left(-\frac{t^2}{T^2} - jw_c t\right) \tag{8.46}$$

where T is pulse duration; $w_c = 2\pi f_c$, f_c is carrier frequency, and all the N elements of the TR array receive the scattering signal and take M sampling points, then the SF-MDM for the received signal in the frequency domain, $\mathbf{K}_n(w)$, can be written as:

$$\mathbf{K}_n(\omega) = \begin{bmatrix} k_{1n}(\omega_1) & k_{1n}(\omega_2) & \dots & k_{1n}(\omega_M) \\ k_{2n}(\omega_1) & k_{2n}(\omega_2) & \dots & k_{2n}(\omega_M) \\ \dots & \dots & \dots & \dots \\ k_{Nn}(\omega_1) & k_{Nn}(\omega_2) & \dots & k_{Nn}(\omega_M) \end{bmatrix} \tag{8.47}$$

where $\omega_M - \omega_1$ is the signal bandwidth. The singular value decomposition of $\mathbf{K}_n(w)$ gives

$$\mathbf{K}_n(w) = \mathbf{U}_n(w)\Lambda_n(w)\mathbf{V}_n(w) \tag{8.48}$$

where $\mathbf{U}_n$ is the $N \times N$ order matrix representing the left singular vector, $\mathbf{V}_n$ is the $M \times M$ order matrix representing the right singular vector, and Λ_n is the singular value matrix of order $N \times M$. When there are M_t targets, there exist M_t singular values greater than zero, corresponding singular vectors $\{\mathbf{u}_{sub_1}(w), \dots, \mathbf{u}_{sub_{M_t}}(w)\}$ forming the signal subspace, and the remaining singular vectors $\{\mathbf{u}_{sub_{M_t+1}}(w), \dots, \mathbf{u}_{sub_N}(w)\}$ singular values near zero is regarded as the noise subspace. The target information is embedded in the amplitude and phase of the singular vector of the signal subspace the target. The time-reversed signal that needs to be reversed is:

$$\sum_{i=M_t+1}^{N} s_{TR-n}^{i} = \sum_{i=M_t+1}^{N} \mathbf{u}_{sub_n}^{i} S(w) \tag{8.49}$$

where $S(\omega)$ is spectrum form of $s(t)$. It follows that the TR-MUSIC imaging function is given by

$$\begin{aligned} M_{SF}^{s}(\mathbf{X}_s, w) &= \left(\sum_{i=M_t+1}^{N} |\langle s_{TR-n}^{i}(w), \mathbf{g}(\mathbf{X}_s, w)\rangle|^2\right)^{-1} M_{SF}(\mathbf{X}_s, w) \\ &= \left(\sum_{i=M_t+1}^{N} \left|\int_{\Omega} S^H(w)\left[\mathbf{u}_{sub_n}^{i}\right]^H \mathbf{g}(\mathbf{X}_s, w)dw\right|^2\right)^{-1} \end{aligned} \tag{8.50}$$

which is based on a single space-frequency matrix, and the steering vector is

$$\mathbf{g}(\mathbf{X}_s, w) = [G(\mathbf{X}_s, \mathbf{R}_1, w), ..., G(\mathbf{X}_s, \mathbf{R}_N, w)]^T \tag{8.51}$$

with $G(\mathbf{X}_s, \mathbf{R}_i, w)$ being the background Green's function.

We see that the decomposed left singular vector $\mathbf{u}_i$, $i = 1, \ldots, N$ forms an orthogonal set containing sensor position information, while the right singular vector $\mathbf{v}_i$, $i = 1, \ldots, N$ forms an orthogonal set containing frequency information. It turns out that $\mathbf{K}_n(w)$ represents a space-frequency multi-static data matrix. The decomposition of $\mathbf{K}_n(w)$ is called space-frequency decomposition, and the imaging based on such space-frequency decomposition is called space-frequency TR-MUSIC.

Recalled that Equation (8.51) is the data matrix for the nth element of the TR array to transmit and the rest of the elements to receive the scattering signal. Now, if every element from 1 to N sequentially transmits the signal, while the rest of the elements receive simultaneously the scattering signal, then the full data matrix of becomes

$$\mathbf{K}(\omega) = \begin{bmatrix} \mathbf{k}_{1n}(w_1) & \mathbf{k}_{1n}(w_2) & \ldots & \mathbf{k}_{1n}(w_M) \\ \mathbf{k}_{2n}(w_1) & \mathbf{k}_{2n}(w_2) & \ldots & \mathbf{k}_{2n}(w_M) \\ \ldots & \ldots & \ldots & \ldots \\ \mathbf{k}_{Nn}(w_1) & \mathbf{k}_{Nn}(w_2) & \ldots & \mathbf{k}_{Nn}(w_M) \end{bmatrix} \tag{8.52}$$

with

$$\mathbf{k}_{i,n}(w_1) = [k_{i,1}(w_1), k_{i,2}(w_1), \ldots, k_{i,N}(w_1)], \ (i, n = 1, \ldots, N) \tag{8.53}$$

The singular value decomposition of $\mathbf{K}(w)$ is

$$\mathbf{K}(w) = \mathbf{U}(w)\Lambda(w)\mathbf{V}^H(w) \tag{8.54}$$

where $\mathbf{U}$ is the left singular vector matrix of order $N^2 \times N^2$, Λ is the singular value matrix of order $N^2 \times M$, $\mathbf{V}$ is the right singular vector matrix of order. It follows that the time-reversal signal is

$$\sum_{n=1}^{N} \sum_{i=M_t+1}^{N} s_{TR-n}^{i}(w) = \sum_{n=1}^{N} \sum_{i=M_t+1}^{N} \mathbf{u}_{sub_n}^{i}(w) S(w) \tag{8.55}$$

and imaging function for the full data matrix is of the form

$$\begin{aligned} M_{SF}^{m}(\mathbf{X}_s, w) &= \left(\sum_{n=1}^{N} \sum_{i=M_t+1}^{N} |\langle s_{TR-n}^{i}(w), \mathbf{g}(\mathbf{X}_s, w)\rangle|^2 \right)^{-1} M_{SF}^{m}(\mathbf{X}_s, w) \\ &= \left(\sum_{n=1}^{N} \sum_{i=M_t+1}^{N} \left| \int_{\Omega} S^H(w) \left(\mathbf{u}_{sub_n}^{i}\right)^H \mathbf{g}(\mathbf{X}_s, w) dw \right|^2 \right)^{-1} \end{aligned} \tag{8.56}$$

The above matrix is TR-MUSIC based on full data from the elements of the TRE array in playing the signal transmission and reception. We may interpret Equation 8.50 as a single-input multiple-output (SIMO) system, while Equation (8.56) is a multiple-input multiple-output (MIMO) system. A quick comparison of SIMO and MIMO performance for imaging targets in free-space is shown in Figure 8.2. A noise level of 0 dB is added. Both SIMO and MIMO offer better azimuth resolution than range resolution. The SIMO space-frequency TR-MUSIC imaging presents a range shifts, while MIMO space-frequency TR-MUSIC imaging seems to have not such range displacement.

For the imaging of a target in a random medium background, assuming that the target is an isotropic scatterer with a constant scattering amplitude, with the imaging scene shown in Figure 1.6. For N elements of TR array, we have a transmission matrix:

$$\langle \mathbf{K} \rangle = \langle \sigma_s \mathbf{g}\mathbf{g}^{\mathrm{T}} \rangle \tag{8.57}$$

where $\langle\rangle$ is the ensemble average; σ_s is the scattering coefficient, T denotes the matrix transpose, and

$$\mathbf{g} = [G(\mathbf{R}_1, \mathbf{X}_p, w), G(\mathbf{R}_2, \mathbf{X}_p, w), \ldots, G(\mathbf{R}_N, \mathbf{X}_p, w)]^{\mathrm{T}} \tag{8.58}$$

The matrix element of **K** is the transmission function between element i and element j given by a random Green's function as:

$$K_{ij} = \sigma_s G(\mathbf{R}_j, \mathbf{X}_p) G(\mathbf{X}_p, \mathbf{R}_i) \tag{8.59}$$

Then the Decomposition of the Time Reversal Operator (DORT) T is

$$\langle \mathbf{T} \rangle = \langle \mathbf{K}\mathbf{K}^{\dagger} \rangle \tag{8.60}$$

where † denotes the conjugate transpose operator, and the elements of **T** are

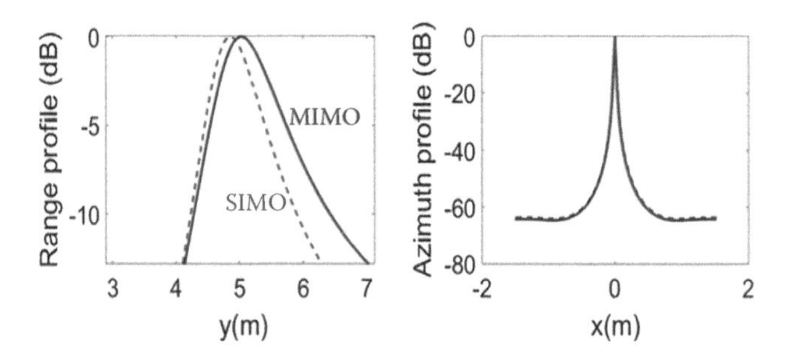

FIGURE 8.2 Comparison of space-frequency TR-MUSIC imaging target in free-space by SIMO and MIMO: range profile and azimuth profile. The noise level was set to 0 dB.

$$\langle T_{mn} \rangle = \sum_{i=1}^{N} \langle G_m^* G_n G_i G_i^* \rangle \tag{8.61}$$

For the strong and weak fluctuations of the wave propagation in a random media, the circular complex Gaussian approximation imposes constraints. Assuming the circular complex Gaussian distribution, the fourth-order moment appearing in Equation (8.61) can be approximated from the second-order moment [25]:

$$\langle G_m^* G_n G_i G_i^* \rangle = \langle G_m^* G_n \rangle \langle G_i G_i^* \rangle + \langle G_m^* G_i \rangle \langle G_n G_i^* \rangle - \langle G_m^* \rangle \langle G_n \rangle \langle G_i \rangle \langle G_i^* \rangle \tag{8.62}$$

The second order moment is recognized as a mutual coherence function:

$$\left\langle G_i G_j^* \right\rangle = \Gamma_{i,j}(w_1, w_2) \tag{8.63}$$

Using the parabolic equation (PE) approximation, we have

$$\Gamma_{i,j} = G_{io}(w_1) G_{jo}(w_2) \exp(-Q_{i,j}) \tag{8.64}$$

where G_{io} is the free-space Green's function:

$$G_{io} = -\frac{i}{4} H_0^{(1)} (k_1 |\mathbf{R}_i - \mathbf{X}_p|),\ k_1 = \frac{w_1}{c} \tag{8.65}$$

where $H_0^{(1)}$ is 0-order Hankle function of first kind. The factor $exp(-Q_{i,j})$ accouts for random medium contribution. One possible solution is given in [26].

Now, the K matrix in full data (MIMO) space-frequency TR-MUSIC writes

$$\mathbf{K}(w_m) = \begin{bmatrix} \mathbf{k}_{1n}(w_m) \\ \mathbf{k}_{2n}(w_m) \\ \cdots \\ \mathbf{k}_{Nn}(w_m) \end{bmatrix} = \begin{bmatrix} k_{11}(w_m) & k_{12}(w_m) & \ldots & k_{1N}(w_m) \\ k_{21}(w_m) & k_{22}(w_m) & \ldots & k_{2N}(w_m) \\ \cdots & \cdots & \cdots & \cdots \\ k_{N1}(w_m) & k_{N2}(w_m) & \ldots & k_{NN}(w_m) \end{bmatrix},\ m = 1, \ldots M \tag{8.66}$$

and

$$\mathbf{T}(w) = \begin{bmatrix} \mathbf{t}_{1n}(w_1) & \mathbf{t}_{1n}(w_2) & \ldots & \mathbf{t}_{1n}(w_M) \\ \mathbf{t}_{2n}(w_1) & \mathbf{t}_{2n}(w_2) & \ldots & \mathbf{t}_{2n}(w_M) \\ \cdots & \cdots & \cdots & \cdots \\ \mathbf{t}_{Nn}(w_1) & \mathbf{t}_{Nn}(w_2) & \ldots & \mathbf{t}_{Nn}(w_M) \end{bmatrix} \tag{8.67}$$

where the elements are

$$\mathbf{t}_{i,j}(w) = [t_{i,j}(w_1),\ t_{i,j}(w_2)\ldots,\ t_{i,j}(w_M)](i, j = 1, \ldots, N) \tag{8.68}$$

Following the same procedure in obtaining the imaging function of Equation 8.52 by singular value decomposition applying to Equation 8.67, we can readily perform

TR-MUSIC imaging for target obscured by random medium. Table 8.1 gives the simulation parameters to compare the performance between the space-space and space-frequency TR-MUSIC. For imageing geometry, please refer to Figure 1.6.

From Figure 8.3, we see that the anti-noise performance of the space-frequency TR-MUSIC is significantly better than that of the space-space TR-MUSIC. When the Gaussian white noise with a signal-to-noise ratio of −10 dB is added, the space-frequency TR-MUSIC has a position offset but imaging quality is retained to a good level; by space-space TR-MUSIC the image is seriously defocused. As the signal-to-noise ratio increases to 0 dB, the space-space TR-MUSIC still cannot achieve the imaging of the target, but the space-frequency TR-MUSIC imaging is improved significantly, and the fact is more so when the signal-to-noise ratio continues to increase. Figure 8.4 shows the profile cuts of the target image T1 under different signal-to-noise ratio levels. From the figure, it can be seen that the azimuth resolution is better than the range resolution by both methods, with space-frequency TR-MUSIC being superior to the space-space TR-MUSIC. It is also clear that the stronger the noise, the greater the range displacement in the space-space TR-MUSIC.

Now, we examine the random media effect on the imaging performance. Figure 8.5 shows the target imaging results under different random media background. When OD = 0.1 and Albedo = 0.1, the space-space TR-MUSIC can achieve accurate imaging of the target; with the increase of the single scattering albedo, although the imaging results appear defocused in the longitudinal direction, the target can still be achieved Imaging. When OD = 1 and Albedo = 0.1, the imaging of the target can be achieved. However, with the increase of optical thickness and the increase of single scattering albedo, the traditional space-space TR-MUSIC cannot achieve focused imaging of the target. It can be seen that when OD = 0.1, it means that the random medium has less influence and is close to free space. When OD = 5, it means that the influence of

TABLE 8.1
Simulation Parameters Setting of Space–Space and Space–Frequency TR-MUSIC Imaging

Parameters	Symbol	Value
Pulse duration	T	4 ns
Carrier frequency	f_c	3 GHz
Number of elements	N	11
Element spacing		0.05 m
Pixel spacing		0.025×0.025 m^2
	d_1	1.2 m
Medium depth	d_2	1.3 m
	d_3	2.5 m
Target location	T1	T1(0 m, 5 m)
Albedo		0.1/0.5/1
Optical depth	OD	0.1/1/5

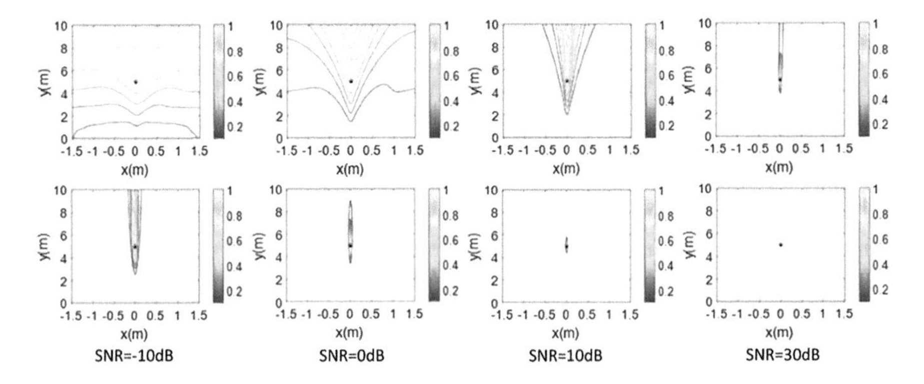

FIGURE 8.3 Comparison of imaging results under different noise levels: top row: space-space TR-MUSIC; bottom row: space-frequency TR-MUSIC.

random media is greater. The greater the optical thickness, the greater the single-scatter albedo, indicating that the random media has a greater impact on the target imaging.

Similarly, we examine the imaging performance of space-frequency TR-MUSIC under the influence of a random medium. It can be seen from Figure 8.6 that when the OD is 0.1, although with the increase of Albedo, the image point becomes larger and the focusing effect becomes worse. But compared with the space-space TR-MUSIC, its change is smaller. When OD = 1, with the increase of Albedo, the vertical focusing performance of the image becomes significantly worse. When OD = 5, albedo = 0.5, and 1, the space-frequency TR-MUSIC has a serious de-focusing phenomenon in both azimuth direction and range direction; its imaging performance is still superior to the space-space TR-MUSIC.

To further verify the space-frequency TR-MUSIC, the imaging performance with the same set of parameters but at a carrier frequency of 24 GHz is shown

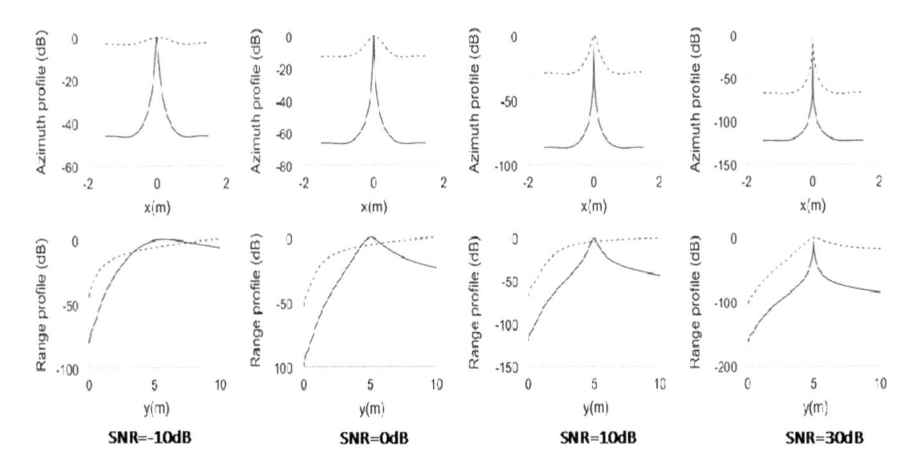

FIGURE 8.4 Comparison of space-space TR-MUSIC (red dash-line) and space-frequency TR-MUSIC (blue-line) imaging resolution under different signal to noise levels.

in Figure 8.7. Comparing it with Figure 8.6 shows that when OD = 0.1, 1, the effect of the random media is weak. As the frequency increases, the wavelength becomes shorter, and the penetration through the random medium is shallower, resulting in a significant deterioration of the distance resolution; when OD = 5, Albedo = 1, the influence of the random medium is stronger and further deteriorates the imaging performance.

A short remark about the imaging performance of TR-MUSIC may be drawn at this point. From the above simulations, both optical thickness and single-scatter albedo are the two main factors that affect the target imaging performance in the presence of random media. Compared to azimuth resolution, range resolution is deteriorated to a greater extent by random media. When the scattering thickness is large, the wave propagation is more affected by the random medium, and more energy is scattered, so that the influence from the random medium is enhanced, resulting in a poor target imaging.

8.3 SYNTHETIC APERTURE IMAGING

8.3.1 Signal Model

SAR echo signal in the time domain is a result of scattered field $E_{s_\tau}(\tau, \eta)$, also called the reflectivity field, convolving with the radar system impulse response, or

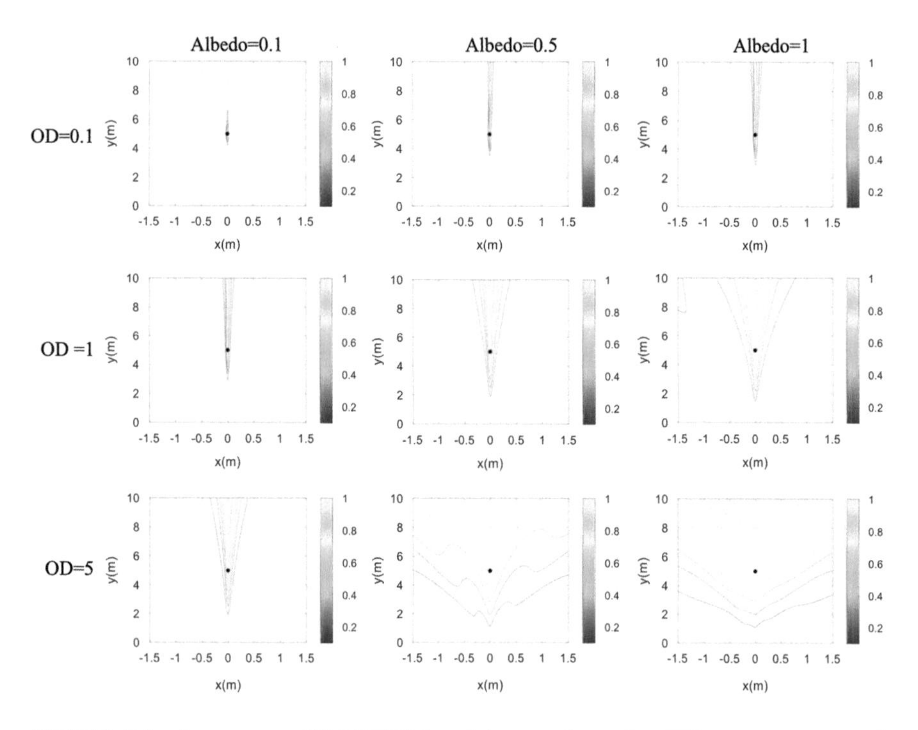

FIGURE 8.5 Space-space TR-MUSIC imagining results under different parameters of random media (f_c=3 GHz).

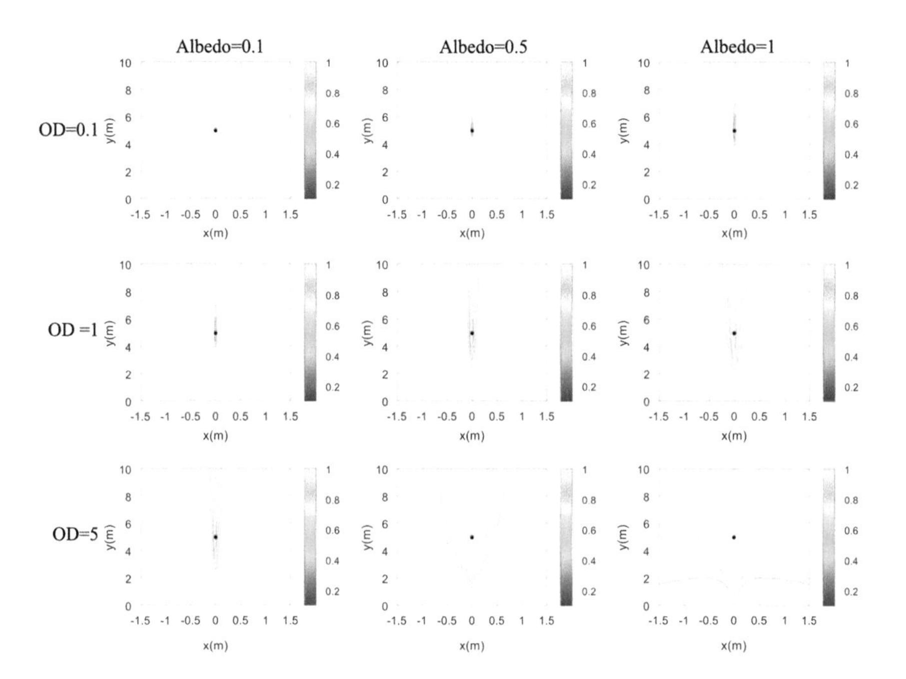

FIGURE 8.6 Space-frequency TR-MUSIC imagining results under different parameters of random media (f_c = 3 GHz).

point spread function, $\text{PSF}_\tau(\tau, \eta)$, and is mathematically expressed as, assuming a pulse radar:

$$\begin{aligned} s_0(\tau, \eta) &= E_{s_\tau}(\tau, \eta) \otimes \text{PSF}_\tau(\tau, \eta) \\ &= E_{s_\tau}(\tau, \eta) \otimes \\ &\left[p_r\left(\tau - \frac{2R(\eta)}{c}\right) g_a(\eta - \eta_c) \exp\left\{-j4\pi f_c \frac{R(\eta)}{c}\right\} \exp\left\{j\pi a_r \left[\tau - \frac{2R(\eta)}{c}\right]^2\right\}\right] \end{aligned} \tag{8.69}$$

where τ, η represent the fast time and slow time, respectively; $\otimes$ is convolution operator in the time domain; R_c is the SAR range to the center of footprint and varies with the slow time; a_r is the chirp rate, f_c is the carrier frequency.

In Equation 8.69, p_r is a pulse waveform, and g_a denotes the azimuthal antenna pattern with a typical form:

$$g_a(\eta - \eta_c) \cong \text{sinc}^2\left\{\frac{0.886 \cdot \theta_{diff}(\eta - \eta_c)}{\theta_{az}}\right\} \tag{8.70}$$

where θ_{diff} denotes the angle difference between beam center and instantaneous target angle; θ_{az} is the antenna beamwidth at azimuthal direction; η_c stands for the azimuth crossing time at azimuth beam center.

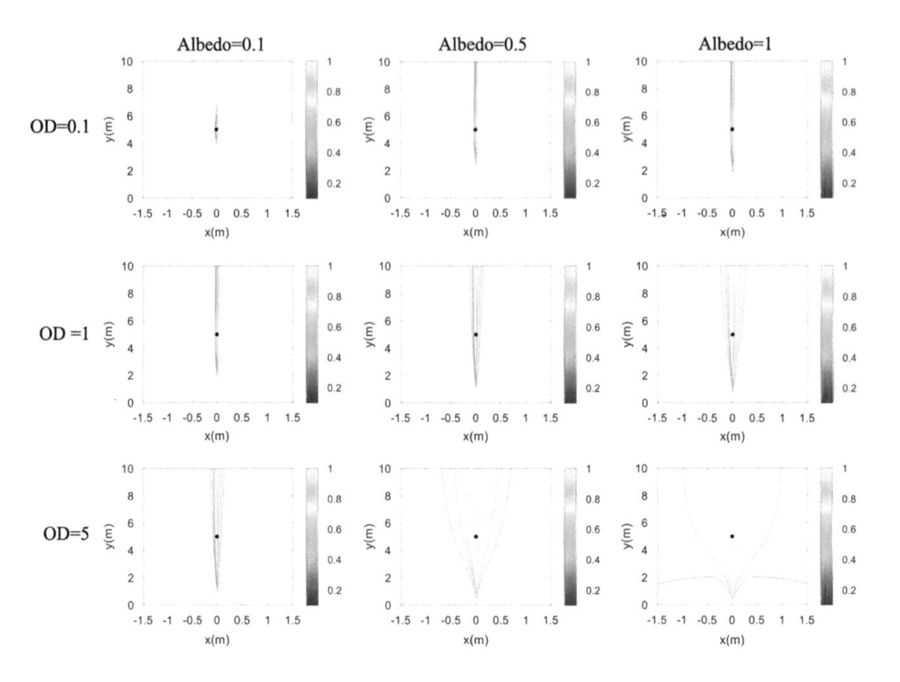

FIGURE 8.7 Same as Figure 8.6 but at $f_c = 24$ GHz.

Equivalently, the echo signal at frequency domain with linear frequency modulation in the chirp signal can be given by taking the Fourier transform of Equation 8.69 in fast time τ as:

$$S_0(f_\tau, \eta) = E_{s_f}(f_\tau, \eta) \times PSF_{f_\tau}(f_\tau, \eta) = E_{s_f}(f_\tau, \eta) \times \left[P_r(f_\tau) g_a(\eta - \eta_c) \times \exp\left\{ -j\frac{4\pi(f_c + f_\tau)R(\eta)}{c} \right\} \exp\left\{ -j\pi\frac{f_\tau^2}{a_r} \right\} \right] \tag{8.71}$$

As for estimating the scattered field from the imaging target, be it distributed or single, there exist numerous fast computational algorithms [27]. For the purpose of demonstration, the ray-tracing approach, combining with the ray tracing, the Physical Optics (PO) and Physical Theory of Diffraction (PTD) methods [22] is given. As we stated in the last section, estimation of the scattered field in the time of SAR transmitting the incident signal toward the target can be by full-wave solution or the approximate solution such as Born, Rytov, PE approximations, among other choices, as discussed previously. In approximate solution, for the more complex targets, the high order Taylor expansion is preferable for solving the surface current density over which the scattered field is obtained by integration, so that the multiple scattering is taken into account. Furthermore, an analytical diffraction solution is applied to the wedge structure that is very common in geometrically complex targets. Accounting for the multiple scattering or bounces, Equation 8.71 can be rewritten as a coherent sum of the individual bounce ($m = 1, 2, \ldots M_b$):

$$S_0(f_\tau, \eta) = \sum_{m=1}^{M_b} \vec{E}_{sf}(f_\tau, \eta) \times \left[P_r(f_\tau) g_a(\eta - \eta_c) \times \exp\left\{ -j\frac{4\pi(f_0 + f_\tau)R_m(\eta)}{c} \right\} \exp\left\{ -j\pi\frac{f_\tau^2}{a_r} \right\} \right] \tag{8.72}$$

where R_m is the slant range of the mth bounce; $\vec{E}_{sf}$ is scattered field in frequency domain. When simulating the echo signal, either Equation 8.69 or Equation 8.71 can be applied, at least mathematically. Computationally, it is not so, as can be seen from the comparison between time domain and frequency domain given in Table 8.2, where M denotes the grid number at azimuth; N is the range bin number; P is the discrete sample of reflectivity map in single azimuth line; Q is the convolution kernel size that is with regard to the chirp signal, and R is the number of bouncing. Note that the ray tracing is desirable in both time and frequency domains.

In the time domain, it requires to compute the convolution for each azimuth line with scattered field (reflectivity field) within the area of instantaneous footprint at certain azimuth slow time. Based on the fast Fourier transformation, the time complexity can be reduced to $\mathbf{O}(\log_2(P+Q))$ and for all of the azimuth positions, M, can be extended to $\mathbf{O}(M(P+Q)\log_2(P+Q))$. Indeed, the multi-bounce is hidden in the pre-processing using the reflectivity field combination. On the other hand, in the frequency domain, we only need to consider the pixel-wise multiplication and the multi-bouncing, and the time complexity is $\mathbf{O}(MNR)$. To this end, it is clear that SAR echo signal simulation in the frequency domain is a better choice [28].

8.3.2 SAR Path Trajectory

For a more realistic echo simulation, SAR's path trajectory that perturbs the Doppler estimation must be considered. For tracking the SAR moving path, a series of co-ordinates transformations are necessary. The geo-location with longitude/latitude/height (LLH) coordinate, known as Geocentric coordinate is attained to produce the geo-reference SAR image, which can be projected onto the ground map. As for SAR's attitudes, pitch/roll/yaw (PRY) coordinate, which generates the squint angle, and determines the antenna pointing vector and the antenna pattern [29]. For example, the pitch angle induces the squint angle that subsequently changes the Doppler centroid. The vibration of the SAR motion also gives rise to bias and noise attached to the attitudes within the path trajectory tunnel, where, the bias is measurable, while the noise is immeasurable. In order to simulate these vibration effects, the coordinate transformations between the SAR observation geometry and the local geometry must be included.

As far as the path trajectory is concerned, one must consider the state vector including the position $[X, Y, Z]$, the tangential velocity $[V_x, V_y, V_z]$, and the attitude $[\alpha, \beta, \gamma]$, all of them must be recorded as the sensor parameters. The basic idea of relationship of each coordinate is below. First of all, we need to define the SAR movement.

TABLE 8.2
Comparison of Time and Frequency Domain for a Single Range Bin

Item	Time Domain	Frequency Domain
Multi-bounce	Finding the bouncing numbers with mesh grid, $\mathbf{T} = M$	Finding the bouncing number with mesh grid, $\mathbf{T} = M$
Multi-level combination	Merging the reflectivity fields for all of complex target, $\mathbf{T} = R$	—
Convolution	2D convolution with reflectivity map, $\mathbf{T} = (PQ)\log_2(PQ)$	—
Multiplication	—	Pixel-wise multiplication for each bouncing level, $\mathbf{T} = MNR$
Time complexity	$\mathbf{O}(n^2\log_2 n^2)$	$\mathbf{O}(n^2)$

From the SAR observation geometry in Figure 8.8, it is known that the slant range $R(\eta)$ varies along the slow time η. Ideally, it can be rewritten as

$$R(\eta) = \sqrt{R_0^2 + (v\eta - v\eta_0)^2} \approx R_0 + \frac{(v\eta - v\eta_0)^2}{2R_0} \tag{8.73}$$

where η_0 corresponds to R_0, the shortest range. Indeed, $R(\eta)$ may be seen as a measurable slant range under the vibration- free motion. In reality, there is no vibration-free motion, where the slant range becomes $R'(\eta)$, which now can be expressed in Line-of-sight (LOS)/parallel/perpendicular (LPP) coordinate [28,29]:

$$\begin{aligned} R'(\eta) &= \sqrt{(R_0 - d_{LOS}(\eta, r))^2 + (v\eta - vt_0)^2 + d_\perp^2(\eta)} \\ &\approx R_0 + \frac{(v\eta - v\eta_0)^2}{2R_0} - d_{LOS}(\eta, r) + \frac{d_{LOS}\eta(\eta, r)}{2R_0^2}(v\eta - v\eta_0)^2 \\ &+ \frac{d_\perp^2(\eta)}{2R_0} + \frac{d_{LOS}(\eta, r)}{2R_0^2} d_\perp^2(\eta) \\ &\triangleq R(\eta) + \Delta R \end{aligned} \tag{8.74}$$

where d_{LOS} is the distance along the line of sight, and $d_\perp$ is the cross product of $d_\parallel$ and d_{LOS}, with $d_\parallel$ being the distance along the instantaneous tangential velocity direction [29].

The differential slant range, ΔR, in Equation 8.74 is given by:

$$\Delta R = -d_{LOS}(\eta, r) + \frac{d_{LOS}(\eta, r)}{2r^2}(v\eta - v\eta_0)^2 + \frac{d_\perp^2(\eta, r)}{2r} + \frac{d_{LOS}(\eta, r)}{2r^2} d_\perp^2(\eta, r) \tag{8.75}$$

where the first two terms on the right-hand side are influenced by d_{LOS}, the third term is determined by $d_\perp$, and the last term is the coupling term, the higher-order term.

The ENU coordinate representing the local position in east/north/up directions, respectively, can be expressed in the LLH coordinate with $[\phi, \theta, h]$ denoting the longitude, attitude, and height, respectively [28,29]:

$$\mathbf{P_{enu}} = \begin{bmatrix} x_e \\ y_n \\ z_u \end{bmatrix} = \begin{bmatrix} (\varphi - \varphi_0)(E_b + h)\cos\theta \\ (\theta - \theta_0)(E_a + h) \\ h - h_0 \end{bmatrix} \tag{8.76}$$

where $[\theta_0, \phi_0, h_0]$ denotes the origin position; E_a and E_b are the semi-major and semi-minor axis length of earth. With the transformation matrices, $\mathbf{M_1}$ and $\mathbf{M_2}$ given in[19], the ENU coordinate can be transformed to the LPP coordinate:

$$\mathbf{P_d} = [d_{LOS}, d_{\|}, d_{\perp}]^T = \mathbf{M_2 M_1 P_{enu}} \tag{8.77}$$

where $[d_{LOS}, d_{\|}, d_{\perp}]$ is in LPP coordinate. The noisy position and velocity can be initialized within the $d_{\perp}$ and d_{LOS}. It follows that to take into account the path trajectory, the SAR's state vector, $\mathbf{P_{enu}}$, can be derived from Equation 8.76.

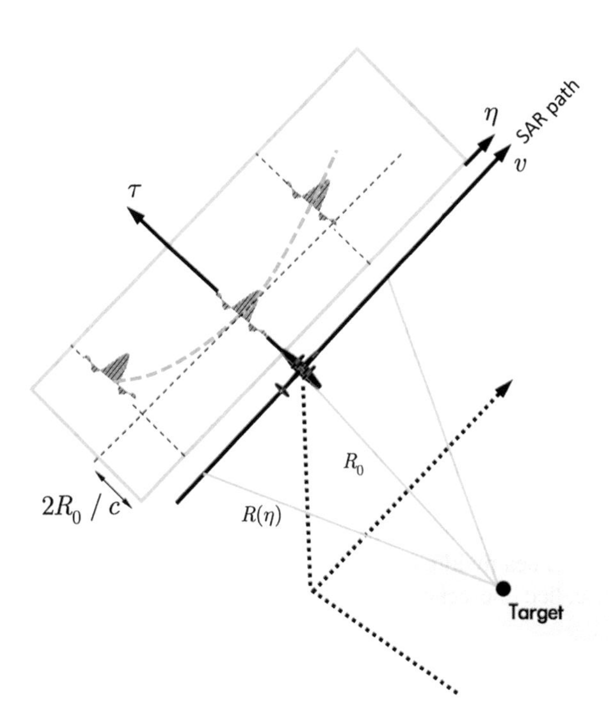

FIGURE 8.8 A side-looking strip map SAR observation geometry in slow-fast time coordinate.

8.3.3 Antenna Beam Tracking

In SAR image simulation, both radiometric and geometric fidelities must be preserved as much as possible. In the radiometric aspect, when computing the scattered field from a target being imaged, we must consider the azimuth antenna angle, defined in Figure 8.9, where η_c is the center azimuth position, and η_1, η_2 denote the initial and final positions of SAR, respectively, within the course of synthetic aperture. For arbitrary antenna pointing direction, Φ measures the azimuth angle covering the target region in a single footprint, while he angle Φ' represents the azimuth angle from the antenna beam center to target's center. It should be noted that in ray tracing, we have to deal with multiple rays within an illuminated target.

In Circle and Spotlight SAR data acquisition modes (see Figure 1.4), only a single footprint is involved because the region of observation is limited to be within a footprint. But it is not so in Stripmap mode, which uses push broom to obtain a wider range of data taken in the azimuth direction. To be more realistic in simulation, modifications are needed for Stripmap mode. As the SAR moves, at arbitrary azimuth position, the azimuthal antenna scanning angle is bounded by [28]:

$$\Phi^i_{\text{scan}} = \min(\Phi^i_{3\text{dB}}, \Phi^i_{\text{target}}), \quad i \in [\text{azimuth, range}] \tag{8.78}$$

where $\Phi^i_{3\text{dB}}$ and Φ^i_{target} are the covering angles of antenna 3 dB beam-width and target, respectively, and need to be estimated for computing the scattered filed. Notice that Φ^i_{scan} is also affected by the squint angle effect, θ_{sq}. Similarly, the covering angles should be estimated along the slant range, so that the image simulation can be feasible for the arbitrary size of the target.

8.3.4 Simulation Examples

Referring to Figure 8.8, suppose that the SAR system moves along the y-direction with ground range in x-direction. Then, the scattered field received by SAR traveling along the y-direction, at far-field range R, can be expressed as

$$\mathbf{E}_s(R, t, \eta) = \frac{e^{-j\omega t} e^{-ikR}}{4\pi R} \int \mathbf{J}(\mathbf{r}') e^{i\mathbf{k}\cdot\mathbf{r}'} e^{-\frac{(x'-x_c)^2\cos^2\theta_s}{R_0^2\beta^2}} e^{-\frac{(y'-y_c)^2}{R_0^2\beta^2}} dS' \tag{8.79}$$

where $\mathbf{J}(\mathbf{r}')$ are surface current density, with a two-dimensional Gaussian antenna gain pattern with full beamwidth β, centered at a resolution cell (x_c, y_c). In SAR, the received signal, called the echo signal, is a coherent sum of all scattered fields received at R:

$$\mathbf{E}_s(t, \eta) = \int_0^{L_s} \mathbf{E}_s(t, R(\eta))dy \tag{8.80}$$

where L_s is the synthetic aperture length.

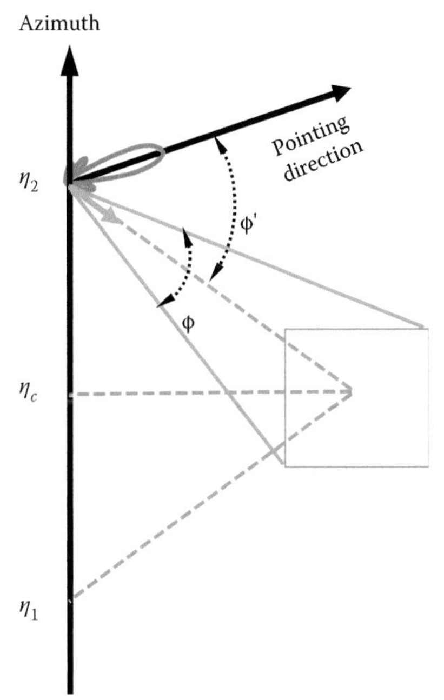

FIGURE 8.9 The azimuth antenna angle with respect to the SAR moving and looking the target within the course of synthetic aperture.

Noted that in transmitting signal and receiving signal, the wave polarizations can be chosen. After demodulation, the received signal of Equation 8.80 is then Fourier transformed and is ready for image focusing processing. Figure 8.10 illustrates a functional block diagram of a SAR image simulation. The simulation processing flow is adopted from [28]. Generally, the inputs include the platform, radar parameter settings, and target computer-aided design (CAD) model. Then, the scattering fields are calculated within different platform positions, and the echo signal can be generated in two dimensions, followed by an imaging algorithm to achieve the final focused image. A refined omega-K algorithm was chosen to perform SAR image focusing [28,29].

The numerical simulations are based on a practical experimental configuration with a total synthetic length of 1 m and a maximum slant range of 1.8 m. The system height from the ground plane is 1.5 m with a look angle of 45°. The antenna beam width is from a typical standard gain horn antenna. The signal carrier frequency was set at 36.5 GHz with a 10 GHz bandwidth. Selected specifications are shown in Table 8.3. More details are given in [30].

Now, simulations of metal, dielectric, and coatings (metal coated with Teflon) for two targets are used. Two spheres were separated along the azimuth direction, with a spacing of 0 mm, 5 mm, 10 mm, 15 mm, 20 mm, and 25 mm. The focused images in Figure 8.11 present the interaction among the targets. The case of two metal spheres features not only the two targets but also their mutual effects. The interactions are stronger with smaller displacement spacing and become weaker as the spacing increases. Due to the effects of transmission, scattering, and multi-path interactions, the results show more complex imagery in the case of the dielectric object.

It is clear that multiple phase delays occur in the range direction, which includes the different levels of interactions with the spacing changes. In the case of coated spheres, only parts of the power is scattered back because of the presence of the inner metal sphere (see Figure 8.12), and the interaction is weaker. By testing different types of material, different degrees of interaction among targets can be analyzed in the simulations.

More complex target arrangements in Figure 8.13 are simulated in three dimensions by three spheres (both the metal and dielectric) placed both along the range and azimuth directions with no spacing in the scene center. In the case of multiple targets, the interactions are weaker in the range than in the azimuth direction. The different level of interactions in the azimuth direction is due to mutual effects among targets. The electromagnetic characteristics in the dielectric objects are more complex, and mutual interactions and self-interactions are shown at the same time. The different interaction degrees are illustrated that reduce target recognition ability. The results in Figure 8.13 highlight the difference between the point-target model and the full-wave method.

From the system aspect of the simulation, the system bandwidth is an important parameter in radar system design and is relative to the range resolution in the focused image. As the radar bandwidth increases, the electromagnetic interaction of a target can be resolved to varying degrees depending on the radar bandwidth. Hence, electromagnetic characteristics according to the variation in bandwidth in different materials are discussed. In the case of a single metal object, the radar

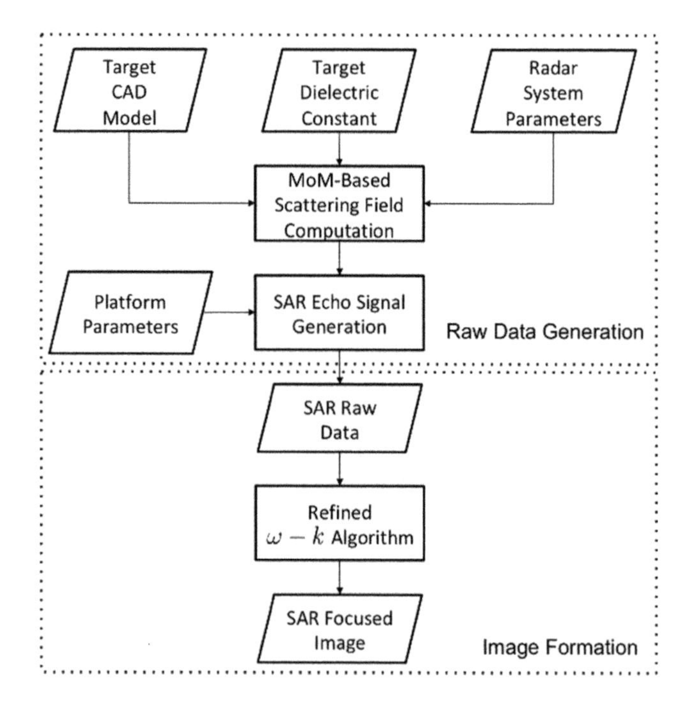

FIGURE 8.10 A Full-wave based SAR image simulation flowchart.

TABLE 8.3
Simulation Parameters for a Full-Wave Based SAR

Parameter	Value	Unit
Carrier Freq.	36.5	GHz
Bandwidth	10	GHz
Sensor Height	1.5	m
Target Location	1.1	m
Sensor Position Interval	1	cm
Look Angle	45	degree
Synthetic Length	1	m
Antenna Beam Width	17.5	degree
Sphere Radius	1.5	cm
Azimuth Angle (Incident)	0	degree
Azimuth Angle (Scattering)	0 and 180	degree
Scattering Angle (Bistatic)	45	degree

bandwidth changes from 1 GHz to 10 GHz are shown in Figure 8.14. Because of single scattering behavior in a single metal target, the results show the bandwidth changing with little influence. Only the range resolution changes with the variation in bandwidth.

Three different types of material (namely, Teflon, glazed ceramic, and GaAs) are used for analyzing the effect of bandwidth change for a single dielectric sphere. Due to the rich electromagnetic information in dielectric objects, the scattering effect based on system bandwidth is more pronounced. When the system bandwidth is small, the phenomenon of multiple scattering cannot be expressed because of the low range resolution. As the range resolution corresponds to the system bandwidth close to the size of the observation targets, the results show that one can make a distinction among different materials. When the system bandwidth reaches 5 GHz, that is, when the resolution is close to the object size level, the different target dielectric constants and the multiple scattering phenomena can be illustrated in the various levels of results. The focused images presented in Figure 8.15 demonstrate that system bandwidth is profoundly significant, as is well known, for imaging dielectric objects relative to the metal targets.

The effect of the system bandwidth on two targets is also discussed. Two different kinds of metal and dielectric objects are simulated with no spacing between targets. As shown in Figure 8.16, when the system bandwidth is lower, the ability to differentiate between metal and dielectric objects is poor, but their mutual interaction can still be illustrated. Moreover, the bandwidth changes only affect the resolution variation because of electromagnetic wave nonpenetration for the two metal objects. On the other hand, bandwidth alternations strongly affect the multipath imaging of a dielectric material. Also, more complex objects are used for

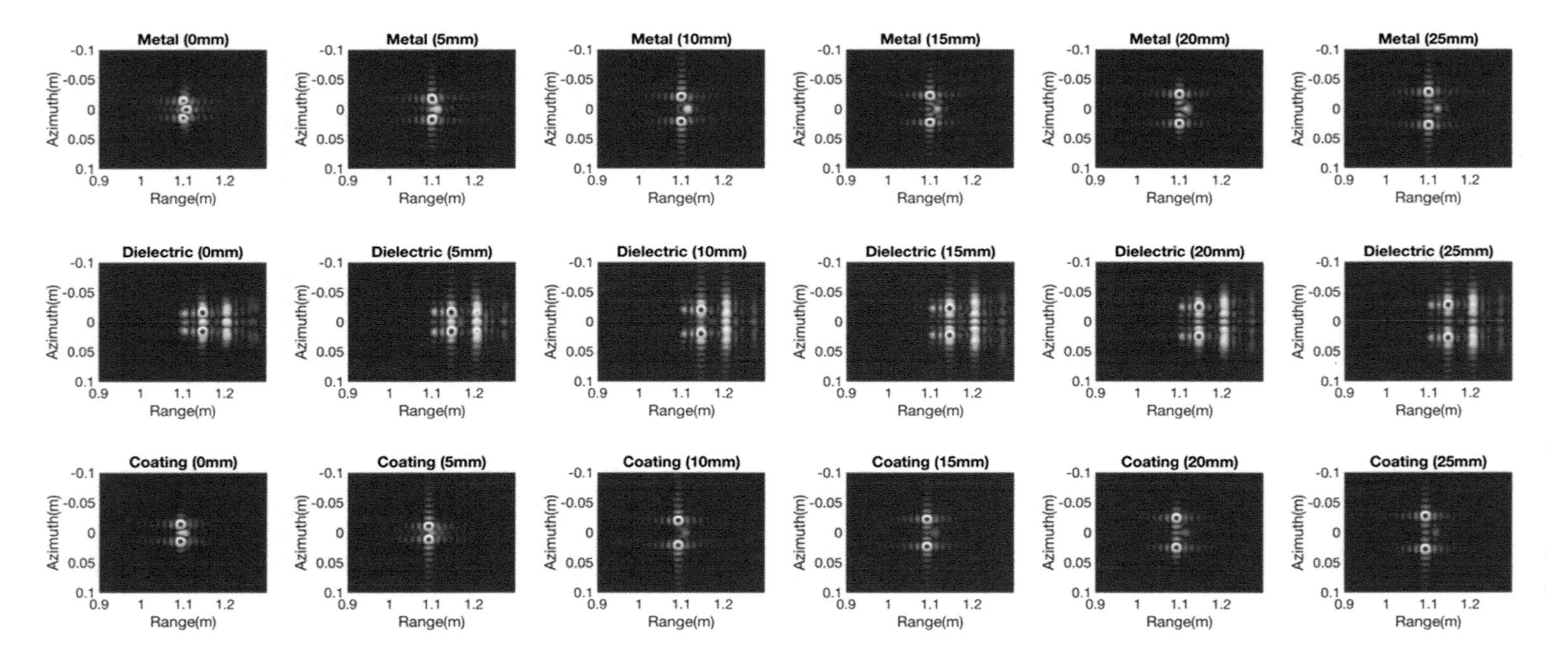

FIGURE 8.11 Images of metal sphere, dielectric sphere, and metal sphere coated with Teflon with a spacing of 0 mm, 5 mm, 10 mm, 15 mm, 20 mm and 25 mm. (Top row: metal, middle row: dielectric and bottom row: metal coated with Teflon).

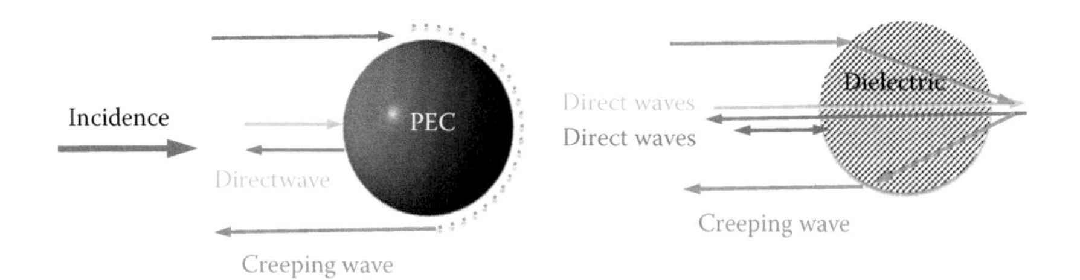

FIGURE 8.12 Physical interpretation of multiple phase delays for PEC sphere (left) and dielectric sphere (right).

analysis of the system bandwidth. Both metal and dielectric materials with three-by-three spheres are placed along the range and azimuth directions with no spacing between objects. In the focused images of the metal spheres, the targets with mutual interaction in the azimuth direction can be identified, and the targets in the distance image cannot be recognized as three objects because of lower resolution in the range direction. As the system bandwidth increases, the mutual interactions in the different levels are presented with the corresponding resolution. For the dielectric materials shown in Figure 8.17, different degrees of electromagnetic interaction are delivered with the variation in system bandwidth.

We now present results of numerical simulation and experimental measurement. The experiments were set in an anechoic chamber and two metal spheres are placed in the scene center with spacings of 0 mm, 5 mm, 10 mm, 15 mm, 20 mm, and 25 mm. An N5224A PNA microwave network analyzer is used as a transmitter and receiver in this experiment. The system applied the motion controller to acquire data with a 1cm interval in the total 1 m synthetic length. After collecting the raw data, focusing is carried out to obtain the focused images. In Figure 8.18, the results from the full-wave method compare well with the experimental results. The approach preserves the electromagnetic wave interactions between two spheres, and the interaction is reduced with spacing increases. Various degrees of interactions are

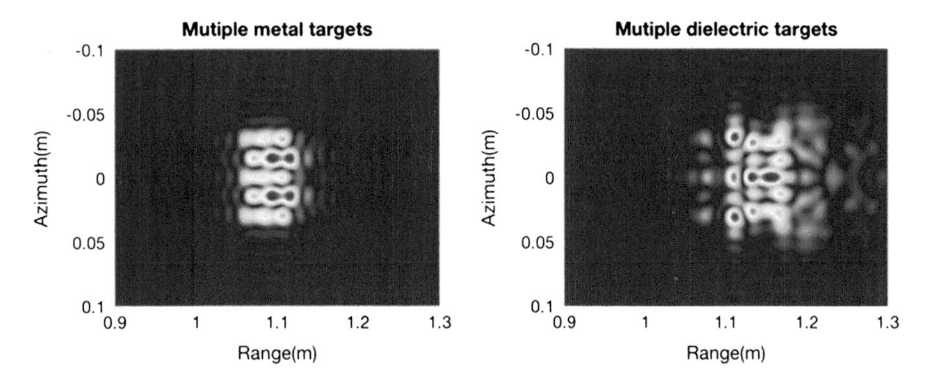

FIGURE 8.13 Image of a cluster of three by three metal spheres (left) and dielectric spheres (right).

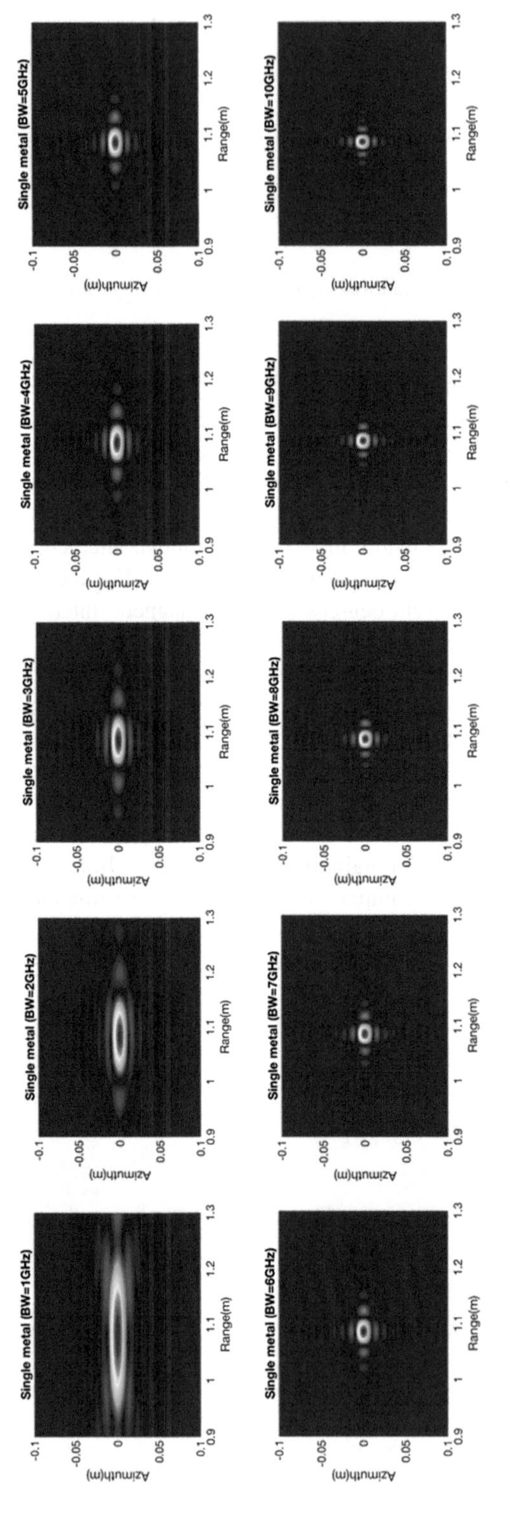

FIGURE 8.14 Image of a metal sphere with SAR bandwidth varying from 1 to 10 GHz (from top to bottom).

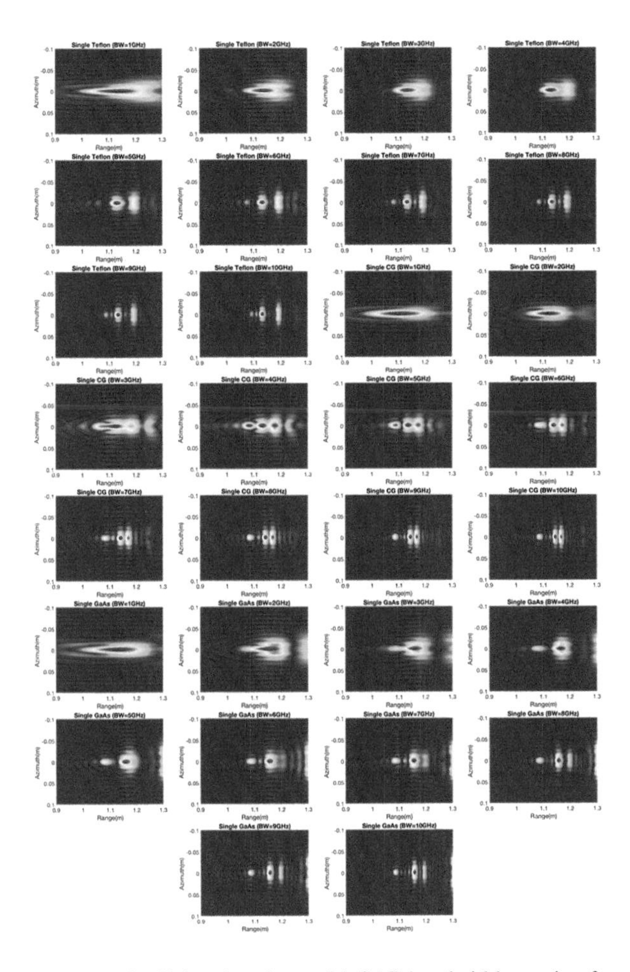

FIGURE 8.15 Images of a dielectric sphere with SAR bandwidth varying from 1 to 10 GHz (top: Teflon; middle: glazed ceramic; bottom: GaAs).

present with the spacing changes between the targets, and the interaction intensity is reduced with increasing spacing, as physically expected.

8.4 MUTUAL COHERENCE FUNCTION

In acquiring the scattered field from the target, be it deterministic or random, we may devise it by forming diversity of mutual coherence function (MCF) in frequency, angular, space, polarization, etc. One such example is the coherency vector of radiated fields received by two antennas with displacement $\vec{d}$ given by [31]:

$$\mathbf{V}_{qp}(\vec{d}) = \left\langle \mathbf{E}_q^s(\vec{r}) \otimes \mathbf{E}_p^{s*}(\vec{r} + \vec{d}) \right\rangle \tag{8.81}$$

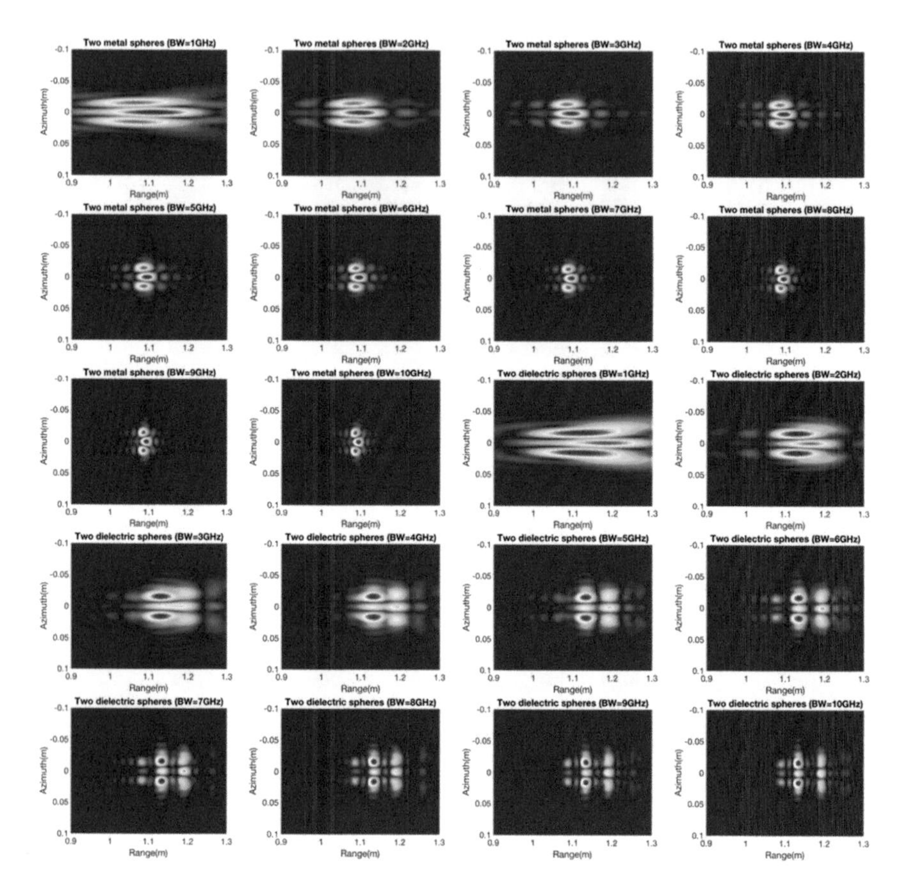

FIGURE 8.16 Image of two metal spheres (top two rows) and two dielectric spheres (bottom two rows) with SAR bandwidth varying from 1 to 10 GHz.

where $\langle\ \rangle$ is ensemble average; $\otimes$ is the outer product operator; * is a complex conjugate operator; p, q represents polarization. The polarimetric coherency vector in Equation 8.81 actually yields a compact expression that provides insight into interferometric SAR. By Cittert–Zernike theorem [32], the Fourier transform relationship of a brightness distribution with a polarimetric coherency vector can be established. A set of integral and differential equations for the correlation function of a wave in a random distribution of discrete scatterers [1]. Theses general integral and differential equations for spatial as well as temporal correlation function allow to observe the effects of the constant velocity as well as the fluctuating velocity. The MCFs for the scattered wave both for two frequencies and two scattering angles for one-dimensional rough surfaces were derived in [33–39]. The Kirchhoff approximation was used to obtain the scattered field. The scattering cross-section in the form of the two-frequency is available to conduct a series of simulation. The numerical calculation of the analytical results was compared with experimental data and Monte Carlo simulations showing good agreement. An analytic expression of

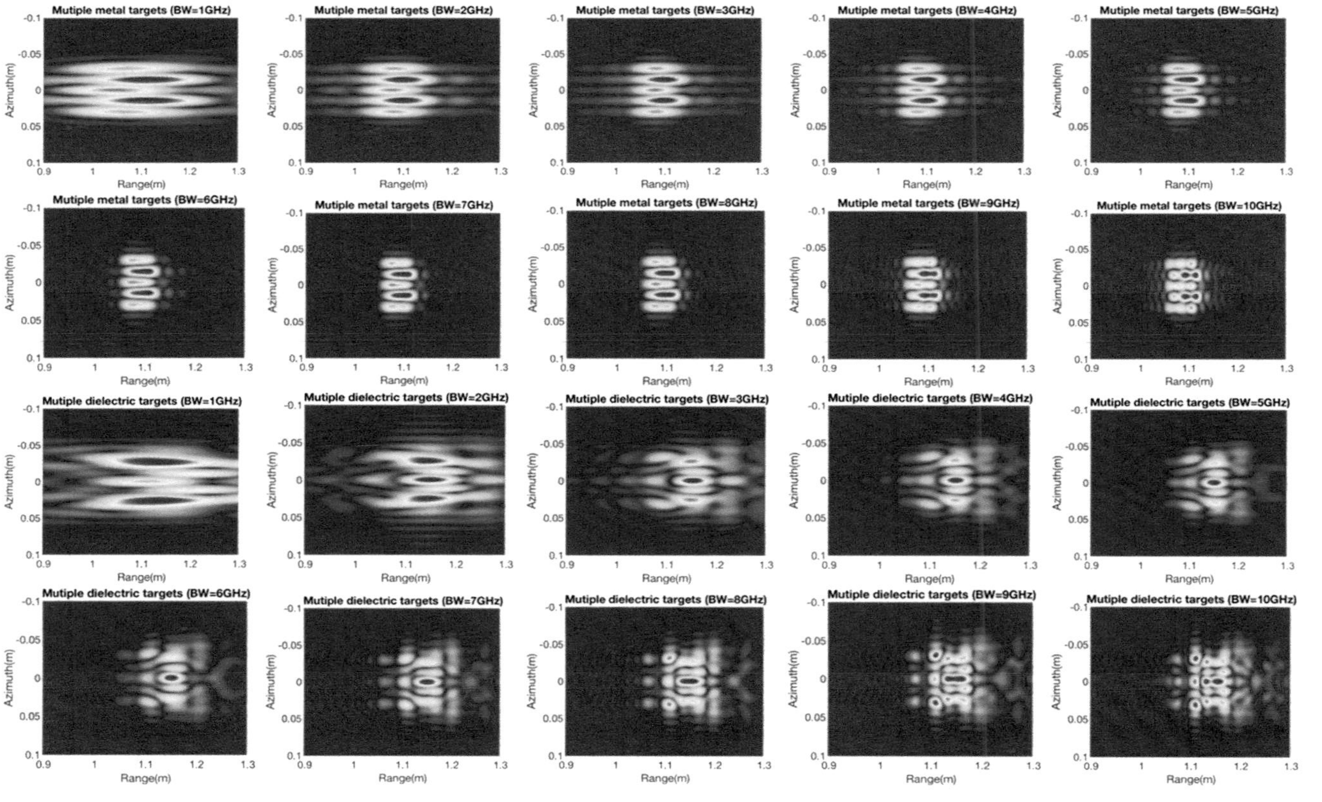

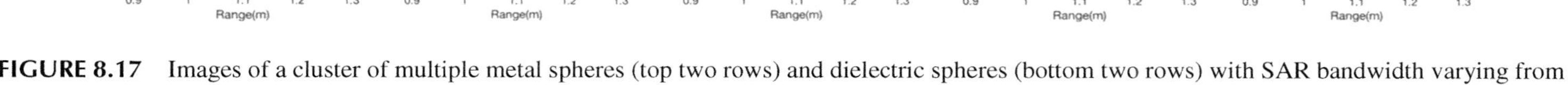
FIGURE 8.17 Images of a cluster of multiple metal spheres (top two rows) and dielectric spheres (bottom two rows) with SAR bandwidth varying from 1 to 10 GHz.

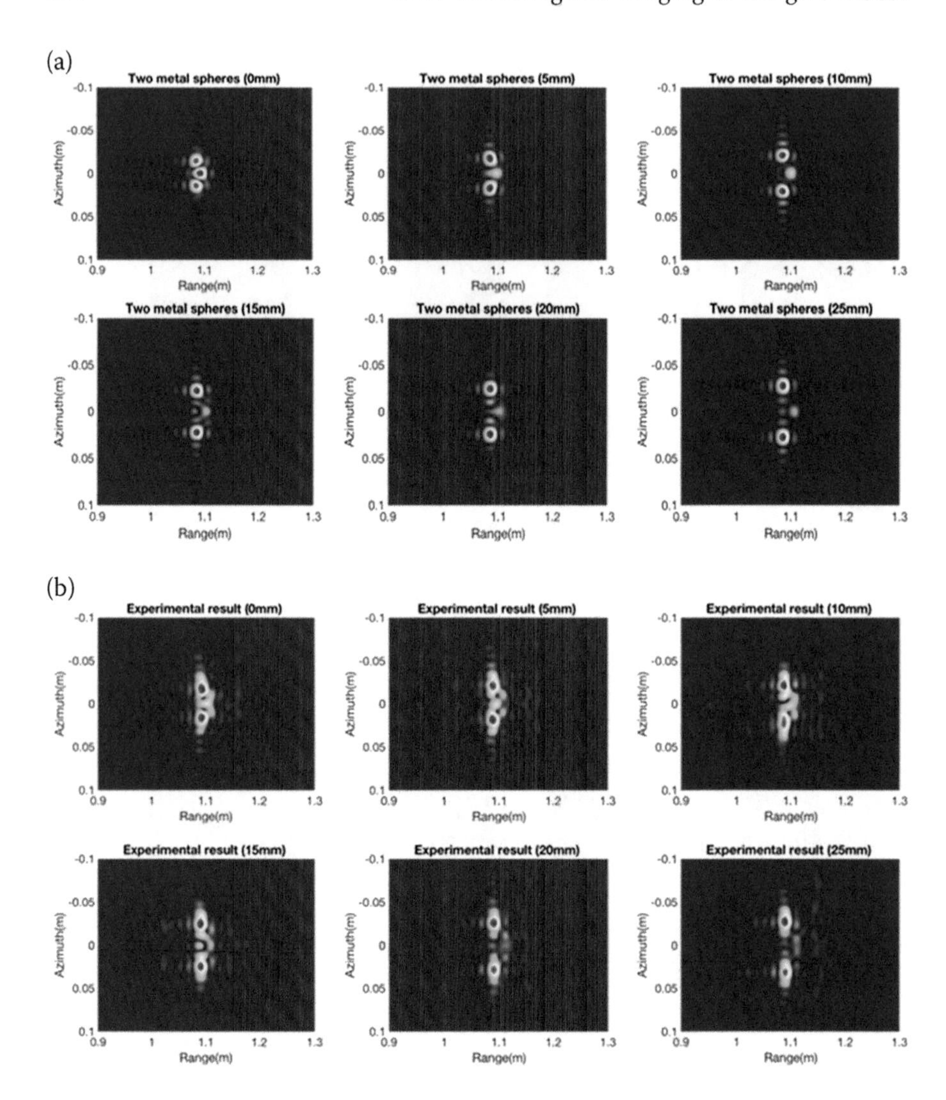

FIGURE 8.18 Images of two metal spheres (top row: simulated; bottom row: experimental) with spacing of 0 mm, 5 mm, 10 mm, 15 mm, 20 mm, and 25 mm (from left to right).

the two-frequency mutual coherence function (MCF) was derived in [40] for a two-dimensional random rough surface. The scattered field was calculated by the Kirchhoff approximation. Their numerical simulations show that the two-frequency MCF is greatly dependent on the root-mean-square (RMS) height, while less dependent on the correlation length. Recalled that the first-order Kirchhoff approximation used in these papers does not include the effect of multiple scattering from different parts of the surface. Using the second-order Kirchhoff approximation (KA) with angular and propagation shadowing functions, the angular correlation function (ACF) of scattering amplitudes was developed to surfaces with large radii of curvature and high slopes of the order of unity [41]. An expression for the two-frequency mutual coherence function was also derived for studying waves propagating close to the ground, based on the parabolic wave equation model, which was solved by the path integral method [39]. The irregular surface height was assumed non-Gaussian distribution. Numerical schemes were developed for simulating the detection of buried objects embedded in rough surfaces [35–37].

For two-angle and two-frequency mutual coherence functions are formed by varying the transmitting or receiving angles or frequency, respectively (see Figure 8.19), and are mathematically given by

$$\mathbf{V}_{pq}(\theta_1, \theta_2) = \left\langle \mathbf{E}^s_{pq}(\theta_1)\mathbf{E}^{s*}_{pq}(\theta_2) \right\rangle \tag{8.82}$$

$$\mathbf{V}_{pq}(f_1, f_2) = \left\langle \mathbf{E}^s_{pq}(f_1)\mathbf{E}^{s*}_{pq}(f_2) \right\rangle \tag{8.83}$$

In either way, the following baseline or memory line must be met:

$$k_1 \sin\theta_{i1} - k_2 \sin\theta_{i2} = k_1 \sin\theta_{s1} - k_2 \sin\theta_{s2} \tag{8.84}$$

where k_1, k_2 form two frequencies, and incident pair θ_{i1}, θ_{i2} or scattered pair θ_{s1}, θ_{s2} form the two angles.

As is noted, the polarizations p, q can be applied so that additional polarization diversity can be devised.

For the purpose of demonstration, we choose the following parameters: carrier frequency at 1.25 GHz, incident angles: 20° and 60°, surface roughness: $k\sigma = 0.5\lambda$, $kl = 2\lambda$, soil permittivity: $\varepsilon_r = 10 - j0.05$. We also set the incident and scattered azimuthal angles to 0°. To establish the mutual coherence functions in angle and frequency, the scattered angle varies from 90° to –90°, while the frequency increases a step of 0.01 GHz from a center frequency of 1.25 GHz and up to 2.25 GHz. Figures 8.20 displays the MCF on the angular-frequency plane at incident angles of 20°. For all four polarizations, we see that the MCF responses bear quite a similar pattern, but the cross polarizations, HH and VH, have much small magnitude; perhaps it is due to the lack of sources for multiple scattering for this roughness scale. Figures 8.21 and 8.22 show the MCF responses to the soil moisture changes from 0.2 to 0.4. It indicates that the MCF is strongly dependent on the moisture content.

8.5 BISTATIC SAR IMAGING

From Chapter 5, we have seen that bistatic scattering offers benefits of increasing the dynamic range of the returned signal strength and prompting higher sensitivity to the surface parameters. Hence it is demanded to acquire bistatic data. However, bistatic observation poses a high degree of freedom to configure the transmitter and receiver, to which each has elevation and azimuth angles to set. The topic itself covers a vast of subjects to explore. This subsection only discusses some basics in bistatic imaging, particularly in SAR.

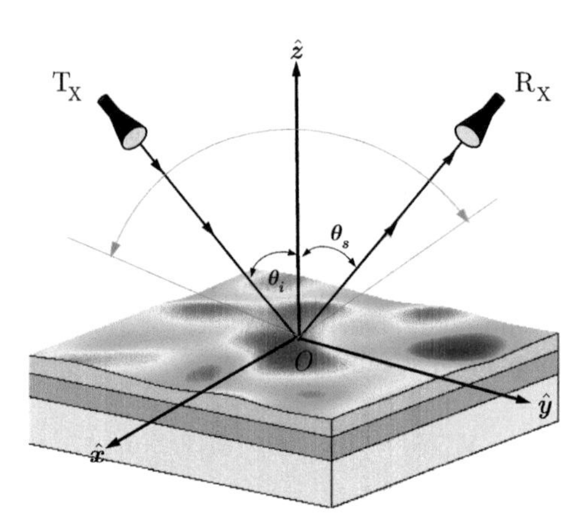

FIGURE 8.19 Correlated scattered fields acquired at two angles or two frequencies.

8.5.1 Bistatic SAR Scattering Property

A bistatic SAR imaging system separates the transmitter and receiver to achieve benefits such as exploitation of additional information contained in the bistatic reflectivity of targets, increased radar cross-section, and increased bistatic SAR data information content with regard to feature extraction and classification. Currently, similar to the monostatic SAR simulation, the point-target is used in bistatic SAR imaging, which is an efficient way to develop and compare algorithms. However, this approach is unable to achieve the advantages of bistatic SAR to extract target information in different aspect angles. Hence, to evaluate the proposed method, the results both in monostatic and bistatic mode are simulated for the special case of equal velocity vectors for the transmitter and receiver. Figure 8.23 presents the different scattering behavior under the same system parameters Table 8.3 with two directions of scattering azimuthal angles at 0° and 180°.

Using the same system parameters but with different observation angles, the intensity of the first bounce point in the bistatic results is the same as that in the monostatic: Increasing with increasing dielectric constant. However, the first bounce place moves backward in the bistatic simulation relative to the monostatic system. The effect of multiple scattering inside a sphere is significant in monostatic mode, but the behavior decays rapidly in bistatic mode. It is clear that the location of multiple scattering moves forward in the bistatic system [32].

Two targets in bistatic mode are also simulated with two spheres separated along the azimuth direction with spacing of 0 mm, 5 mm, 10 mm, 15 mm, 20 mm, and 25 mm. The look angle is 45° with azimuth angle at 0° and 180° in two observation modes. The results in Figure 8.24 show the interaction between two targets in both monostatic and bistatic observations; diversified scattering information can be obtained through different observation angles. For the simulation results of the two objects, the intensity of mutual interaction is lower in bistatic mode than monostatic. However, the intensity of creeping waves is enhanced in the bistatic simulation over the monostatic system.

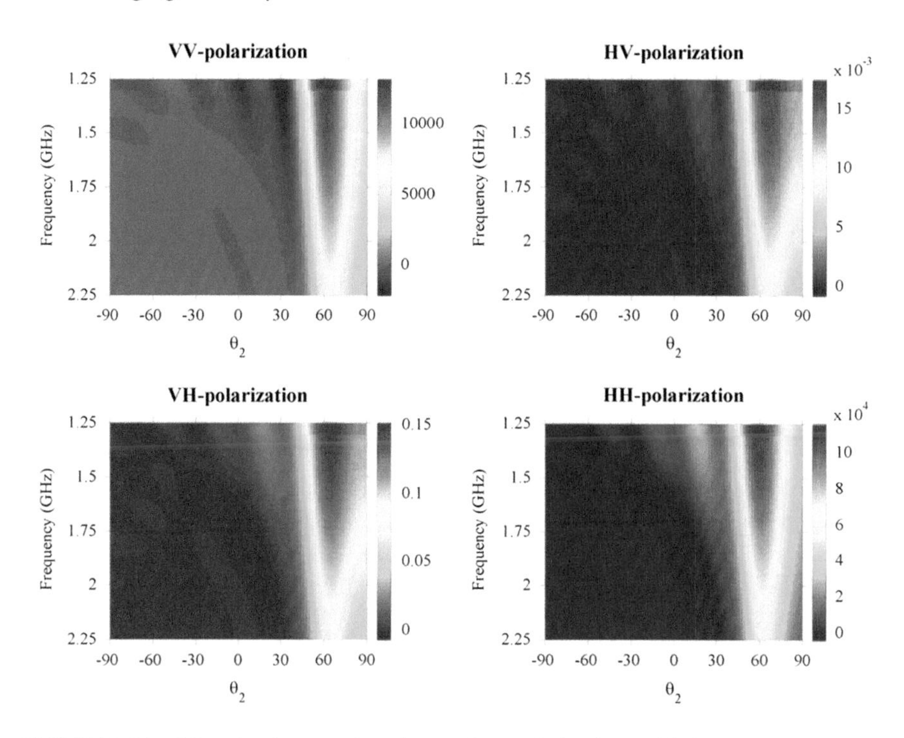

FIGURE 8.20 Mutual coherence functions at four polarizations with 60° of incident angle.

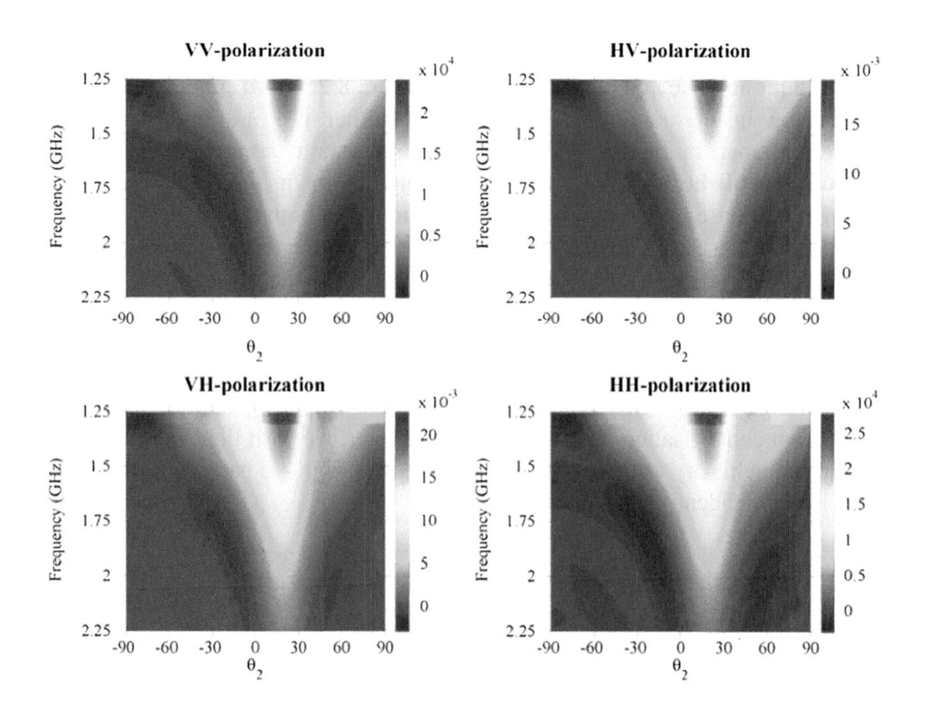

FIGURE 8.21 Response of mutual coherence function to soil moisture: $m_v = 0.2$.

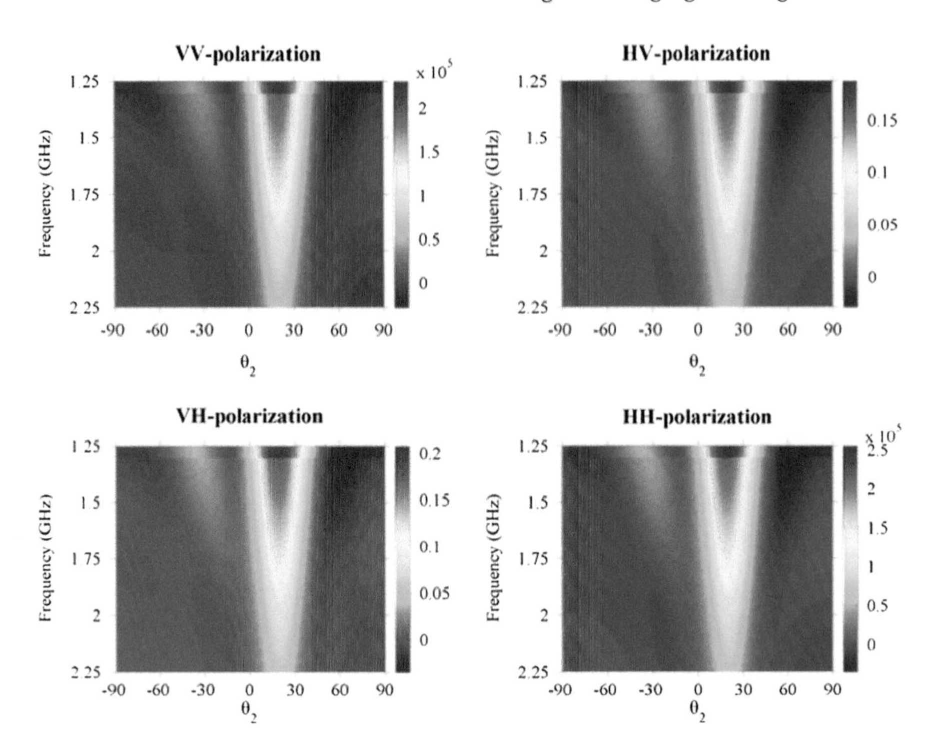

FIGURE 8.22 Response of mutual coherence function to soil moisture: $m_v = 0.4$.

Two-by-two metal targets placed along the range and azimuth directions are shown in monostatic and bistatic modes in Figure 8.25. Two spheres are set with no spacing in the azimuth direction, and another two are five wavelengths (0.0205 m) apart in the range direction. The monostatic case shows that original targets can still be identified with strong interaction in their centers. The interaction response is enhanced with the interaction in both the range and azimuth directions. As a result of the enhanced mutual interaction intensity, only the interaction is shown in the focused image, and the target recognition ability is reduced.

Because of two separate carrier platforms, the performance analysis of the bistatic SAR imaging becomes more complicated than of monostatic SAR, in terms of bistatic range history, two-dimensional resolution, Doppler parameter estimation, and motion compensation and so on. In this chapter, we limit the evaluation of the bistatic range history and two-dimensional resolutions in the backward zone and in the forward incident plane zone.

8.5.2 Bistatic Imaging Geometry and Signal Model

Now, we discuss the geometric property of backward and forward bistatic SAR based on the imaging geometry as depicted in Figure 8.26 [42]. In the imaging geometry, θ_T and φ_T are the incidence angle and transmitted azimuth angle and φ_R the received incidence angle and received azimuth angle, with subscripts T and R

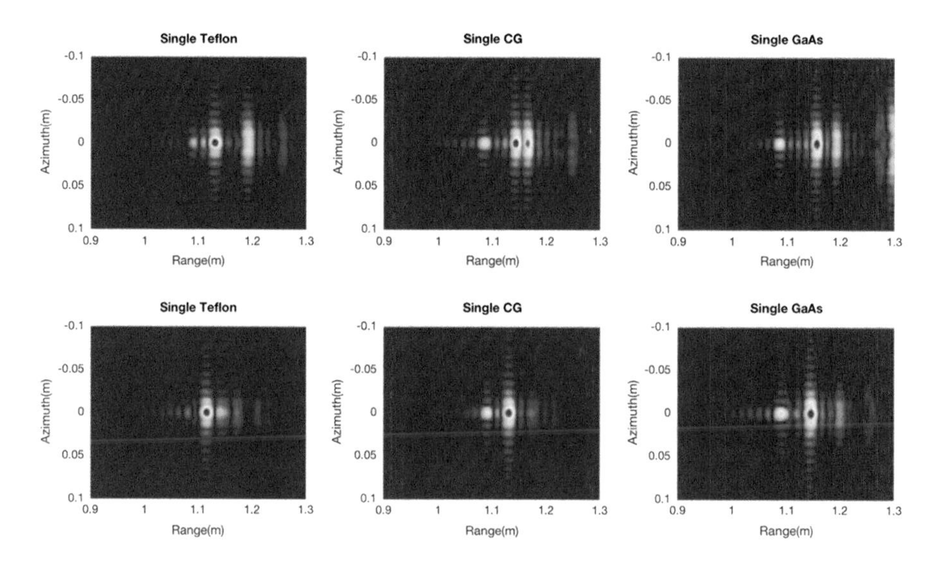

FIGURE 8.23 Image of a single dielectric sphere for the monostatic versus bistatic (left: Teflon; center: ceramic glaze; right: GaAs). (top row: monostatic; bottom row: bistatic).

denoting the transmitter and the receiver; the imaging space perhaps can be roughly divided into two zones, assuming a positive azimuth angle is counter-clockwise from the x-axis; the forward imaging zone is the area with the azimuth angle in the range of $\varphi_R \in (0°, 90°) \cup (270°, 360°)$ and another part belongs to the backward imaging zone. $\mathbf{P}_T(\eta)$ and $\mathbf{P}_R(\eta)$ are the instantaneous position vectors; $\mathbf{V}_T$ and $\mathbf{V}_R$ are the velocity vectors; $\mathbf{u}_T(\eta)$ and $\mathbf{u}_R(\eta)$ are the unit vectors in the direction from target P to the transmitter and receiver, respectively, at η, with $\mathbf{w}_T(\eta)$ and $\mathbf{w}_R(\eta)$ denoting the angular velocities of the transmitter and receiver, respectively, at the time η; β is the velocity angle between the transmitter and the receiver velocity vectors. Notice that the monostatic backward imaging and bistatic forward specular imaging are located at $\theta_T = \theta_R$, $\varphi_R = 180°$ and $\theta_T = \theta_R$, $\varphi_R = 0°$, respectively.

In the stripmap mode bistatic SAR, for analysis, some hypotheses are made in our study: first, we consider that the transmitter and the receiver sweep a continuous strip synchronously during the entire observation time, and second, the stop-and-go model is adopted. Finally, the ground plane of the imaging scenario is flat. If the curvature of the earth is considered, the whole scene can be divided into sub blocks so that the following analysis is still valid. Suppose that the transmitter sends a pulsed signal with duration time T_p and the carrier frequency f_c, defined as:

$$s(\tau) = w_r(\tau)\exp[j(2\pi f_c \tau + \pi a_r \tau^2)] \tag{8.85}$$

where τ is the range time, a_r is the chirp rate and w_r is a rectangular gate function with width T_p. The demodulated baseband signal from a point target having a constant scatter amplitude A_0 is of the form

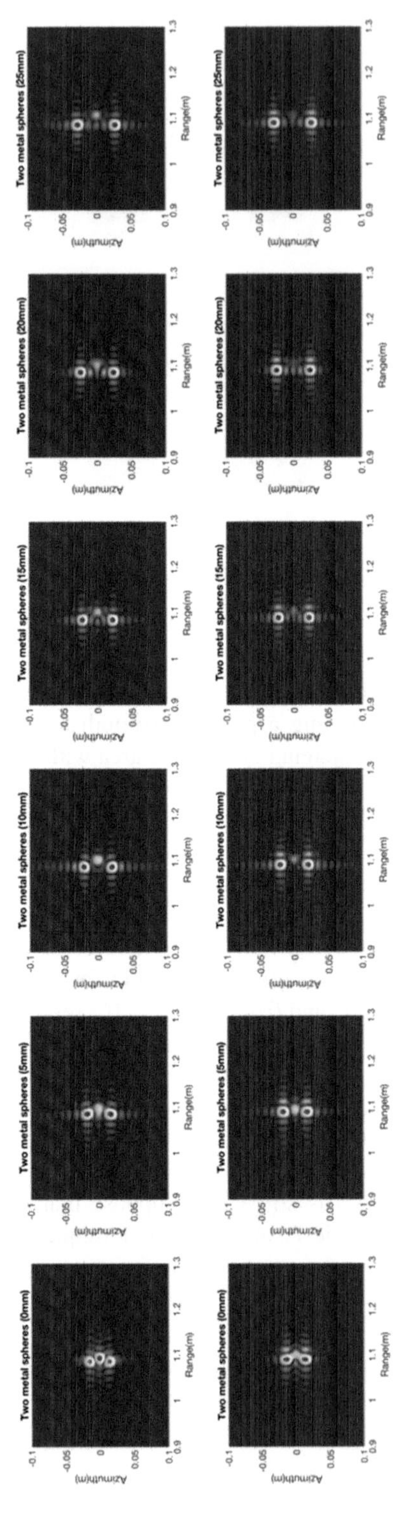

FIGURE 8.24 Case simulations of two metal spheres in the monostatic and bistatic modes with spacing of 0 mm, 5 mm, 10 mm, 15 mm, 20 mm, and 25 mm (up: monostatic and bottom: bistatic).

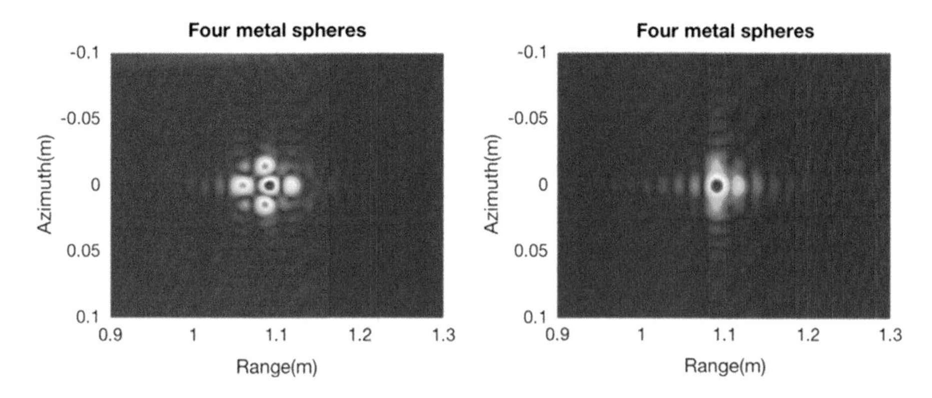

FIGURE 8.25 Images of a cluster of two by two metal spheres in backward and forward modes (left: monostatic; right: bistatic).

$$s_r(\tau, \eta) = A_0 w_r\left(\tau - \frac{2R_{bi}(\eta)}{c}\right) w_a(\eta) \exp\left\{\left(-j\frac{4\pi f_c R_{bi}(\eta)}{c}\right) + j\pi a_r\left(\tau - \frac{2R_{bi}(\eta)}{c}\right)^2\right\} \tag{8.86}$$

where c denotes the speed of light, η is the cross-range time, w_a is the antenna pattern in the cross-range direction and $R_{bi}(\eta)$ is the bistatic range, which is the sum of the ranges from the transmitter and the receiver to the target.

We now consider the two-dimensional ground resolution in a general configuration of bistatic SAR as shown in Figure 8.26. For the targets with a constant arrival time satisfy an iso-range surface $t(\eta) = \frac{\|\mathbf{P}_T(\eta) - \mathbf{P}\| + \|\mathbf{P}_R(\eta) - \mathbf{P}\|}{c}$, upon projecting to the iso-range gradient vector, one can obtain the general form of the bistatic ground range resolution [43,44].

$$\rho_{gr} = \frac{\kappa c}{B\mathbf{P}_{\mathbf{Z}_s}^{\perp}\mathbf{u}_T(\eta) + \mathbf{P}_{\mathbf{Z}_s}^{\perp}\mathbf{u}_R(\eta)} \tag{8.87}$$

where $\kappa = 0.886$ when the antenna patterns and ranging waveform can be approximated by the rectangle pulse function; B is the signal bandwidth; $\mathbf{P}_{\mathbf{Z}_s}^{\perp}$ is the ground projection matrix given by

$$\mathbf{P}_{\mathbf{Z}_s}^{\perp} = \begin{bmatrix} 1 & 0 & 0 \\ 0 & 1 & 0 \\ 0 & 0 & 0 \end{bmatrix} \tag{8.88}$$

and $\mathbf{u}_T(\eta)$ and $\mathbf{u}_R(\eta)$ are given by

$$\mathbf{u}_T(\eta) = \frac{\mathbf{P}_T(\eta) - \mathbf{P}}{\mathbf{P}_T(\eta) - \mathbf{P}} \tag{8.89}$$

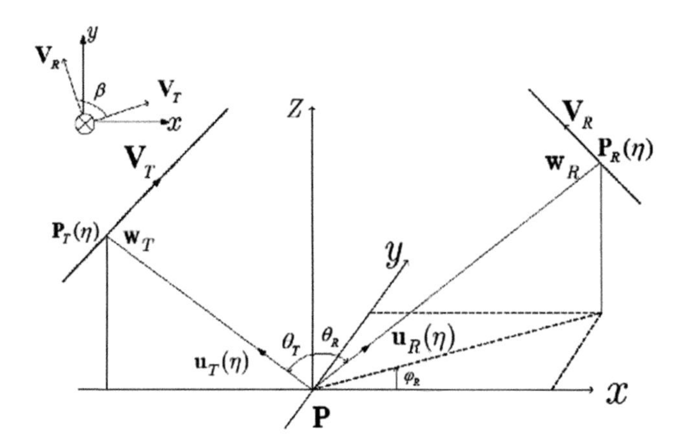

FIGURE 8.26 Imaging geometry of bistatic SAR.

$$\mathbf{u}_R(\eta) = \frac{\mathbf{P}_R(\eta) - \mathbf{P}}{\mathbf{P}_R(\eta) - \mathbf{P}} \tag{8.90}$$

Referring to Figure 8.26, for $\mathbf{P}_T(\eta)$, the main factors to determine the ground range resolution are the receiver motion parameters, including θ_R and φ_R. Based on the concept of the wavenumber vector or K-space, the bistatic azimuth resolution is calculated by

$$\rho_{ga} = \frac{\lambda}{T_a(x, y)\mathbf{P}_{z_s}^{\perp}\mathbf{w}_T(\eta) + \mathbf{P}_{z_s}^{\perp}\mathbf{w}_R(\eta)} \tag{8.91}$$

where the λ is the wavelength and T_a is the synthetic aperture time of the target at (x, y); $\mathbf{w}_T(\eta)$ and $\mathbf{w}_R(\eta)$ are the angular velocities of the transmitter and receiver, respectively, at the time η, which are given by

$$\mathbf{w}_T(\eta) = \frac{[\mathbf{I} - \mathbf{u}_T(\eta)\mathbf{u}_{T'}(\eta)]\mathbf{v}_T}{\mathbf{P}_T(\eta) - \mathbf{P}} \tag{8.92}$$

$$\mathbf{w}_R(\eta) = \frac{[\mathbf{I} - \mathbf{u}_R(\eta)\mathbf{u}_{R'}(\eta)]\mathbf{v}_R}{\mathbf{P}_R(\eta) - \mathbf{P}} \tag{8.93}$$

with $\mathbf{I}$ the 3×3 identity matrix. If $\mathbf{P}_T(\eta)$ and $\mathbf{v}_T$ are known, for azimuth resolution, the main influences are the receiver motion parameters and the velocity angle, including θ_R, φ_R and β.

It is known that the ground range resolution and the azimuth resolution directions can be non-orthogonal in bistatic SAR mode [45]. From Figure 8.27, the resolution direction angle between the direction of the range gradient and that of the Doppler gradient can be calculated as

$$\psi = \text{accos}(\Theta \cdot \Xi) \tag{8.94}$$

where Θ and Ξ are the unit direction vectors along the range resolution and azimuth resolution, respectively, given by

$$\Theta = \frac{\mathbf{P}_{\mathbf{z}_s}^{\perp}(\mathbf{u}_T(\eta) + \mathbf{u}_R(\eta))^{\mathrm{T}}}{\mathbf{u}_T(\eta) + \mathbf{u}_R(\eta)} \tag{8.95}$$

$$\Xi = \frac{\mathbf{P}_{\mathbf{z}_s}^{\perp}(\mathbf{w}_T(\eta) + \mathbf{w}_R(\eta))^{\mathrm{T}}}{\mathbf{w}_T(\eta) + \mathbf{w}_R(\eta)} \tag{8.96}$$

The intercept imaging area of bistatic SAR is determined by the ground range resolution, the azimuth resolution, and the resolution direction angle, given by [43–45]

$$A_{cell} = \frac{\rho_{gr}\rho_{ga}}{|\sin \psi|} \tag{8.97}$$

8.5.3 Bistatic Range History

An important difference in the forward bistatic SAR to that in the backward bistatic SAR is the existence of ghost effect, which will be explained by the range-history analysis as follows. At cross-range time η, the bistatic range $R_{bi}(\eta)$ is the summarization of the distances of the transmitter and the receiver to the target (x, y, z):

$$\begin{aligned} R_{bi}(\eta) = &\sqrt{(x - x_{T\eta})^2 + (y - y_{T\eta})^2 + (z - H)^2} \\ &+ \sqrt{(x - x_{R\eta})^2 + (y - y_{R\eta})^2 + (z - H)^2} \end{aligned} \tag{8.98}$$

It is clear that the iso-bistatic range $R_{bi}(\eta)$ surface in 3D space forms an ellipsoid with the transmitter and receiver at two foci. As the baseline between the transmitter and receiver decreases, the ellipsoid surface approaches a spherical surface. The intersection curve of the iso-bistatic range $R_{bi}(\eta)$ surface with the ground can be written as, setting $z = 0$:

$$\begin{aligned} R_{bi}(\eta) = &\sqrt{(x - x_{T\eta})^2 + (y - y_{T\eta})^2 + H^2} \\ &+ \sqrt{(x - x_{R\eta})^2 + (y - y_{R\eta})^2 + H^2} \end{aligned} \tag{8.99}$$

where $(x_{T\eta}, y_{T\eta}, H)$ and $(x_{R\eta}, y_{R\eta}, H)$ are the positions of the transmitter and receiver, respectively, as a function of cross-range time η. Though the iso-bistatic range forms an ellipsoid surface for the target P at any given cross-range time η, there exists a difference in the backward bistatic and forward bistatic modes that

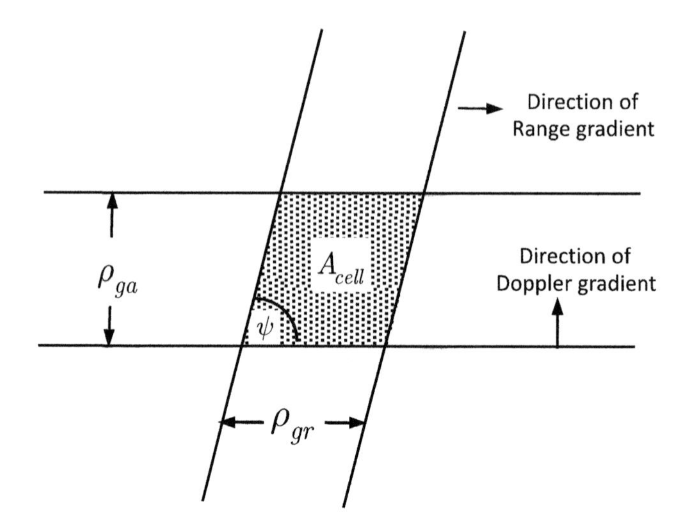

FIGURE 8.27 The ground resolution cell of bistatic SAR.

should be noted. In the backward bistatic, only one intersection curve is confined in the imaging scene. In contrast, in the forward incident plane bistatic mode, there are two intersecting curves in the imaging scene. The above scenario ranges are depicted in Figure 8.28. We see that the point P' has the same bistatic range as the target point P in the forward vertical profile, which does not occur in the backward bistatic mode. If the point P' has dual bistatic range histories with the target point during the whole observation time, the point P' will induce a "false or ghost" in the focused imaging.

To further explore the properties of the bistatic range histories of the point $P'(x_{P'}, y_{P'})$ and target point $P(x_P, y_P)$, the iso-bistatic range and iso-bistatic Doppler frequency during the entire synthetic aperture time are depicted in Figure 8.29. Note that for the backward bistatic mode (see Figure 8.29a), as both platforms move synchronously, all intersecting curves cross at one point, the target position. For the forward bistatic modes (see Figure 8.29b), the intersecting curves meet in two points P and P', due to equal bistatic range histories. The two range histories, Doppler histories, are equal, creating dual but identical targets in the data domain, so that a "ghost target" may appear in the focused image. It is the difference of the projection rule that the difference between the backward and forward bistatic SAR.

In the forward bistatic mode, we can locate the position of the ghost P' of a certain target P with the conditions: for plane assumption, $Z_R = 0$, and for ellipsoid, $R_{bi}(\eta_i) = R_{bi\_P}(\eta_i)$; where R_{bi} is the bistatic range to the point in the imaging scene, R_{bi_P} is the bistatic range to the target point and η_i is the instantaneous azimuth time, slow time, during the target exposure period. In plane surface assumption, it means only the points on the earth plane are considered. The ellipsoid case means that the bistatic range of ghost image is equal to the bistatic range of target point in the whole observation time, which is often selected to increase the prediction accuracy of the position of ghost image.

8.5.4 Examples

For the purpose of simulation and to be more practical, we adopt the system parameters from SAOCOM-CS mission [46–48], as given in Table 8.4. In SAOCOM-CS mission, the bistatic incidence angle range is 20.7~38.4°. In our simulation, the incidence angle from the transmitter is selected as a central incidence angle 29.55°.

In what follows, the imaging property of backward and forward bistatic SAR will be analyzed. This system works in the receiver incidence angle between 21° and 57° with 7 beams in total (3° on the left and right sides of the beam center) and its corresponding ground range resolution is shown in Figure 8.30.

We see that the ground range resolution in the backward mode is similar to monostatic SAR and that is beneficial for imaging. In addition, only in the forward scattering zone, the ground range resolution deteriorates and the phenomenon is more obvious as the incidence angle differences between the transmitter and the receiver become small, particularly near the specular region. In the forward specular bistatic, the transmitter and the receiver are symmetrical about the center region within the imaging scene. This symmetry causes the two opposite direction vectors in the x-axis to counteract each other and the resolution becomes extremely poor. As the angular difference between the transmitter and receiver becomes larger, the ground range resolution changes are mitigated and improve considerably. Thus, conclusions can be drawn that the ground range resolution in the backscattering zone is superior to that in the forward backscattering zone with a certain incidence angle. To improve the ground range resolution, the receiver incidence angle should be selected away from the transmitter incidence angle and the azimuth angle should be increased as much as possible. In addition, special attention should be paid to avoid the approximately symmetrical imaging configuration in the forward bistatic SAR.

Figure 8.31 shows the azimuth resolution with respect to the velocity direction angle β and azimuth angle φ_R. It is seen that the azimuth resolution has a slight change when $\beta = 0°$, since the two SAR platforms move in the same direction with

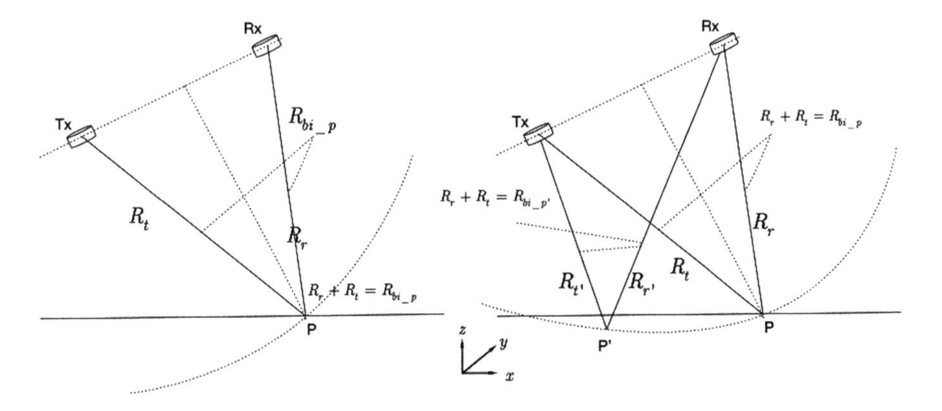

FIGURE 8.28 The illustration of the spherical and ellipsoid surface of the backward bistatic and forward bistatic.

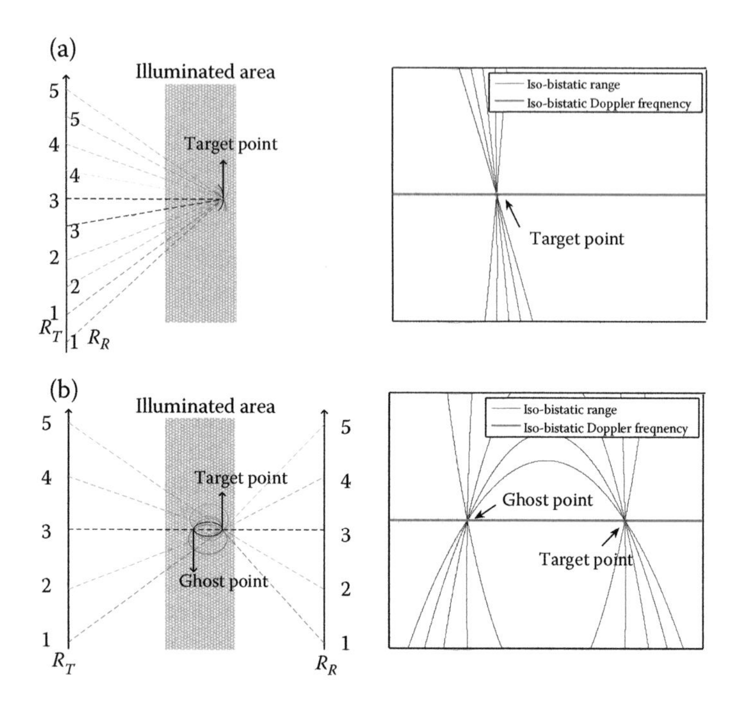

FIGURE 8.29 The range histories of the backward bistatic (a) and forward bistatic (b) at target P during full aperture and the numbers 1–5 represent different positions of the transmitter and receiver at the azimuth direction. The right column is the zoom of the target imaging scene of the left column.

a parallel track. When the velocity angle is close to 180°, the azimuth resolution diverges, because the angular velocity vectors $\mathbf{w}_T(\eta)$ *and* $\mathbf{w}_R(\eta)$ are in opposition, which therefore is not recommended in bistatic SAR. The influence of the velocity direction angle is dominant and the receiver incidence angle adds to the effect. Conclusions that the opposite velocity vectors are not desirable even though the azimuth resolution becomes better as the receiver incidence angle changes. Except in the case of velocity angle near to 180°, there is not much difference for the azimuth resolution for backward and forward bistatic SAR.

In most practical cases, the velocity angle is set zero, which means a parallel track. The ground range and azimuth resolutions, resolution direction angle, and the ground resolution cell area with respect to the azimuth angle and the receiver incidence angle, with $\theta_T = 29.55°$ and $\beta = 0°$ are plotted in Figure 8.32. Figure 8.32a shows that the azimuth resolution changes slightly in the whole scattering zone when $\beta = 0°$ but it suffers degradation when φ_R is near 90° and 270° because the sum of angular velocity units $\mathbf{w}_T(\eta)$ *and* $\mathbf{w}_R(\eta)$ is smaller. Figure 8.32b shows that the directions of the two resolutions are almost collinear in some regions in the forward scattering zone which causes defocus in the final image. Figure 8.32c also indicates the same trend; it can be seen that the areas with an orthogonal resolution direction angle are not strictly in conformity with the area with the smallest ground

TABLE 8.4
Key Simulation Parameters for Bistatic Imaging

System/Motion	Parameter	Symbol	Value
System parameters	Chirp bandwidth	B	45 MHz
	Processed Doppler bandwidth	B_a	1050 Hz
	Center frequency	f_c	1275 MHZ
	Wavelength	λ	23 cm
	Integration time	T_a	10 s
	Transmitter peak power	P_T	3.1 kW
	Antenna gain of transmitter	G_T	55 dB
	Antenna gain of receiver	G_R	50 dB
	Receiver noise temperature	T_R	300 K
	Receiver noise figure	F_n	4.5 dB
	Propagation losses	L	3.5 dB
	Duty cycle	D_c	0.05
Motion parameters	Orbit height	H	619.6 km
	Transmitter incidence angle	θ_T	29.55°
	Transmitter azimuth angle	φ_T	0°
	Flight velocity	v	7545 m/s

resolution cell and it is also influenced by the two ground resolutions. In some regions in the forward bistatic mode, the resolution cell is totally lost and should be avoided when designing the imaging geometric parameters.

In the forward quasi-specular bistatic mode, the azimuth resolution is fine while the ground range resolution deteriorates rapidly, implying that the forward specular bistatic is not preferable for forward imaging in view of the resolution. In this

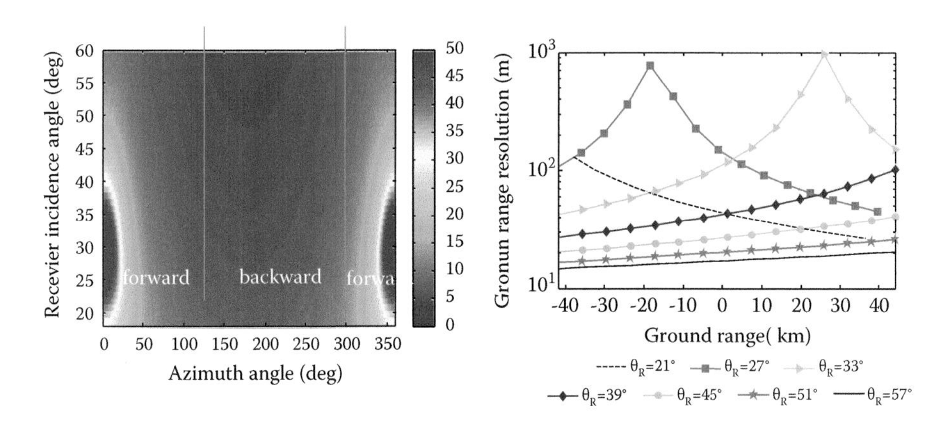

FIGURE 8.30 (Left): The ground range resolution with respect to the received azimuth angle φ_R. (Right): The ground range resolution in different beams.

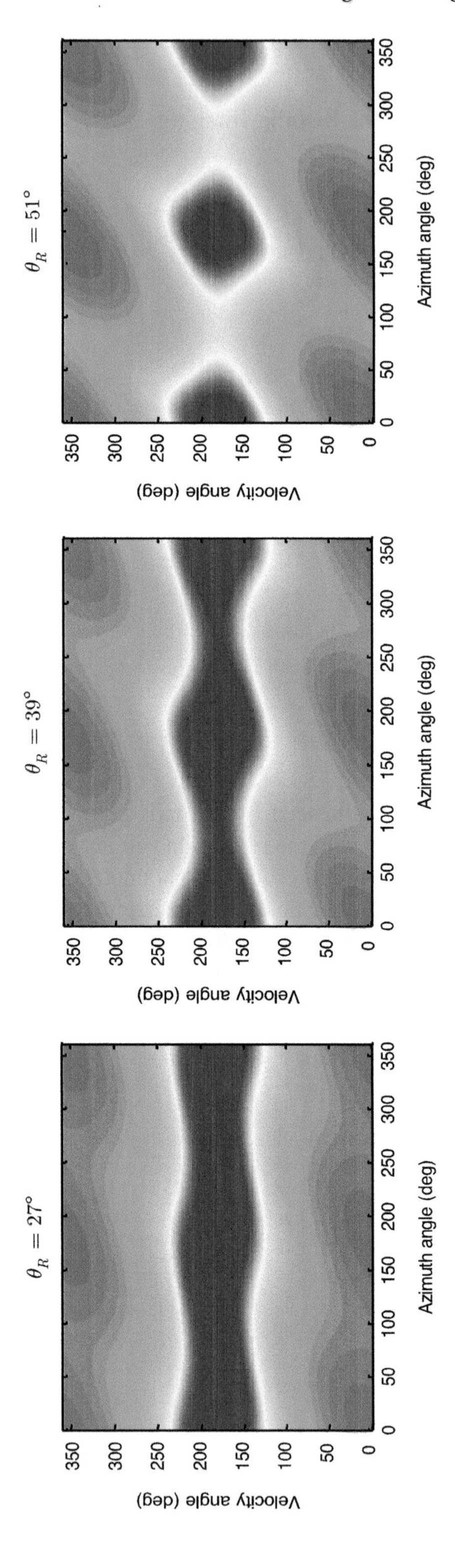

FIGURE 8.31 The azimuth resolution with respect to velocity angle β and received azimuth angle φ_R . with $\theta_T = 29.55°$.

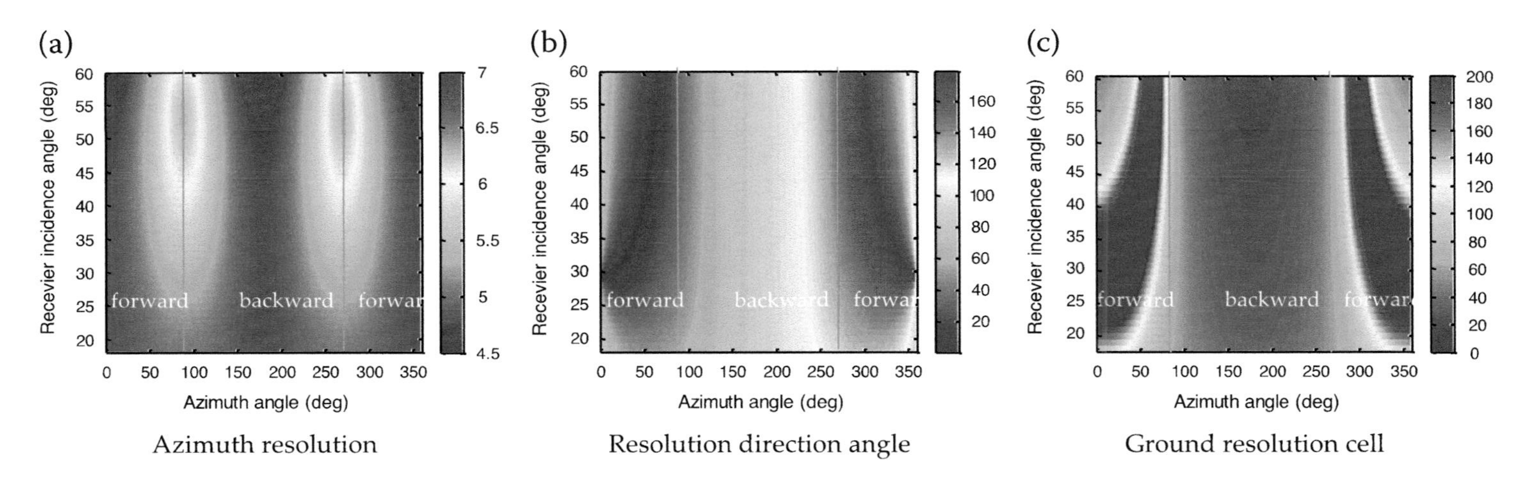

FIGURE 8.32 The resolution analysis with respect to θ_R . and φ_R . . (a) Azimuth resolution; (b) Resolution direction angle; (c) Ground resolution cell with $\theta_T = 29.55°$.

simple demonstration, the geometric properties and power considerations of the backward and forward bistatic SAR are analyzed in a formation mode. In addition, the focus is on the forward bistatic configuration which has been proven to be beneficial for remote sensing applications. As predicted by the range history analysis and verified by the back-projection imaging simulation results, the bistatic range history phenomenon that introduces "ghost targets" exists when the imaging area is across the specular region. That drawback should be avoided by limiting the illumination area beside the specular region rather than across it. By enlarging the difference in the incidence angles of the transmitter (θ_T) and receiver (θ_R), the ghost-free area can be broadened for practical observations. From the view of realizing a sufficiently high resolution, the quasi-specular forward imaging geometry should also be avoided, as the ground range resolution deteriorates badly in this case. The azimuth resolution variation is relatively insignificant as long as the velocity directions of the transmitter and receiver are not notably different. Then, based on the considerations of the range ambiguity effect, the ground resolution cell area size, remarks on the forward bistatic SAR imaging geometry design can be concluded: a sufficiently large difference in the transmitting and receiving incidence angle (θ_T and θ_R, $\theta_R > \theta_T$) should be guaranteed near the incident plane, for ghost-free imaging and achieving a good balance between fine resolution and obtaining richer scattering features from earth surfaces.

Bistatic SAR is a fast-growing and active research area for its high potential of providing much richer information about the targets, including the geometric and dielectric properties. As we also demonstrated in Chapters 4 and 5, for remote sensing of soil moisture, bistatic scattering offers a much wider dynamic range and higher sensitivity in response to moisture content changes. The bistatic scattering measurement, if properly configured, also help to decouple the responses of surface roughness and moisture content, namely, the strong coupling between the geometric and dielectric properties can be better separated. More research efforts, however, should be devoted to better capture a more complete picture of the bistatic scattering pattern.

REFERENCES

1. Ishimaru, A., *Wave Propagation and Scattering in Random Media*, Academic Press, New York, 1978.
2. Tsang, L., Kong, J. A., and Shin, R. T., *Theory of Microwave Remote Sensing*, Wiley, New York, 1985.
3. Fung, A. K., *Microwave Scattering and Emission Models and Their Applications*, Artech House, Norwood, MA, 1994.
4. Ulaby, F. T. and Long, D. G., *Microwave Radar and Radiometric Remote Sensing*, University of Michigan Press, Ann Arbor, MI, 2014.
5. Chew, W. C., *Waves and Fields in Inhomogeneous Media*, IEEE Press, Piscataway, NJ, 1995.
6. Chew, W. C., Jin, J. M., Michielssen, E., and Song, J., *Fast and Efficient Algorithms in Computational Electromagnetics*, Artech House, Norwood, MA, 2000.
7. Tsang, L., Kong, J. A., Ding, K. H., *Scattering of Electromagnetic Waves*: Theories and Applications, Wiley, New York, 2000.
8. Jing, J. M., *Theory and Computation of Electromagnetic Fields*, 2nd Edition, Wiley-IEEE Press, New York, 2015.

9. Graglia, R. D., Peterson, A. F., *Higher-Order Techniques in Computational Electromagnetics*, SciTech Publishing, Raleigh, NC, 2016.
10. Tsang, L., Ding, K. H., Huang, S. W., and Xu, X. L., Electromagnetic computation in scattering of electromagnetic waves by random rough surface and dense media in microwave remote sensing of land surfaces, *Proceedings of the IEEE*, 101(2), 255–279, 2013.
11. Tsang, L., Liao, T.-H., Tan, S., Huang, H., Qiao, T., and Ding, K. H., Rough surface and volume scattering of soil surfaces, ocean surfaces, snow, and vegetation based on numerical Maxwell model of 3D simulations, *IEEE Journal of Selected Topics in Applied Earth Observations and Remote Sensing*, 10(11), 4703–4720, 2017.
12. Tsang, L., Cha, J. H., and Thomas, J. R., Electric Fields of spatial Green's functions of microstrip structures and applications to the calculations of impedance matrix elements, *Microwave and Optical Technology Letters*, 20(2), 90–97, 1999.
13. Li., Q., Chan, H., and Tsang, L., Monte-Carlo simulations of wave scattering from lossy dielectric random rough surfaces using the physics-based two-gird method and canonical grid method, *IEEE Transactions on Antennas and Propagation*, 47(4), 752–763, 1999.
14. Arfken, G., *Mathematical Methods for Physicists*, 3rd Edition, Academic Press, Orlando, FL, 1985.
15. Chen, X. D., *Computational Methods for Electromagnetic Inverse Scattering*, Wiley-IEEE Press, Hoboken, NJ, 2018.
16. Keller, J. B., Accuracy and validity of the Born and Rytov pproximations. *Journal of the Optical Society of America*, 59(8), 1003, 1969.
17. Marks, D. L., A family of approximations spanning the Born and Rytov scattering series, *Optics Express*, 14(19), 8837, 2006.
18. Levy, M., *Parabolic Equation Me*scattering media on super*thods for Electromagnetic Wave Propagation*. IET, London, 2000.
19. Zhang, P., Bai, L., Wu, Z., and Guo, L., Applying the parabolic equation to tropospheric groundwave propagation: A review of recent achievements and significant milestones, *IEEE Antennas and Propagation Magazine*, 58(3), 31–44, 2016.
20. Devaney, A. J., Time reversal imaging of obscured targets from multistatic data, *IEEE Transactions on Antennas and Propagation*, 53(5), 1600–1610, 2005.
21. Devaney, A. J., *Mathematical Foundations of Imaging, Tomography and Wavefield Inversion*, Cambridge University Press, Cambridge, UK, 2012.
22. Ishimura, A., *Electromagnetic Wave Propagation, Radiation, and Scattering: From Fundamentals to Applications*, 2nd Edition, Wiley-IEEE Press, Hoboken, NJ, 2017.
23. Ishimaru, A., Jaruwatanadilok, S., & Kuga, Y., Time reversal effects in random scattering media on super-resolution, shower curtain effects, and backscattering enhancement. *Radio Science*, 42(6), RS6S28. 2007. doi:10.1029/2007RS003645
24. Ishimaru, A., Jaruwatanadilok, S., and Kuga, Y., Imaging through random multiple scattering media using integration of propagation and array signal processing, *Waves in Random & Complex Media*, 22(1): 24–39, 2012.
25. Goodman, J. W., 2nd Edition, *Statistical Optics*, Wiley, New York, 2015.
26. Chan, T., Jaruwatanadilok, S., Kuga, Y., and Ishimura, A., Numerical study of the time-reversal effects on super-resolution in random scattering media and comparison with an analytical model, *Waves in Random & Complex Media,* 18(4), 627–639, 2008.
27. Cumming, I., and Wong, F., *Digital Signal Processing of Synthetic Aperture Radar Data: Algorithms and Implementation*, Artech House, Norwood, MA, 2004.
28. Chiang C. Y., Chen K. S. Simulation of complex target RCS with application to SAR image recognition 3rd International Asia-Pacific Conference on Synthetic Aperture Radar (APSAR), Seoul, Korea, 1-4, 26–30 September 2011

29. Chen, K. S., *Principles of Synthetic Aperture Radar: A System Simulation Approach.* CRC Press: Boca Raton, FL, 2015.
30. Ku, C. S., Chen, K. S., Chang, P. C., and Chang, Y. L., Imaging simulation of synthetic aperture radar based on full wave method, *Remote Sensing*, 10(9), 1404, 2018.
31. Piepmeier, J. R., and Simon, N. K., A polarimetric extension of the van Cittert-Zernike Theorem for use with microwave interferometers, *IEEE Geoscience and Remote Sensing Letters*, 1(4), 300–303, 2004.
32. Born, M., and Wolf, E., *Principles of Optics*, 7th Edition, Cambridge University Press, New York, 1999.
33. Hong, S. T., and Ishimaru, A., Two-frequency mutual coherence function, coherence bandwidth, and coherence time of millimeter and optical waves in rain, fog, and turbulence. *Radio Science*, 11, 551–559, 1976.
34. Ishimaru, A., Ailes-Sengers, L., Phu, P., and Winebrenner, D., Pulse broadening and two-frequency mutual coherence function of the scattered wave from rough surfaces, *Waves Random Media*, 4, 139–148, 1994.
35. Kuga, Y., Le, T. C. C., Ishimaru, A., and Ailes-Sengers, L., Analytical, experimental, and numerical studies of angular memory signatures of waves scattered from one-dimensional rough surfaces, *IEEE Transactions on Geoscience and Remote Sensing*, 34(6), 1300–1307, 1996.
36. Tsang, L., Zhang, G., and Pak, K., Detection of a buried object under a single random rough surface with angular correlation function in EM wave scattering, *Microwave and Optical Technology Letters*, 11(6), 300–304, 1996.
37. Zhang, G., Tsang, L., and Kuga, Y., Studies of the angular correlation function of scattering by random rough surfaces with and without a buried object, *IEEE Transactions on Geoscience and Remote Sensing*, 35(2), 444–453, 1997.
38. Guo, L. X., Wu, Z. S., Study on the two-frequency scattering cross section and pulse broadening of the one-dimensional fractal sea surface at millimeter wave frequency. *Progress in Electromagnetic Research*, 37, 221–234, 2002.
39. Wu, K., Two-frequency mutual coherence function for electromagnetic pulse propagation over rough surfaces, *Waves in Random and Complex Media*, 15(2), 127–143, 2005.
40. Ren, Y., Guo, L., Two-frequency mutual coherence function and its applications to pulse scattering by random rough surface, *Science China Physics, Mechanics, & Astronomy,* 51, 157–164, 2008.
41. Le, C. T., Kuga, Y., and Ishimaru, A., Angular correlation function based on the second-order Kirchhoff approximation and comparison with experiments, *Journal of Optical Society of America A*, 13(5), 1057–1067, 1996.
42. Li, T., Chen, K. S., and Jin, M., Analysis and simulation on imaging performance of backward and forward bistatic SAR, *Remote Sensing*, 10, 1676, 2018.
43. Cardillo, G. P., On the use of the gradient to determine bistatic SAR resolution, *Proceedings of Antennas and Propagation Society International Symposium*, Dallas, TX, USA, 7–11 May; 1032–1035, 1990 10.1109/APS.1990.115286.
44. Moccia, A., and Renga, A., Spatial resolution of bistatic synthetic aperture radar: Impact of acquisition geometry on imaging performance, *IEEE Transactions on Geoscience and Remote Sensing*, 49(10), 3487–3503, 2011.
45. Qiu, X. L., Ding, C. B., Hu, D. H., *Bistatic SAR Data Processing Algorithms*, John Wiley Sons, New York, 2013.
46. Gebert, N., Dominguez, B. C., Davidson, M. W. J., Martin, M. D., and Silvestrin, P., SAOCOM-CS—apA passive companion to SAOCOM for single-pass L-band SAR interferometry. *In Proceedings of 2014 10th European Conference on Synthetic Aperture Radar*, Berlin, Germany, 1–4, 3–5 June 2014.

47. Bordoni, F., Younis, M., Cassola, M. R., Iraol, P. P., Dekker, P. L., Krieger, G. SAOCOM-CS SAR imaging performance evaluation in large baseline bistatic configuration. *Proceedings of 2015 IEEE International Geoscience and Remote Sensing Symposium*, Milan, Italy, 26–31 July 2015; 2107–2110, 10.1109/IGARSS.2015.7326218.
48. Gebert, N., Dominguez, B. C., Martin, M. D., Salvo, E. D., Temussi, F., Giove, P.V., Gibbons, M., Phelps, P., and Griffiths, L., SAR Instrument Pre-development Activities for SAOCOM-CS. *Proceedings of 2016 11th European Conference on Synthetic Aperture Radar (EUSAR)*, Hamburg, Germany, 1–4, 6–9 June 2016.

APPENDIX 8A

Explicit forms of $\vec{I}^r$ and $\vec{F}^r$

For simple notation, let

$$\bar{\bar{\mathbf{G}}}_t^d \equiv \begin{bmatrix} \hat{x}\hat{x}G_{xx} & \hat{x}\hat{y}G_{xy} & \hat{x}\hat{x}G_{xz} \\ \hat{y}\hat{x}G_{yx} & \hat{y}\hat{y}G_{yy} & \hat{y}\hat{z}G_{yz} \\ \hat{z}\hat{x}G_{zx} & \hat{z}\hat{y}G_{zy} & \hat{z}\hat{z}G_{zz} \end{bmatrix} \quad (8.A1)$$

$$\nabla \times \bar{\bar{\mathbf{G}}}_t^d = \begin{bmatrix} \hat{x}\hat{x}D_{xx} & \hat{x}\hat{y}D_{xy} & \hat{x}\hat{x}D_{xz} \\ \hat{y}\hat{x}D_{yx} & \hat{y}\hat{y}D_{yy} & \hat{y}\hat{z}D_{yz} \\ \hat{z}\hat{x}D_{zx} & \hat{z}\hat{y}D_{zy} & \hat{z}\hat{z}D_{zz} \end{bmatrix} \quad (8.A2)$$

After a series of vector manipulations, we can obtain

$$\begin{aligned} I_x^r &= \hat{x}\cdot\vec{I}^r = \hat{x}\cdot(\vec{s}\times\vec{E}^i) + \hat{x}\cdot\vec{s}\times\iint(\nabla\times\bar{\bar{\mathbf{G}}}\cdot\vec{I}^{r'} + j\omega\mu\bar{\bar{\mathbf{G}}}\cdot\vec{F}^{r'})dx'dy' \\ &= -E_{iy} + \iint\{[(-\xi_y D_{zx} - D_{yx}) + \xi_x'(-\xi_y D_{zz} - D_{yz})]I_x^{r'} \\ &\quad + [(-\xi_y D_{zy} - D_{yy}) + \xi_y'(-\xi_y D_{zz} - D_{yz})]I_y^{r'} \\ &\quad + j\omega\mu[(-\xi_y G_{zx} - G_{yx}) + \xi_x'(-\xi_y G_{zz} - G_{yz})]F_x^{r'} \\ &\quad + j\omega\mu[(-\xi_y G_{zy} - G_{yy}) + \xi_y'(-\xi_y G_{zz} - G_{yz})]F_y^{r'}\}dx'dy' \end{aligned} \quad (8.A3)$$

$$\begin{aligned} I_y^r &= \hat{y}\cdot\vec{I}^r = \hat{y}\cdot(\vec{s}\times\vec{E}^i) + \hat{y}\cdot\vec{s}\times\iint(\nabla\times\bar{\bar{\mathbf{G}}}\cdot\vec{I}^{r'} + j\omega\mu\bar{\bar{\mathbf{G}}}\cdot\vec{F}^{r'})dx'dy' \\ &= E_{ix} + \xi_x E_{iy} + \iint\{[(D_{xx} + \xi_x D_{zx}) + \xi_x'(D_{xz} + \xi_x D_{zz})]I_x^{r'} + [(D_{xy} + \xi_x D_{zy}) \\ &\quad + \xi_y'(D_{xz} + \xi_x D_{zz})]I_y^{r'} + j\omega\mu[(G_{xx} + \xi_x G_{zx}) + \xi_x'(G_{xz} + \xi_x G_{zz})]F_x^{r'} \\ &\quad + j\omega\mu[(G_{xy} + \xi_x G_{zy}) + \xi_y'(G_{xz} + \xi_x G_{zz})]F_y^{r'}\}dx'dy' \end{aligned} \quad (8.A4)$$

$$F_x^r = \hat{x}\cdot\vec{F}^r = \hat{x}\cdot(\vec{s}\times\vec{H}^i) + \hat{x}\cdot\vec{s}\times\iint\{-j\omega\varepsilon(\bar{\bar{\mathbf{G}}}\cdot\vec{I}^{r'}) + (\nabla\times\bar{\bar{\mathbf{G}}}\cdot\vec{F}^{r'})\}dx'dy'$$
$$= -H_{iz} - \beta H_{iy} + \iint\{-j\omega\varepsilon[(-\xi_y G_{zx} - G_{yx}) + \xi_x'(-\xi_y G_{zz} - G_{yz})]I_x^{r'}$$
$$- j\omega\varepsilon[(-\xi_y G_{zy} - G_{yy}) + \xi_y'(-\xi_y G_{zz} - G_{yz})]I_y^{r'} + [(-\xi_y D_{zx} - D_{yx}) + \xi_x'(-f_y D_{zz} - D_{yz})]F_x^{r'}$$
$$+ [(-\xi_y D_{zy} - D_{yy}) + \xi_y'(-\xi_y D_{zz} - D_{yz})]F_y^{r'}dx'dy' \quad (8.A5)$$

$$\begin{aligned} F_y^r &= \hat{y}\cdot\vec{F}^r \\ &= \hat{y}\cdot(\vec{s}\times\vec{H}^i) + \hat{y}\cdot\vec{s}\times\iint\{-j\omega\varepsilon(\bar{\bar{\mathbf{G}}}\cdot\vec{I}^{r'}) + (\nabla\times\bar{\bar{\mathbf{G}}}\cdot\vec{F}^{r'})\}dx'dy' \\ &= -H_{iy} - \xi_y H_{iy} + \iint\{-j\omega\varepsilon[(G_{xx} + f_x G_{zx}) + \xi_x'(G_{xy} + \xi_x G_{zy})]I_x^{r'} \\ &\quad - j\omega\varepsilon[(G_{xy} + \xi_x G_{zy}) + \xi_y'(G_{xz} + \xi_x G_{zz})]I_y^{r'} + \\ &\quad [(D_{xx} + \xi_x D_{zx}) + \xi_x'(D_{xy} + \xi_x D_{zy})]F_x^{r'} \\ &\quad + [(D_{xy} + \xi_x D_{zy}) + \xi_y'(D_{xz} + \xi_x D_{zz})]F_y^{r'}dx'dy \end{aligned} \quad (8.A6)$$

Note that

$$\hat{n}\cdot\vec{I}^r = 0 \rightarrow I_z^r = \xi_x I_x^r + \xi_y I_y^r \quad (8.A7)$$

$$\hat{n}\cdot\vec{F}^r = 0 \rightarrow F_z^r = \xi_x F_x^r + \xi_y F_y^r \quad (8.A8)$$

It follows that the normal components of $\vec{I}^r$ and $\vec{F}^r$ are written as, respectively

$$I_n^r = (-\xi_x E_{ix} - \xi_y E_{iy} + E_{iz}) + \int\{[-(\xi_x D_{xx} + \xi_y D_{yx} - D_{zx} + \xi_x\xi_x' D_{xz} + \xi_y\xi_x' D_{yz}$$
$$- \xi_x' D_{zz}]I_x^{r'} + [\xi_x D_{xy} + \xi_y D_{yy} - D_{zy} + \xi_x\xi_x' D_{xz} + \xi_y\xi_x' D_{yz} - \xi_x' D_{zz}]I_y^{r'}$$
$$-j\omega\varepsilon[\xi_x G_{xx} + \xi_y G_{yx} - G_{zx} + \xi_x\xi_x' G_{xz} + \xi_y\xi_x' G_{yz} - \xi_x' G_{zz}]F_x^{r'}$$
$$+j\omega\varepsilon[\xi_x G_{xy} + \xi_y G_{yy} - G_{zy} + \xi_x\xi_y' G_{xz} + \xi_y\xi_y' G_{yz} - \xi_y' G_{zz}]F_y^{r'}\}dx'dy' \quad (8.A9)$$

$$F_n^r = (-\xi_x H_{ix} - \xi_y H_{iy} + H_{iz}) + \iint\{(-i\omega\varepsilon)[\xi_x G_{xx} + \xi_y G_{yx} - G_{zx} + \xi_x\xi_x' G_{xz}$$
$$+ \xi_y\xi_x' G_{yz} - \xi_x' G_{zz}]I_x^{r'} + (-j\omega\varepsilon)[\xi_x G_{xy} + \xi_y G_{yy} - G_{zy} + \xi_x\xi_y' G_{xz} + \xi_y\xi_y' G_{yz}$$
$$- \xi_y' G_{zz}]I_y^{r'} - [\xi_x D_{xx} + \xi_y D_{yx} - D_{zx} + \xi_x\xi_x' D_{xz} + \xi_y\xi_x' D_{yz} - \xi_x' D_{zz}]F_x^{r'}$$
$$+[\xi_x D_{xy} + \xi_y D_{yy} - D_{zy} + \xi_x\xi_y' D_{xz} + \xi_y\xi_y' D_{yz} - \xi_y' D_{zz}]F_y^{r'}\}dx'dy' \quad (8.A10)$$

APPENDIX 8B

Explicit forms of $\bar{\bar{\mathbf{G}}}_t^r$ and $\nabla \times \bar{\bar{\mathbf{G}}}_t^r$

To write explicit forms of $\bar{\bar{\mathbf{G}}}_t^r$ and $\nabla \times \bar{\bar{\mathbf{G}}}_t^r$, let $X = x - x'$, $Y = y - y'$, $Z = z + z' + 2d_1$, with $\Delta = e^{-i2k_{tz}d_1}$, and $\rho = \sqrt{X^2 + Y^2}$, $\varphi = \tan^{-1}(Y/X)$. After some mathematical manipulations, we have

$$
\begin{aligned}
\bar{\bar{\mathbf{G}}}_t^r(X, Y, Z) = & \frac{j}{4\pi}\hat{x}\hat{x}\int_0^\infty dk_\rho k_\rho \frac{1}{k_{tz}} e^{ik_{tz}Z}\left[R_h\Delta\left(J_1'(k_\rho\rho)\sin^2\varphi + \frac{J_1(k_\rho\rho)}{k_\rho\rho}\cos^2\varphi\right)\right.\\
& \left. - R_v\Delta\frac{k_{tz}^2}{k_t^2}\left(J_1'(k_\rho\rho)\cos^2\varphi + \frac{J_1(k_\rho\rho)}{k_\rho\rho}\sin^2\varphi\right)\right]\\
& + \frac{j}{4\pi}\hat{x}\hat{y}\int_0^\infty dk_\rho k_\rho \frac{1}{k_{tz}} e^{ik_{tz}Z}\left[-R_h\Delta - R_v\Delta\frac{k_{tz}^2}{k_t^2}\right]\left[J_1'(k_\rho\rho) - \frac{J_1(k_\rho\rho)}{k_\rho\rho}\right]\cos\varphi\sin\varphi\\
& + \frac{j}{4\pi}\hat{y}\hat{x}\int_0^\infty dk_\rho k_\rho \frac{1}{k_{tz}} e^{ik_{tz}Z}\left[-R_h\Delta - R_v\Delta\frac{k_{tz}^2}{k_t^2}\right]\left[J_1'(k_\rho\rho) - \frac{J_1(k_\rho\rho)}{k_\rho\rho}\right]\cos\varphi\sin\varphi\\
& + \frac{j}{4\pi}\hat{y}\hat{y}\int_0^\infty dk_\rho k_\rho \frac{1}{k_{1z}} e^{ik_{1z}Z}\left[R_h\Delta\left(J_1'(k_\rho\rho)\cos^2\varphi + \frac{J_1(k_\rho\rho)}{k_\rho\rho}\sin^2\varphi\right)\right.\\
& \left. -R_v\Delta\frac{k_{1z}^2}{k_1^2}\left(J_1(k_\rho\rho)\cos^2\varphi + \frac{J'_1(k_\rho\rho)k_{tz}^2}{k_t^2}\sin^2\varphi\right)\right]\\
& -\frac{i}{4\pi}\hat{x}\hat{z}\int_0^\infty dk_\rho k_\rho \frac{1}{k_{tz}} e^{ik_{tz}Z}\left[R_v\Delta J_1(k_\rho\rho)\left(\frac{-k_{tz}}{k_t^2}\right)k_\rho\cos\varphi\right]\\
& -\frac{j}{4\pi}\hat{z}\hat{x}\int_0^\infty dk_\rho k_\rho \frac{1}{k_{tz}} e^{ik_{tz}Z}\left[R_v\Delta J_1(k_\rho\rho)\left(\frac{k_{tz}}{k_t^2}\right)k_\rho\cos\varphi\right]\\
& -\frac{j}{4\pi}\hat{y}\hat{z}\int_0^\infty dk_\rho k_\rho \frac{1}{k_{tz}} e^{ik_{tz}Z}\left[R_v\Delta J_1(k_\rho\rho)\left(\frac{-k_{tz}}{k_t^2}\right)k_\rho\sin\varphi\right]\\
& -\frac{j}{4\pi}\hat{z}\hat{y}\int_0^\infty dk_\rho k_\rho \frac{1}{k_{tz}} e^{ik_{tz}Z}\left[R_v\Delta J_1(k_\rho\rho)\left(\frac{k_{tz}}{k_t^2}\right)k_\rho\sin\varphi\right]\\
& + \frac{j}{4\pi}\hat{z}\hat{z}\int_0^\infty dk_\rho k_\rho \frac{1}{k_{tz}} e^{ik_{tz}Z}\left[R_v\Delta J_0(k_\rho\rho)\left(\frac{k_\rho^2}{k_t^2}\right)\right]
\end{aligned}
\tag{8.B1}
$$

$$\nabla \times \bar{\bar{\mathbf{G}}}_t^r(X, Y, Z) = \frac{j}{4\pi}\hat{x}\hat{x}\int_0^\infty dk_\rho k_\rho \frac{1}{k_{tz}} e^{ik_{tz}z}\Bigg[R_h\Delta\frac{k_{tz}}{k_t}\left(J_1'(k_\rho\rho) - \frac{J_1(k_\rho\rho)}{k_\rho\rho}\right)\cos\varphi\sin\varphi$$

$$+ R_v\Delta\frac{k_{tz}}{k_t}\left(J_1'(k_\rho\rho) - \frac{J_1(k_\rho\rho)}{k_\rho\rho}\right)\cos\varphi\sin\varphi\Bigg]$$

$$+ \frac{j}{4\pi}\hat{x}\hat{y}\int_0^\infty dk_\rho k_\rho \frac{1}{k_{tz}} e^{ik_{tz}z}\Bigg[- R_h\Delta\frac{k_{tz}}{k_t}\left(J_1'(k_\rho\rho)\cos^2\varphi - \frac{J_1(k_\rho\rho)}{k_\rho\rho}\sin^2\varphi\right)$$

$$+ R_v\Delta\frac{k_{tz}}{k_t}\left(J_1(k_\rho\rho)\frac{J_1(k_\rho\rho)}{k_\rho\rho}\cos^2\varphi - J_1'(k_\rho\rho)\sin^2\varphi\right)\Bigg]$$

$$-\frac{j}{4\pi}\hat{x}\hat{z}\int_0^\infty dk_\rho k_\rho \frac{1}{k_{tz}} e^{ik_{tz}z}\left[R_v\Delta J_1(k_\rho\rho)\frac{k_\rho}{k_t}\sin\varphi\right]$$

$$+ \frac{j}{4\pi}\hat{y}\hat{x}\int_0^\infty dk_\rho k_\rho \frac{1}{k_{tz}} e^{ik_{tz}}\Bigg[R_h\Delta\left(J_1(k_\rho\rho)\frac{J_1(k_\rho\rho)}{k_\rho\rho}\cos^2\varphi - J_1'(k_\rho\rho)\sin^2\varphi\right)\frac{k_{tz}}{k_t}$$

$$-R_v\Delta\left(J_1'(k_\rho\rho)\cos^2\varphi - \frac{J_1(k_\rho\rho)}{k_\rho\rho}\sin^2\varphi\right)\frac{k_{tz}}{k_t}$$

$$+ \frac{j}{4\pi}\hat{y}\hat{y}\int_0^\infty dk_\rho k_\rho \frac{1}{k_{tz}} e^{ik_{tz}z}\Bigg[- R_h\Delta\frac{k_{tz}}{k_t}\left(J_1'(k_\rho\rho) - \frac{J_1(k_\rho\rho)}{k_\rho\rho}\right)\cos\varphi\sin\varphi$$

$$-R_v\Delta\frac{k_{tz}}{k_t}\left(J_1'(k_\rho\rho) - \frac{J_1(k_\rho\rho)}{k_\rho\rho}\right)\cos\varphi\sin\varphi\Bigg]$$

$$-\frac{j}{4\pi}\hat{y}\hat{z}\int_0^\infty dk_\rho k_\rho \frac{1}{k_{tz}} e^{ik_{tz}z}\left[- R_v\Delta J_1(k_\rho\rho)\frac{k_\rho}{k_t}\cos\varphi\right]$$

$$-\frac{j}{4\pi}\hat{z}\hat{x}\int_0^\infty dk_\rho k_\rho \frac{1}{k_{tz}} e^{ik_{tz}z}\left[- R^{TM}\Delta J_1(k_\rho\rho)\frac{k_\rho}{k_t}\cos\varphi\right]$$

$$-\frac{j}{4\pi}\hat{z}\hat{y}\int_0^\infty dk_\rho k_\rho \frac{1}{k_{tz}} e^{ik_{tz}z}\left[R_v\Delta J_1(k_\rho\rho)\frac{k_\rho}{k_t}\cos\varphi\right]$$

(8.B2)

In Equations 8.B1 and 8.B2, J_n is Bessel function of order n and J'_n is derivative of Bessel function of order n. It is noted that the $\hat{z}\hat{z}$ component vanishes.

9 Computational Electromagnetic Imaging of Rough Surfaces

9.1 ROUGH SURFACES FABRICATION BY 3D PRINTING

For both the experimental target fabrication and numerical discretization, digital samples for both Gaussian and exponentially correlated rough surfaces are generated. Unlike the computer generation, it has been difficult to fabricate an exponentially correlated experimental sample by a general-purpose milling machine due to the rich high-frequency roughness. In this regard, the 3D printing technique may be applied for the fabrication. This technique allows one to directly manufacture actual samples based on the computer-generated digital sample with specific parameters, as shown in Table 9.1. More specifically, the FDM (Fused Deposition Modeling) technique is utilized for the fabrication. In this 3D printing process very thin lines of fused materials are continuously printed to construct a structure, and a 100% filling rate can be achieved to form a solid. In this way the shaping accuracy of about 0.5 mm can be achieved, which is approximately $\lambda/20$ in the Ka-band and sufficient for the imaging study. The main advantage of this technique is that it permits the usage of lossy material, to generate samples simulating the half-space rough surface scattering scenario, such as the moist soil ground. In this work, the measured permittivity is 6.22-j2.86 at 32 GHz, and that value was used in the computations of the scattering field. Figure 9.1 shows the CAD model and fabricated samples for Gaussian (a) and exponentially (b) correlated rough surface; top panel: CAD model; bottom panel: Fabricated sample. As should be noted, the considered sample has rich high frequency roughness, which generates numerous local scatters, leading to the strong speckle effects.

To study the speckle properties, we need a large number of computer-generated surface samples. An exponential correlation function for the random rough surface is taken since it describes better the natural ground surface [1]. In this context, the following procedures are excised: first, a very large rough surface mesh is generated on grids, then the digital sample for each realization (400*400 meshing points) is cut from that large mesh. In the cutting, one sample is neighboring and overlapping to next with 100 discretize interval (100*interval size~31 mm) distant, which is close to the footprint size in diameter in the numerical simulations. For the speckle study, 1600 (40*40) samples are obtained for one to achieve an enough number of

TABLE 9.1
Parameters of the Rough Surface Samples

	Size	RMS Height σ	Correlation Length l
Measured Sample	250 mm in diameter	4 mm	48 mm
(in λ @ 32 GHz)	(26.7 in diameter)	(0.427)	(5.12)
Numerical Samples	125 mm × 125 mm	2 mm, 4 mm, 8 mm, 12 mm	48 mm
(in λ @ 32 GHz)	(13.3 × 13.3)	(0.214, 0.427, 0.854, 1.281)	(5.12)

realizations. The digital samples are then input into the FDTD simulator one by one for computing scattering in the 3D computational domain.

In the context of radar imagery, understanding the speckle properties is imperative for both image de-noising and applications such as land-cover classification and parameter retrieval [1–4]. Radar speckle arises due to the coherent sum of numerous distributed scattering contributions [4]. A general assumption is the "fully developed" speckle model, which is applicable when the resolution cell size is much larger than the correlation length (l) of the ground surface [4]. This requirement

FIGURE 9.1 Geometry Configuration of the 3D model and the fabricated sample of the Gaussian (a) and exponentially (b) correlated rough surface. Top: CAD model; Bottom: Fabricated sample.

may be met in low-to-medium image resolution. The fully developed speckle model is based on the following two hypotheses: (1) total scattering is contributed by independent scatters; (2) there is a sufficiently large number of scatters contributing to a resolution cell so that a complex Gaussian distribution is followed. For VHR radar images, these assumptions are no longer valid.

Radar backscattering and its speckle statistic variation with the resolution cell size or antenna footprint has been a subject of interest in rough surface scattering [5–9]. In [5], the convergence of backscattering coefficients versus footprint size was studied by 1-D method of moment solution. By theoretical models and indoor SAR experiments in EMSL [6,7], the dependence of polarimetric backscattering characteristics from Gaussian correlated rough surfaces on the SAR resolution cell size was examined, while the SAR images of rough soil surfaces were measured in EMSL [8]. That work includes results and analysis on the backscattering coefficients and image probability density functions (PDF), versus different resolution cell size. As the radar image resolution reaches evolves from tens of meters to approximately one meter, or even a smaller size, this topic becomes more important currently for land observations. To model high-resolution speckle properties, Di Martino et al, proposed the predicting method for the equivalent number of scatterers within the resolution cell [9], considering stochastic stationary rough surface description, and more advanced, fractal surface models. In this work, we address the topic of very high-resolution (VHR) radar speckle statistics using indoor SAR experiments and supporting full-wave simulations.

Since the prediction for the equivalent number of scatterers has been established in [9], it is used as an evaluating tool in this work. To be specific, the exponentially correlated rough surface is considered, due to its rich high-frequency roughness on the surface-air interface that leads to numerous local scattering contributions. In particular, the 3-D printing process was used to fabricate the sample surface that is used for scattering measurement conducted at Ka-band. In this manner, the image resolution scale close to the rough surface correlation length can be achieved, namely, the image resolution cell size can be at the same scale of the surface geometric undulation. In this view, the scattering mechanism from an exponentially correlated rough surface is similar to that from the sea surface, where the scattering process from short scale roughness is modulated by relatively long scale undulation [10,11]. The prediction for the equivalent number of scatterers per resolution cell from [9] is served as a reference in the analysis. Meanwhile, based on the results from experimental and supporting numerical practices considering different incident directions, the effects of scattering scale on the speckle properties can be observed. Further numerical studies are conducted with different RMS height, for discussions on both the factors of the equivalent number of scatterers and the scattering scale effects.

9.2 EXPERIMENTAL MEASUREMENTS AND CALIBRATION

In the scanning measurement, the scattering signals were captured through the antennas and microwave transceiver all along the scanning track and all over the Ka-band in a microwave anechoic chamber as shown in Figure 9.2. Before we proceed

FIGURE 9.2 Measurements at microwave anechoic chamber.

with the data acquisition, system calibration is required., as we have already introduced in section 3.42 of Chapter 3. The procedures for polarimetric calibration have been available and well documented in numerous papers [12–16]. We present a general polarimetric calibration technique using hybrid corner reflectors. We deploy only one trihedral and two dihedral corner reflectors with different rotation angles for polarimetric calibration. It is simple and makes no assumptions.

In a real system, under non-ideal conditions, such as polarization cross-talk errors and channel imbalance between transmitter and receiver, the polarimetric calibration is required. This is illustrated in Figure 9.3, where E_h^t and E_v^t are the horizontal and vertical polarization components of the incident field, E_h^r and E_v^r are the horizontal and vertical polarization components of the scattered field; T_{hh}, T_{vv}, R_{hh}, R_{vv} represent channel imbalance and T_{hv}, T_{vh}, R_{hv}, R_{vh} stand for cross-talk. The wave propagation in transmitting and receiving causes the target scattering matrix **S** distorted to a measured scattering matrix. The goal of the polarimetric calibration is to invert the target scattering matrix **S** from the measured one to obtain a minimum difference. Propagation matrices and noise matrix are acquired from the known and reference targets. We detail the approach to obtaining the matrix elements using the calibration or reference targets as follows.

Referring to Figure 9.3, in a polarimetric SAR, the measured scattering matrix $\mathbf{S}^m$ and true scattering matrix $\mathbf{S}^t$ are related by:

$$\mathbf{S}^m = \mathbf{R}\mathbf{S}^t\mathbf{T} + \mathbf{N} \tag{9.1}$$

where $\mathbf{S}^m$ and $\mathbf{S}^t$ are the true and measured 2×2 scattering matrices, respectively, **T** and **R** represent the transmitting and receiving distortion matrices, respectively, and **N** accounts for random noise. To simplify the numerical calculation, via

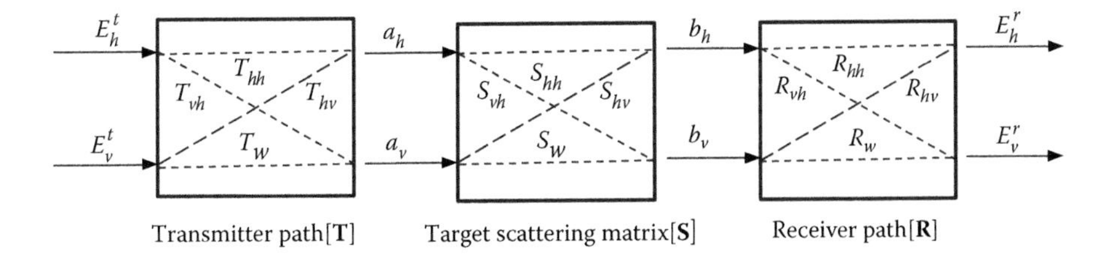

FIGURE 9.3 The process for measuring the target scattering matrix.

absorbing all possible elements of $\mathbf{R}_{ij}$ and $\mathbf{T}_{ij}$ into one matrix without considering the noise contribution involving the scattering reciprocity theorem, Equation 9.1 can be further expressed as follows:

$$\mathbf{S}^m = \mathbf{M}\mathbf{S}^t \tag{9.2}$$

where

$$\mathbf{M} = \begin{bmatrix} m_{11} & m_{12} & m_{13} & m_{14} \\ m_{21} & m_{22} & m_{23} & m_{24} \\ m_{31} & m_{32} & m_{33} & m_{34} \\ m_{41} & m_{42} & m_{43} & m_{44} \end{bmatrix} \tag{9.3}$$

The matrix $\mathbf{M}$ is called the calibration matrix, which contains all distortion in the transmitter and receiver of the polarimetric SAR system. Using eight m matrices in place of the eight matrices, with eight unknowns are given in Equation 9.4:

$$\begin{bmatrix} S_{vv}^m \\ S_{hh}^m \\ S_{vh}^m \\ S_{hv}^m \end{bmatrix} = \begin{bmatrix} m_{11} & \frac{m_{32}m_{42}}{m_{22}} & \frac{m_{33}m_{42}}{m_{22}} & \frac{m_{32}m_{11}}{m_{33}} \\ \frac{m_{31}m_{41}}{m_{11}} & m_{22} & \frac{m_{33}m_{41}}{m_{11}} & \frac{m_{31}m_{22}}{m_{33}} \\ m_{31} & m_{32} & m_{33} & \frac{m_{32}m_{31}}{m_{33}} \\ m_{41} & m_{42} & \frac{m_{42}m_{41}}{m_{44}} & m_{44} \end{bmatrix} \begin{bmatrix} S_{vv}^t \\ S_{hh}^t \\ S_{vh}^t \\ S_{hv}^t \end{bmatrix} \tag{9.4}$$

A total of eight complex unknowns in the calibration matrix need to be solved related to the co-polarizations and cross-polarizations. Therefore, we need at least two calibration targets for co-polarization and one for cross-polarization, forming eight independent complex equations for eight unknowns. In viewing six unknowns for co-polarization, and two unknowns for cross-polarization, passive corner reflectors (CR) are sufficient. They provide more convenient tracking and identification of error sources, if they occur, in the whole wave propagation process. Given these constraints, a trihedral and dihedral corner reflector constitute two equations in co-polarizations, a 45°-rotated dihedral corner reflector makes up two equations

TABLE 9.2
Selected Corner Reflectors and Typical Scattering Matrices

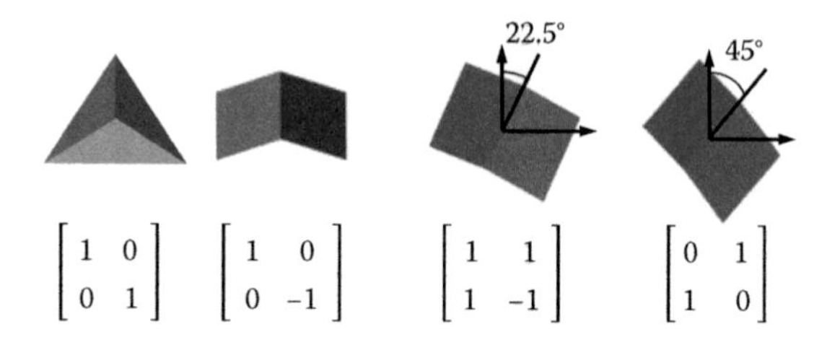

$\begin{bmatrix} 1 & 0 \\ 0 & 1 \end{bmatrix}$	$\begin{bmatrix} 1 & 0 \\ 0 & -1 \end{bmatrix}$	$\begin{bmatrix} 1 & 1 \\ 1 & -1 \end{bmatrix}$	$\begin{bmatrix} 0 & 1 \\ 1 & 0 \end{bmatrix}$

in cross-polarization, and a 22.5°-rotated dihedral corner reflector forms four equations in full polarization. Note that the rotation is about the radar incidence plane. The theoretical scattering matrices for the above-selected reflectors are listed in Table 9.2.

Based on the theoretical scattering matrices of reference targets, one trihedral and one dihedral are used to solve for the six unknown matrices (m_{11}, m_{22}, m_{21}, m_{31}, m_{41}, m_{42}), which are connected with co-polarization. The solutions are given in Equation 9.5:

$$m_{11} = \frac{S_{vv}^{m1} - S_{vv}^{m2}}{2}, \; m_{22} = \frac{S_{hh}^{m1} + S_{hh}^{m2}}{2}$$
$$m_{31} = \frac{S_{vh}^{m1} - S_{vh}^{m2}}{2}, \; m_{32} = \frac{S_{vh}^{m1} + S_{vh}^{m2}}{2} \tag{9.5}$$
$$m_{41} = \frac{S_{hv}^{m1} - S_{hv}^{m2}}{2}, \; m_{42} = \frac{S_{hv}^{m1} + S_{hv}^{m2}}{2}$$

where .. represents the element of the measured scattering matrix of a trihedral corner reflector and similarly S_{qp}^{m2} denotes the element of measured scattering matrix of a dihedral corner reflector. One 22.5°-rotated dihedral is used to work out the two unknowns (m_{33}, m_{44}), which are related to cross-polarization. The solutions read

$$m_{33} = \frac{(m_{31} - m_{32} + S_{vh}^{m3}) \pm \sqrt{(-m_{31} + m_{32} - S_{vh}^{m3})^2 - 4m_{32}m_{31}}}{2},$$
$$m_{44} = \frac{(m_{41} - m_{42} + S_{hv}^{m3}) \pm \sqrt{(-m_{41} + m_{42} - S_{hv}^{m3})^2 - 4m_{42}m_{41}}}{2} \tag{9.6}$$

where S_{qp}^{m3} denotes the element of the measured scattering matrix of a 22.5°-rotated corner reflector. After the **M** matrix is determined, the calibrated scattering matrix can be readily obtained from

$$\mathbf{S}^c = \mathbf{M}^{-1}\mathbf{S}^m \tag{9.7}$$

We employ three reference targets with known measured scattering matrices $\mathbf{S}^{m1}$, $\mathbf{S}^{m2}$ and $\mathbf{S}^{m3}$ for the polarimetric calibration. We can make use of the relationship between measured and theoretical scattering matrices of reference targets to solve the calibration matrix. After the calibration matrix is determined, the scattering matrices of the reference targets can be calibrated. It follows that the scattering matrices of unknown targets can be calibrated. To assess the accuracy of the proposed calibration technique, the calibrated amplitude and phase error is calculated according to

$$\tilde{A} = 20\log_{10}\left(\frac{|\mathbf{S}^c|}{|\mathbf{S}^{ref}|}\right) \tag{9.8}$$

$$\tilde{\phi} = \arg(\mathbf{S}^c) - \arg(\mathbf{S}^{ref}) \tag{9.9}$$

where $\mathbf{S}^c$ represents the calibrated scattering matrices of point targets, and $\mathbf{S}^{ref}$ represents the theoretical scattering matrices of point targets.

Another good measure of calibration quality is the calibrator's polarimetric response, which is a way of visualizing the target scattering properties. We compare the polarimetric responses of the corner reflectors before and after calibration.

9.3 DATA ACQUISITION AND IMAGE FORMATION

To achieve the rough surface imaging in the chamber, the measurements are performed in the Ka-band considering the physical size of the sample prepared in section 9.1. Different bandwidths and synthetic aperture lengths were selected to achieve different spatial resolutions (in ground range), as concluded in Table 9.3. During the image processing, the Kaiser window ($\beta = 2.5$) was utilized along with the frequency and aperture signal, and the resolution cell size in Table 9.3 are realized in the presence of the Kaiser window.

Detailed configuration parameters include: the track span (L) is 800 mm with an interval of 8 mm; the distance between the target scene center and the antenna (T/R)

TABLE 9.3
Imaging Parameters for Different Resolution Cell Size (Ground Range Resolution, Kaiser Windows ($\beta = 2.5$) Applied in Both Domains)

Ground Resolution	Center Frequency	Aperture Length	Bandwidth (GHz)			
			$\theta_i = 30°$	$\theta_i = 40°$	$\theta_i = 50°$	$\theta_i = 60°$
30 × 30 mm^2	32.0 GHz	240 mm	11.00	8.25	6.60	6.15
45 × 45 mm^2		160 mm	7.35	5.50	4.40	4.10
60 × 60 mm^2		120 mm	5.50	4.13	3.30	3.07
75 × 75 mm^2		96 mm	4.40	3.30	2.65	2.45
90 × 90 mm^2		80 mm	3.70	2.75	2.20	2.05

is approximately 1400 mm; the transmitting and receiving antennas are of 20 dBi gain with a half-power beam angle of 17°. In the implementation, the height difference H and distance D can be varied accordingly to achieve the beam direction or incident direction angle θ_i of 30°, 40°, 50°, and even 60°. The target is placed on a low-scattering supporting cylinder made ofthe foam material, which is set on a motor-driven rotation platform. Multiple image acquisitions should be achieved for the speckle statistics. More specifically, in each case of θ_i, the imaging measurements for the rough surface sample are performed 20 times, and after each time turntable is rotated with an equal angular interval. Therefore, 20 images are obtained for a specific θ_i, then speckle statistical analysis can be conducted based on those results.

Figure 9.4 displays radar image of rough surfaces with different resolutions of 1.5 cm, 3.0 cm, 6.0 cm, 9.0 cm in both range and azimuth directions (from top to bottom): left panel: Gaussian correlation; right panel: Exponential correlation. Only VV polarization is shown. The images were focused by the Omega-K algorithm. Finer resolution reveals subtler and detail of the scattering centers, and comparatively, in this regard, exponentially correlated rough surface contains much more fine structures that are responsible for the radar returns, or physically equivalently, the speckle strength. In fact, such phenomena have been numerically studied and confirmed from the field observations.

After the images from the rough surface were processed, calibrated, and compensated, the amplitude speckle results were obtained for the VHR radar speckle analysis.

Although the experimental study is the most straightforward and reliable approach, it is restricted by the actual rough surface sample in both the aspects of number and size. On the other hand, the full-wave full-3D Finite Difference Time Domain (FDTD) simulator provides complementary and extensible data support. This numerical method has been reported in computing scattering from layered and heterogeneous rough surfaces [17,18]. The FDTD simulator used in this work is developed based on that reported in [19] which is a full-3D simulator, and by performing simulations on a large number of samples (realizations) one can get the backscattering results for speckle analysis as well as averaged scattering coefficients.

Figure 9.5 shows the simulation configuration in each realization, which is a typical scene for the half-space scattering. In general, the tapered incident beam is injected by the Total Field/Scatter Field boundary (TF/SF), below that boundary the incident beam immerges to illuminate the rough surface, and above that the boundary scattered field can be extracted. The Huygens Aperture collects the recorded near-field scattered field and turns that into far-field results like scattering coefficients. On the other hand, the Perfect Matching Layer (PML) is used to truncate the computation domain without introducing disturbing reflections.

Here, as a common practice, a tapered wave with a fixed ground footprint is set for the illumination. In each realization, the incident beam points to the center of the rough surface for different incident angles of 30°, 40°, and 50°. Furthermore, different surface RMS heights are also considered. The discretization cell size in the FDTD simulation is set to $1/30\lambda$ at 32 GHz, which

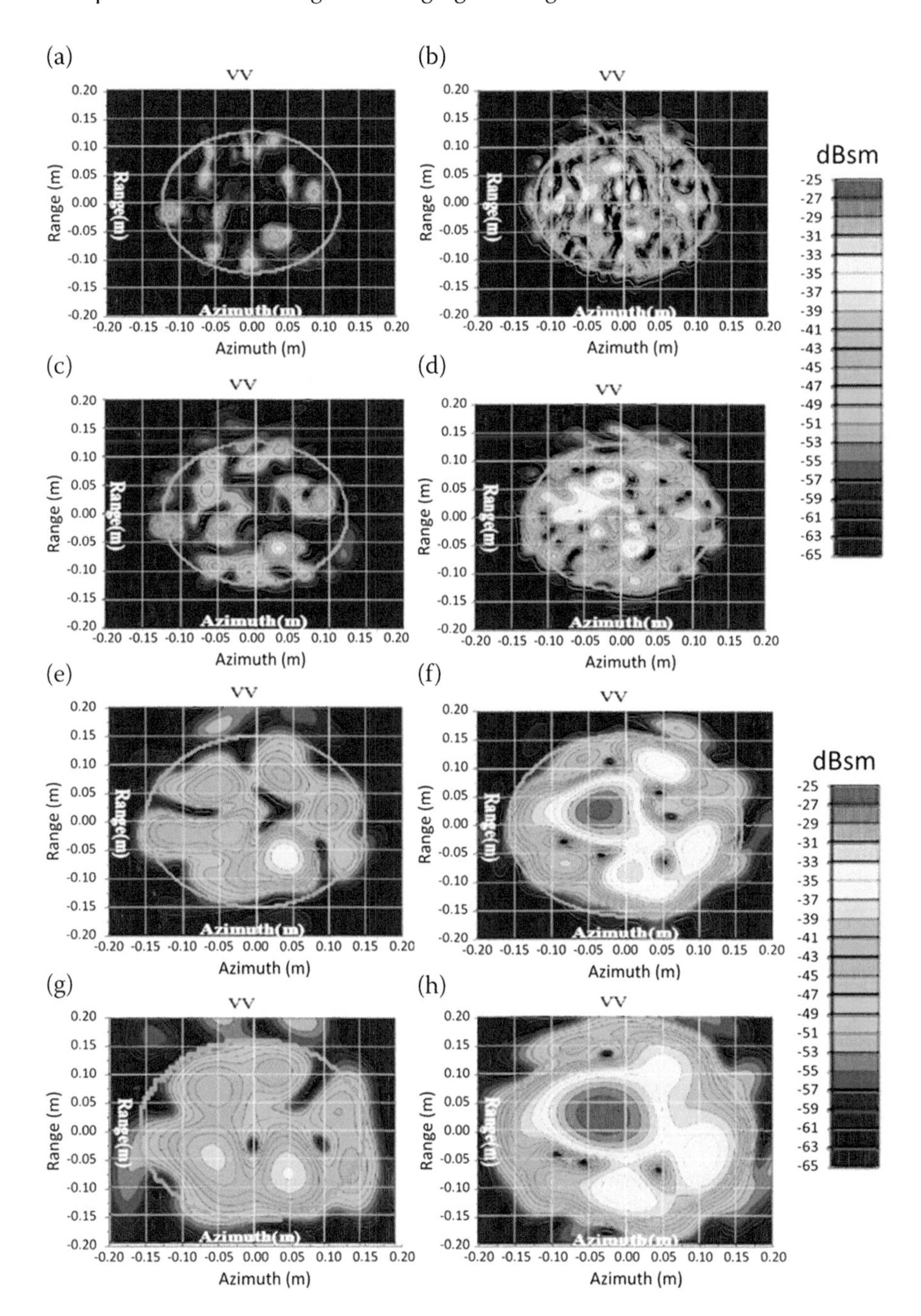

FIGURE 9.4 Radar images of rough surfaces with different resolutions of 1.5 cm, 3.0 cm, 6.0 cm, 9.0 cm in both range and azimuth directions (from top to bottom) @ VV polarization: left panel: Gaussian correlation; right panel: Exponential correlation.

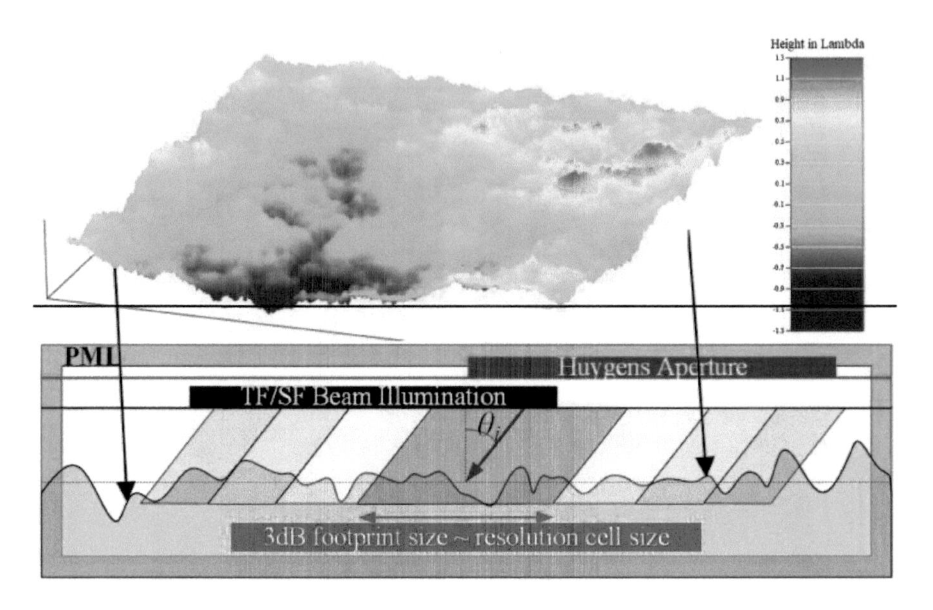

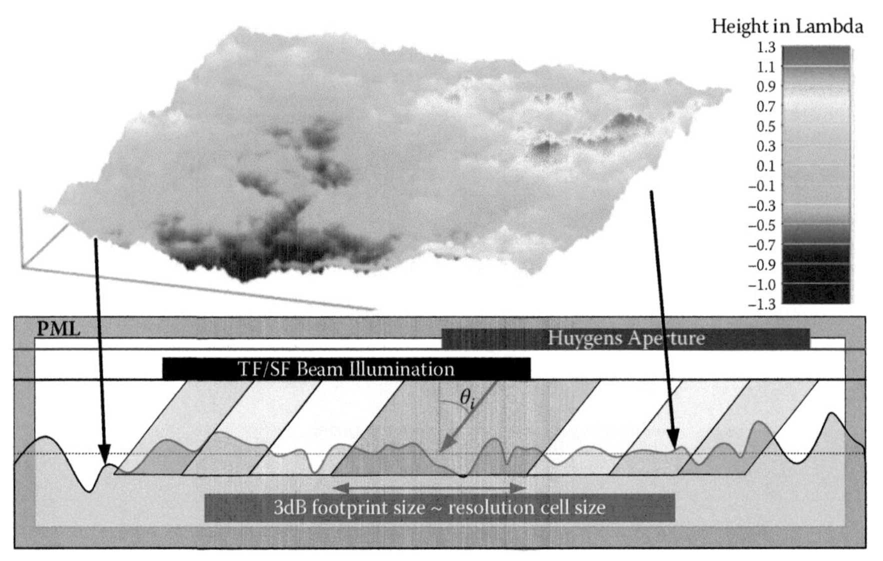

FIGURE 9.5 Configuration of FDTD simulations for rough surface scattering (The presented profile is one of realizations conducted in the numerical simulations).

is sufficient in modeling the high-frequency roughness of the exponentially correlated rough surface.

In general, the aim of this simulation work, which is similar to that reported in [1], is to investigate speckles by counting backscattering far-field amplitude results over a large number of realizations. The difference is that we focused on the VHR situation, which means we used a smaller tapered wave illumination region than those in the general rough surface scattering simulations. Specifically, the simulated resolution cell size is defined by the beam footprint in the numerical studies, realized by setting the footprint size equal to the required value (30 mm in diameter for 3 dB edge power drop). The detailed computation parameters are summarized in Table 9.4, and these simulations are performed to support the conclusions drawn from the VHR radar imaging experiments with a 30 mm resolution as well as further discussions. *ds* denotes the discretizing interval in the computation domain. The simulations were run on a 2-way Intel Xeon workstation with octal channel memories with each realization taking less than 5 min.

We now compare the measured scattering coefficient from the fabricated experimental sample and the model prediction by AIEM model. First, to validate the fabricated sample for the experimental imaging tests, we directly measured its scattering coefficients without the imaging process. Since in the scattering coefficient measurement only one data sample can be obtained through one measurement; apparently, sufficient ensemble samples are needed to estimate the mean value of the scattering coefficient. It turns out that more than 48 sets of backscattering data were acquired by rotating the turntable with a pedestal supporting the rough surface under measuring. Recalled that the rough surface is isotropic. Hence, the rotation of the sample generates independent ensembles as long as the sample is within the same antenna footprint. Then the backscattering coefficient data were derived according to Equations 3.39~3.43 in Chapter 3. Also, the backscattering coefficient at VV polarization derived from the calibrated image of the resolution of 60mm. The measured scattering coefficients are compared with the AIEM model predictions at different θ_i; clearly, a good agreement can be observed in the Figure 9.6.

To validate the FDTD simulation, the computational domain was increased to 1200 × 1200 × 180 Yee cells in the domain size, so as to obtain a converged scattering coefficient. In other words, the rough surface aperture size was increased to $40\lambda \times 40\lambda$. It should be noted that in common scattering simulations for rough surfaces [1], the aperture size of $32\lambda \times 32\lambda$ is electrically large enough. Because of

TABLE 9.4
Computational Parameters for the FDTD Simulations @ 32 GHz

Set 1	3 dB Footprint Size	ds(mm)	Domain Size (Yee cells⁺)	Realization number	θ_i (σ = 4mm)
	30 mm in diameter	0.3123	400 × 400 × 180	1600	30°, 40°, 50°
Set 2	**3 dB Footprint Size**	ds(mm)	**Z Direction Size (Yee cells)**	**Realization number**	σ(mm) θ_i = 30°
	30 mm in diameter	0.3123	160, 180, 220, 280	1600	2,4,8,12

Notes:
+ Yee cell is the mesh cell in the FDTD

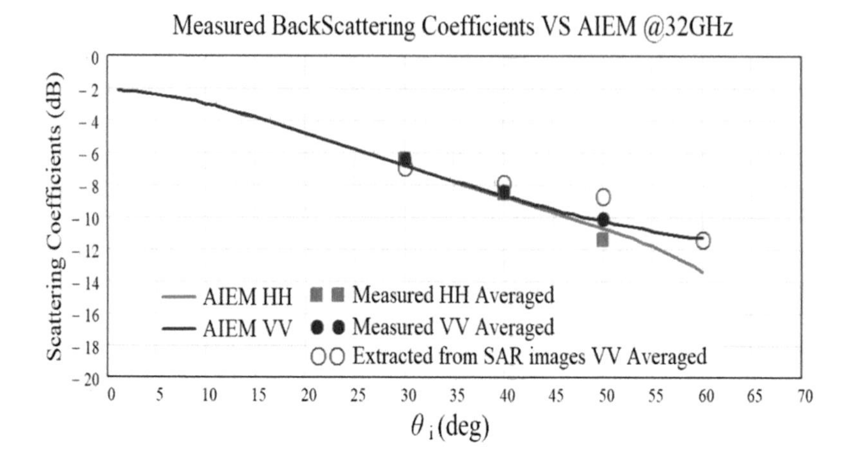

FIGURE 9.6 Comparison between measured scattering coefficients (σ^0) of the exponentially correlated rough surface sample, and the results by AIEM prediction, at 32 GHz.

the enormous computational burden - with one realization requiring approximately 1 hour, only 50 realizations are performed to obtain the averaged scattering coefficients distributions over the upper hemisphere. Again, the AIEM results are utilized as a reference, and two sets of results are compared in Figure 9.7. Although the FDTD results failed in presenting a smooth contour due to the insufficiently large realization number, the results by numerical and analytical methods clearly agree well with each other in the overall angular patterns in the whole bistatic scattering plane.

9.4 IMAGE STATISTICS AND QUALITY

Now, based on the measured images and the simulated backscattering returns, the speckle properties can be obtained. First, the probability density function (PDF) can be computed for the amplitude and compared to the known distribution models, in particular, Rayleigh and K-distribution models. As already noted in Chapter 6, the shape factor coefficients. A larger α will push the K-distribution closer to the Rayleigh distribution that describes the fully-developed speckle. The α value for the K-distribution fitting is obtained from Equation 6.7 up to the fourth order moment of the amplitude or second-order moment of mean normalized intensity. The speckle strength at L-look data can also be readily obtained by Equation 6.4.

We now present the imaging results from the rough surface sample, the image amplitude speckle statistics, and the numerical results. In Figure 9.8, the image examples of the exponentially correlated surface are presented at different resolution scales and different values of θ_i. The scattering hot-spots at a VHR scale are observed to be fused into those at the lower resolution scales. In addition, clearly, as θ_i increases, the RCS of those spots decreases. The presented images are obtained at one of the 20 angular positions, and along with the results at other angular positions, they are used to produce the speckle results.

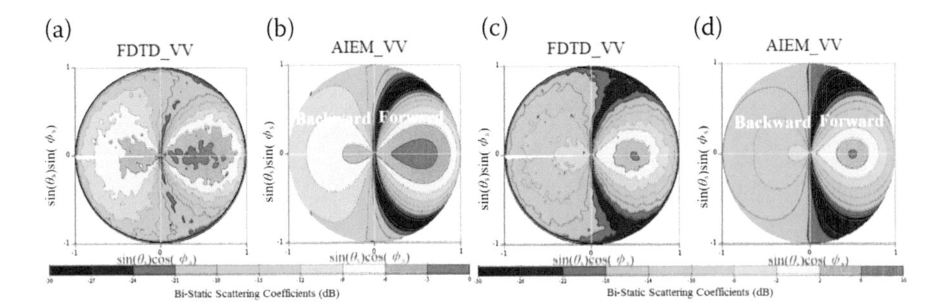

FIGURE 9.7 Comparisons between FDTD simulated upper hemisphere bistatic scattering coefficients ($\sigma°$) and AIEM results, VV, 32 GHz, exponentially correlated rough surface, relative permittivity: 6.22-2.86j; (a) and (b): correlation length l = 48 mm, RMS height σ = 4 mm; (c) and (d): l = 24 mm, σ = 2 mm.

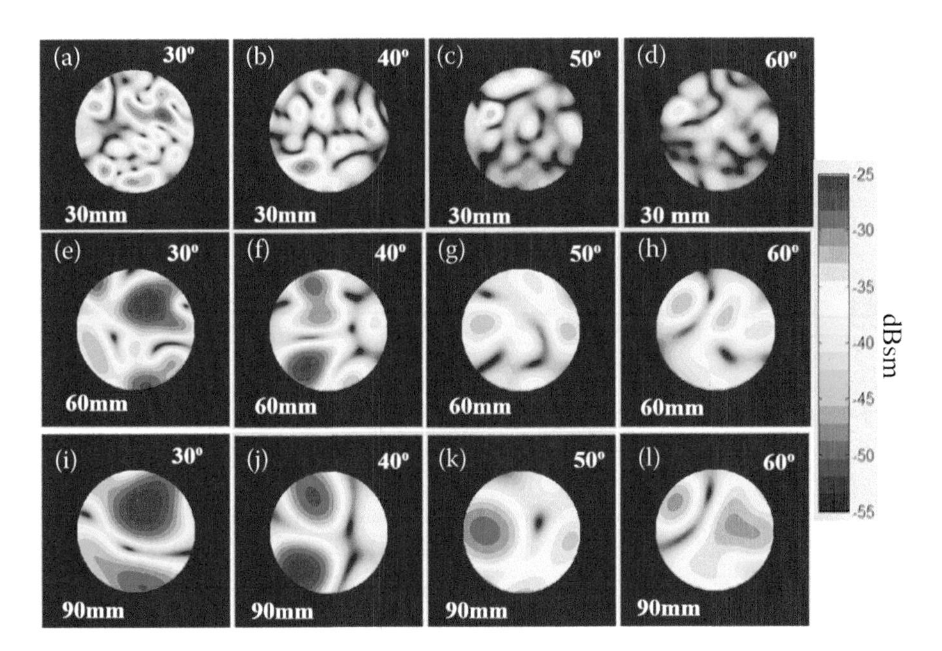

FIGURE 9.8 Image example of the exponentially correlated rough surface sample at different resolutions (30 mm, 60 mm, and 90 mm) and different θ_i (30°, 40°, 50°, and 60°), VV.

In Figure 9.9 and Figure 9.10, the image amplitude PDFs are plotted and compared to the Rayleigh distribution and the fitted K-distribution. In each subfigure of Figure 9.9, the fitted α for the K-distribution is marked, where the resolution cell size is set to 30 mm (in azimuth and ground range), or in correlation length, 0.63l. Clearly, at the smallest θ_i of 30°, the fitted α is the smallest and the PDF curve is with the tail in semi-log plot most away from the Rayleigh curve. This fact is also observed in Figure 9.10, the results in which are with a resolution cell size of 60 mm, or in correlation length, 1.25l.

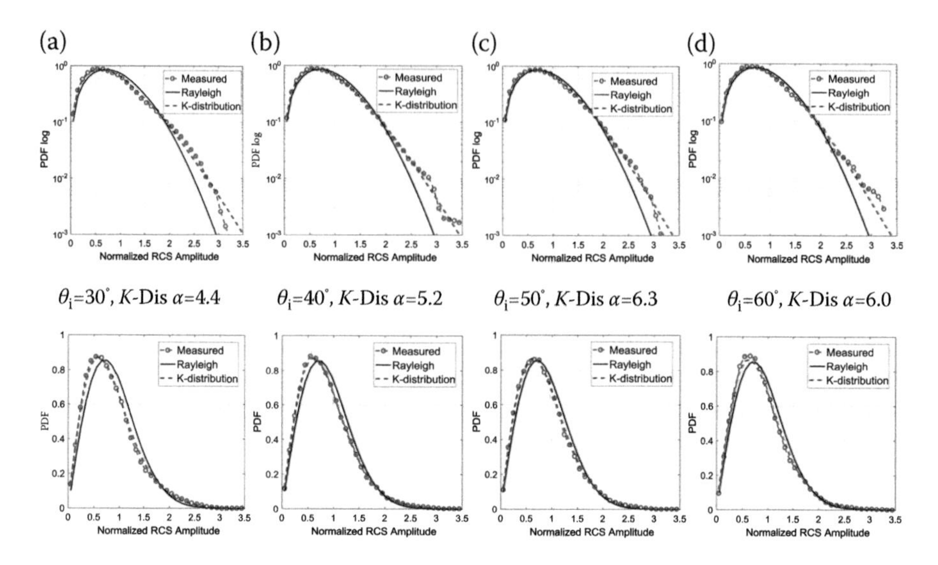

FIGURE 9.9 The normalized amplitude PDF curves of measured images at different θ_i, at the resolution cell size of 30 mm, VV, compared with Rayleigh distribution and K-distribution.

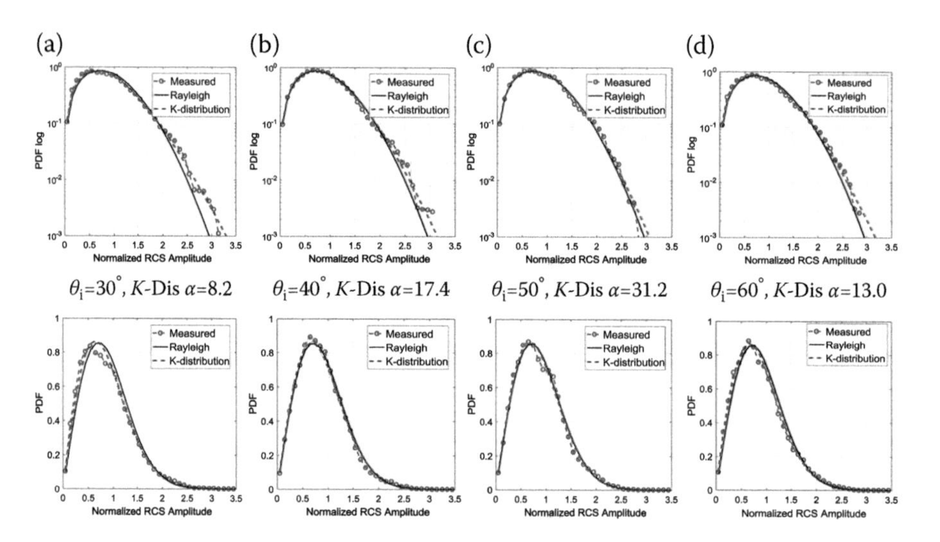

FIGURE 9.10 The normalized amplitude PDF curves of measured images at different θ_i, at the resolution cell size of 60 mm, VV, compared with Rayleigh distribution and K-distribution.

Meanwhile, as the resolution cell size enlarges from 30 mm to 60 mm, the fitted α of K-distribution at each θ_i gets larger and the corresponding PDF is closer to the Rayleigh reference. That agrees with the common sense. On the other hand, it should be noted that the PDFs in the case of resolution cell size 60 mm (Figure 9.10) is not fitting the K-dis as well as those in the case of resolution cell size 30 mm (Figure 9.9). That is due to the limited size of the sample (250 mm in diameter), as it becomes insufficient in providing a good number of scatter returns with increasing resolution cell size.

After the amplitude PDF results based on experimental images are presented, the corresponding averaged RCS σ_{RCS}, the averaged scattering coefficients σ^0, the fitted α of the K-distribution, and the computed γ_a by Equation 6.6 are listed in the Table 9.5 for reference. Further, the computed γ_a in case of different θ_i are plotted via resolution cell sizes in Figure 9.11. The coarse but not perfect trends can be well observed, that the γ_a curves are approaching the theoretical value of $\gamma_a = 0.5227$ with the rising of the resolution cell size. The imperfection in those trends is also due to the limited sample size. Also, in Figure 9.11, the results of computed γ_a results without antenna pattern compensation conducted in the imaging process are also presented, showing the necessity of such a treatment in the chamber imaging experiments for speckle properties.

The simulated scattering coefficients over the upper hemisphere are presented in Figure 9.12, including results by one realization and averaged results by 1600 realizations, for the cases of $\theta_i = 30°$, 40°, and 50° of incidence angles. It can be seen that after averaged over 1600 realizations, the computed bistatic scattering coefficients converge to a smooth distribution.

Then the backscattering amplitude PDF results in cases of $\theta_i = 30°$, 40°, 50°, are compared in Figure 9.13, with a 3 dB footprint size of 30 mm in diameter. It is clear that from $\theta_i = 30°$ to $\theta_i = 50°$, the larger θ_i leads to larger fitted a, and the amplitude

TABLE 9.5
Comparisons of the Average Intensity Values, and Amplitude Speckle Values from the Measured Images at the Resolution Cell Size of 30 and 60 mm

Resolution	θ_i	30°	40°	50°	60°
	AIEM σ^0 (dB)	−6.8	−8.6	−10.4	−11.3
30 mm × 30 mm	Average σ_{RCS} (dBsm)	−37.5	−39.3	−41.4	−43.9
	Average σ^0 (dB)	−6.4	−7.7	−9.0	−10.5
	α for K-dis	4.5	5.2	6.3	6.0
	Computed γ_a	0.598	0.576	0.571	0.564
60 mm × 60 mm	Average σ_{RCS} (dBsm)	−32.0	−33.5	−35.1	−38.8
	Average σ^0 (dB)	−6.9	−7.9	−8.7	−11.4
	α for K-dis	8.2	17.4	31.2	13.0
	Computed γ_a	0.562	0.532	0.534	0.555

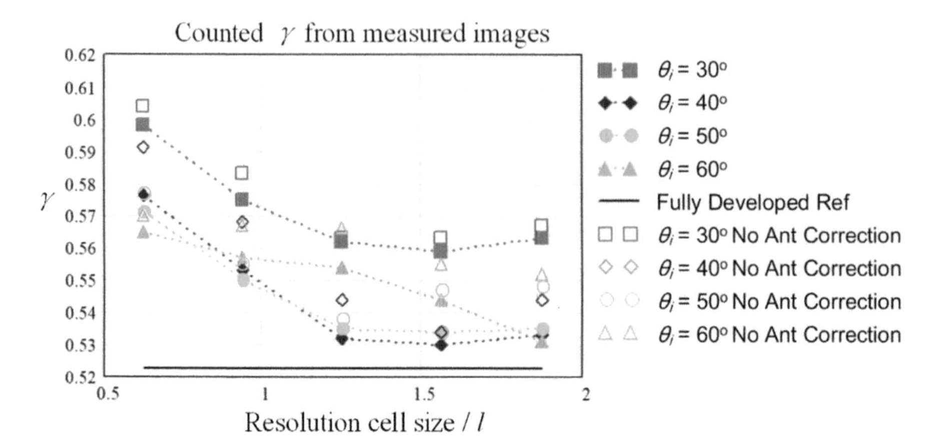

FIGURE 9.11 The computed .. value versus resolution cell size in case of different θ_i from the measured images, VV. "No Antenna Correction" denotes for the γ_a values from images without the antenna pattern correction procedure.

PDF is more close to the Rayleigh distribution. That trend has also been observed in PDF results from measured images (see Figures 9.9 and 9.10). For reference, the averaged backscattering σ^0 fitted α for *K*-distribution, and computed γ_a are concluded in Table 9.7. As can be observed from the results in Tables 9.6 and 9.7, both the experimental imaging and numerical simulation statistics show the same trend, that a larger θ_i (from 30° to 50°) leads to the γ_a closer to the fully developed speckle and thus approaches the theoretical value. That is, as the θ_i gets larger in the moderate region, the observed very high-resolution speckle of the exponentially correlated sample approaches the fully developed speckle.

It is well known that the high-resolution speckle properties are different from the fully developed speckle that in the case of low-resolution sizes. Driven by the trends of high-resolution imaging system development and deployment, it becomes increasingly desired to model the high-resolution speckle for land observations. Starting from the low-resolution basis, the amplitude PDF of Rayleigh is built on the condition that many independent scatters contribute to one resolution cell. It is straightforward to explore new speckle descriptions based on the concept of an equivalent number of scatterers [9,20].

The *K*-distribution, especially for the speckle intensity PDF, has been derived based on the mathematical concept of the coherent sum of a finite number of field returns [20]. In that derivation, each of the independent scatters is assumed to be *K*-distributed in amplitude and uniformly random in phase. Under these assumptions, a close form of *K*-distribution intensity PDF was obtained. For considering image speckle, the number of scatters per resolution N is an important factor, and it is proportional to α of the overall *K*-distribution.

The prediction model for the equivalent number of scatterers per resolution cell was proposed in [9] based on the concept of sub-area dividing, to quantitatively predict the speckle properties of rough surfaces in high-resolution observations, and

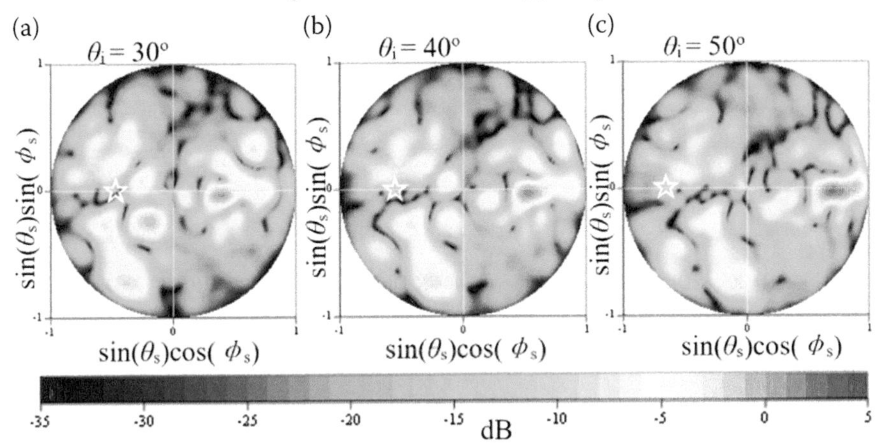

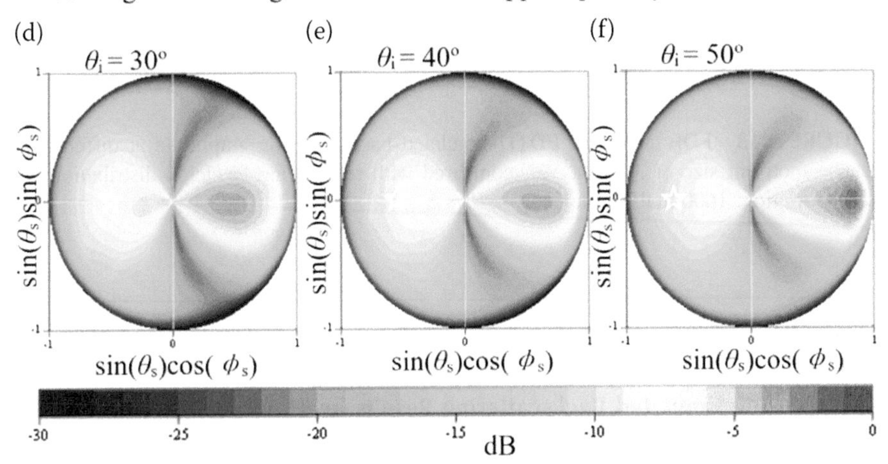

FIGURE 9.12 Results of simulated bistatic scattering coefficients over the upper space, considering different incident angle θ_i. The backscattering angular position is marked by the white star symbol.

to provide a quantitative description linking the equivalent number of scatterers N to the K-distribution, that a larger N generally leads to speckle closer to Rayleigh and a smaller N may lead to a K-distribution further away from the Rayleigh model. A compact formulation is given as in Equation 9.10, yielding the equivalent number (N) of distributed scatters within a resolution cell size with respect to surface correlation function types, parameters, incident direction, and observation frequency [9]:

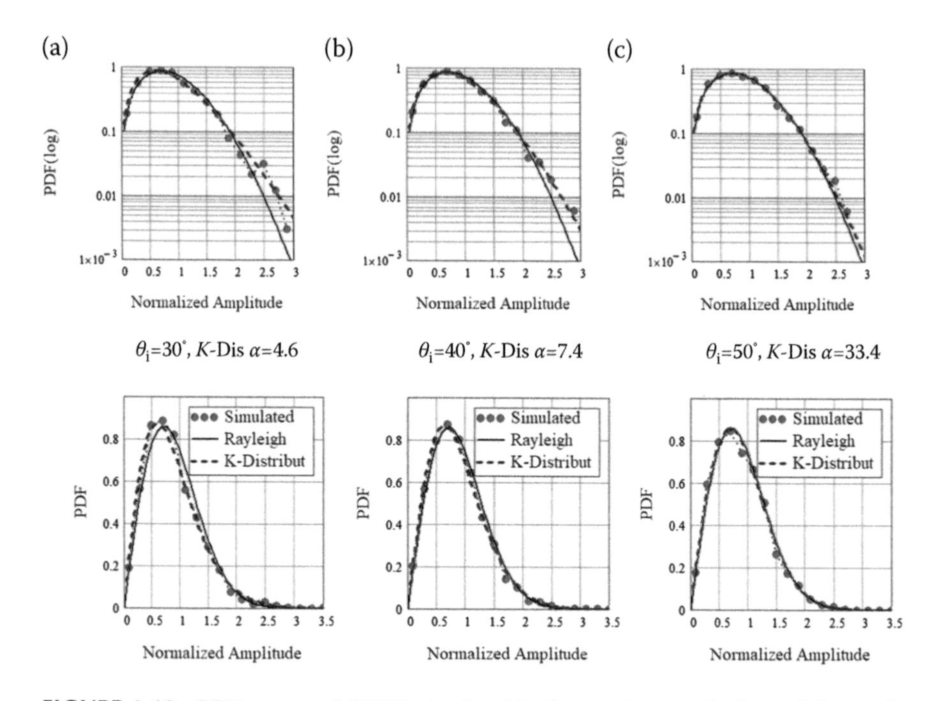

FIGURE 9.13 PDF curves of FDTD-simulated backscattering amplitude at different θ_i, at the footprint size of 30 mm, VV, compared with referencing Rayleigh distribution and K-distribution, 1600 realizations.

TABLE 9.6
Comparisons of the Average Intensity Values, and Amplitude Speckle Values from Simulated Backscattering Results by 1600 Realizations, at the Footprint Size of 30 mm

θ_i	30°	40°	50°
AIEM σ^0 (dB)	−6.8	−8.6	−10.4
Average σ^0 (dB)	−7.3	−9.0	−10.2
α for K-dis	4.6	7.4	33.4
Computed γ_a	0.560	0.547	0.531

$$N = \frac{A_e/\pi \ell^2}{\left[-\ln\left(1-\frac{t}{4k_z^2\sigma^2}\right)\right]^{2/n}} = N\left(\frac{A_e}{\pi \ell^2}, k_z\sigma, n\right) \tag{9.10}$$

TABLE 9.7
Comparison of Amplitude γ_a in the Case of Imaging Resolution Cell/Footprint Size of 30 mm, at Different θ_i, from Measured Images and Simulated Backscattering

Method \ θ_i	30°	40°	50°
Imaging Measurement	0.598	0.576	0.571
Numerical Simulation	0.560	0.547	0.531
Fully Developed speckle	0.5227	0.5227	0.5227

where $k_z = k \cos\theta_i$, l is the correlation length; A_e is the area of a resolution cell (not image pixel, and n is 1 for the exponential correlation, 2 for the Gaussian correlation function, with $t = 1$ [9].

It should be noted that the α of the overall K-distribution is also proportional to α of the sub-scatters (α_s), and a relationship can be concluded as $\alpha = N*\alpha_s$. It is also important that in the rough surface scattering, the α_s namely the K-distribution parameters for the "independent scatter" may vary according to surface roughness parameters and incident angle θ_i. It is possible that in the radar image speckle from rough surfaces if there is only one equivalent scatter in a resolution cell, but the scatter is with Gaussian statistics (α_s is very large), then the overall image speckle properties acts as fully developed. On the other hand, if there is a sufficiently large N so that $N*\alpha_s$ is sufficiently large, then the overall image speckle properties also act nearly as fully developed.

The experimental results demonstrate that larger resolution cell size tends to be more toward Rayleigh distribution, as expected. Physically, when the resolution cell size becomes larger, the equivalent scatterer number N gets larger while α_s keep unchanged; therefore, the speckle distribution follows the Rayleigh model. We focus on the scattering from an exponentially correlated rough surface, especially when the resolution is at the level of correlation length. This is a case that will be encountered in the SAR observations of the land surface, as the resolution performance keeps evolving. On the other hand, since the correlation length (l) of the sea surface is much larger so that the SAR imaging resolution is already at the level of l, the non-Rayleigh speckle phenomenon in sea speckle has been widely noticed and studied for decades [10,11,20–23].

Specifically, the K-distribution has also been developed for modeling sea speckles as a compound PDF in [10]. In that derivation, the radar speckle is viewed as a multiplicative process, where the random process y has Rayleigh PDF, with its power modulated by the process x which is assumed to follow the Gamma distribution. The above mathematical description goes along with the physical understanding of a two-scale model for the rough sea surface scattering at intermediate incidence angles [11]. In this case, the backscatter can be regarded as a collection of the small/middle scale returns from the local surfaces such as the Bragg scattering, while the strength of those returns is modulated by the long scale undulation of the surface through tilting and other mechanisms.

Apparently, both the scattering from short scale roughness and long scale roughness modulation are functional for the total scattering. Actually, one can find the similarity of the scattering process from exponentially correlated rough surface to that from the sea surface, especially when the resolution cell size is close to the correlation length. In this case, the exponentially correlated rough surface contains rich high-frequency roughness leading to scattering sources all over the surface as the scattering from short-scale roughness, meanwhile, the undulations at the level of correlation length (long scale roughness) provide with the relatively long scale modulation carrier. The difference in the mechanisms between the scattering from an exponentially correlated rough surface observed in high resolution, to that from sea surface, is the lacking of middle scale Bragg scattering. Because on exponentially correlated surface there is no wind-driven periodic undulated structure.

From both the experimental and numerical results at the very high-resolution level, it is interesting to note that when the incident angle θ_i gets larger from 30° to 50° (moderate region), the speckle is approaching towards the fully developed Rayleigh description. The answer to this phenomenon, however, can also be found from the knowledge in scattering mechanisms from sea surface [24,25]. In [25], the scattering mechanisms were discussed in the aspects of roughness scales for the two-scale or more precisely multi-scale sea rough surface. Specifically, the wavelength filtering effect states that when the incident angle becomes larger, the dominating factor shifts from the long-scale roughness towards the short-scale roughness [24,25]. And for the vertical polarization, this effect is more notable than that of horizontal polarization [25]. If the effect of long-scale roughness is weakened enough due to the enlarging of θ_i in the moderate region, then the multiplicative process is weakened because that the scattering from short-scale is taking the dominance of the scattering mechanism, as well as the speckle properties. And the scattering from short scale roughness most possibly follows a Rayleigh PDF. It is interesting that, from the results of simulated sea backscattering in [22], one can find that the speckle PDF results are also approaching toward Rayleigh when θ_i get larger in the moderate region.

From the sea surface scattering, the observed very high-resolution speckle variation from an exponentially correlated rough surface in the moderate θ_i region, can be clearly explained: the multiplicative effect of scattering process is weaker when the θ_i get larger from small to moderate, as the effects ofcarrier long-scale roughnessgets weaker. In this case, because the scattering from short scale roughness gets more dominating as the θ_i get larger and itself acts as Gaussian, the overall speckle distribution approaches towards the Rayleigh model.

The widely applied *K*-distribution in modeling non-Rayleigh speckle can be either described based on the concepts of equivalent number of (independent) scatter N [9,20] and the two-scale scattering of multiplicative process [10,11]. To further explore the very high-resolution speckle properties from exponentially correlated rough surface as a significant description for ground roughness, and to discuss the dominating factors, another set of computations are performed for analysis. Specifically, different RMS heights σ are considered: 2 mm (0.21λ), 4 mm (0.43λ), 8 mm (0.85λ), 12 mm (1.28λ@32 GHz) at $\theta_i = 30°$. For each RMS height σ, a total of 1600 realizations are computed for the speckle analysis. In Figure 9.14, the electric fields at 32 GHz in the incident plane cut of the computation domain in one specific realization were presented, as an intuitive exhibition for the difference of scattering process in case of different RMS heights. It seems that when σ = 4 mm, occasional specular reflection may contribute to the backscattering of θ_i = 30°. As shown in the scattering coefficient results of Figure 9.15, the backscattering at σ = 4 mm is larger than those at σ = 2 mm, σ = 8 mm, and σ = 12 mm. In Figure 9.15, it is also interesting to observe that as the σ changes from 2 to 12 mm, the dominant bistatic scattering region gradually moves from the forward region (σ = 2 mm) to the backward region (σ = 12 mm). Also, the averaged computed backscattering coefficients are listed in Table 9.5, and a good agreement with the AIEM results can be observed.

The computed amplitude speckle results are then presented in Figure 9.16 and Table 9.8, showing that: at the lowest and highest considered RMS height values, the backscattering speckle amplitude PDF is very close to the Rayleigh distribution; meanwhile at the moderate σ of 4 mm, the PDF is away from the Rayleigh one. The observed trends of speckle properties are interesting that it can be directly explained by neither the concept of an equivalent number of scatterers N(see Equation 9.10), nor two-scale scattering multiplicative process as the scattering scale factor. The equivalent number of scatterers theory predicts that the N varies from several, to tens and then to thousands, as the RMS height changes from 2 mm to 12 mm, at the resolution level of 30 mm and θ_i = 30°. Keeping in mind that the overall α is proportional to both the N and the statistical property of each independent scatters (α_s), in the theoretic framework of an equivalent number of scatterers $N\alpha_e$ ($N^*\alpha_s$) is the statistical property of an equivalent number of scatterers, and α_s is assumed to be 0.1. The rapid change of the estimated α_s shown in the last line of Table 9.8 as the RMS heights vary from 2 mm to 4 mm, is hard to predict when using the equivalent number of scatters theory if one does not consider full scattering

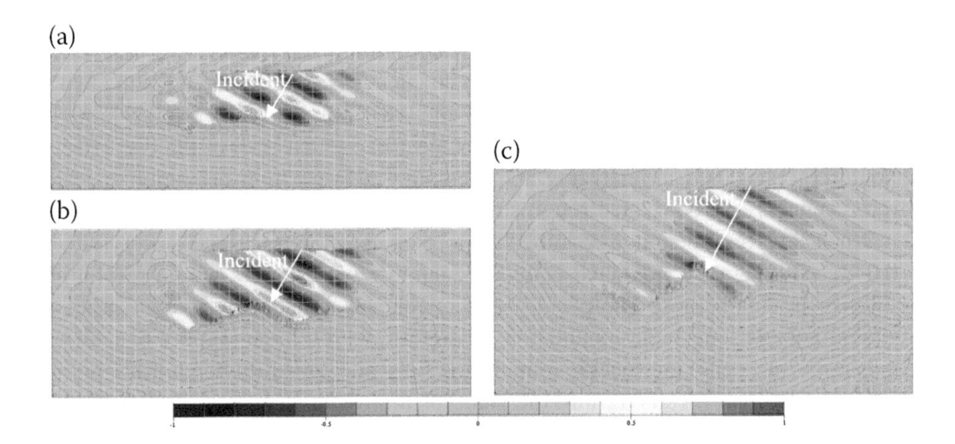

FIGURE 9.14 Recorded total field (E-Field, maximum normalized, real part) in the incident plane cut from the exponentially correlated rough surface simulation, θ_i = 30°. (a) σ = 2 mm, (b) σ = 4 mm, (c) σ = 8 mm.

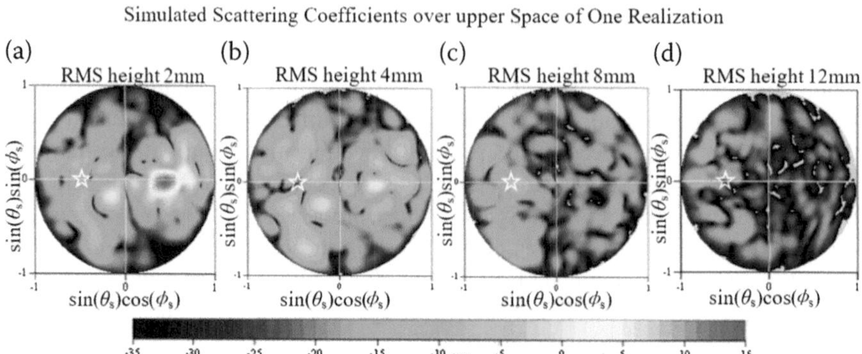

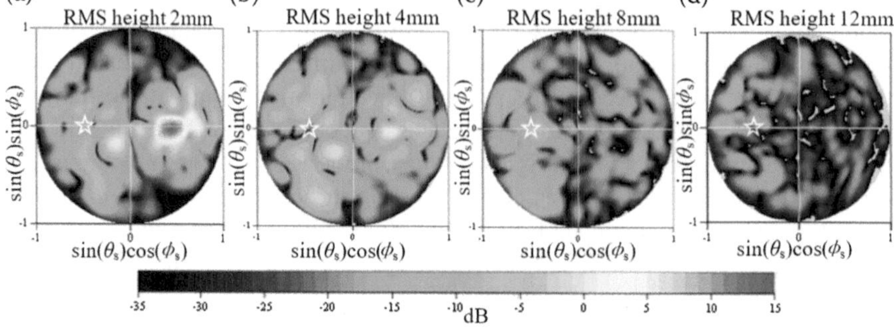

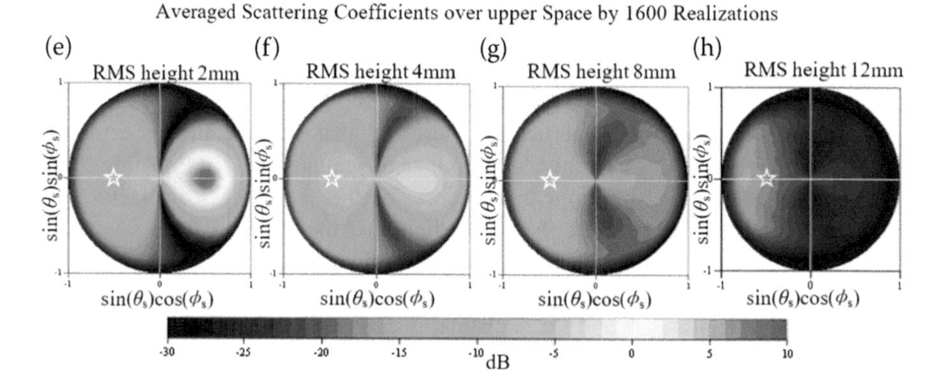

FIGURE 9.15 Results of simulated bistatic scattering coefficients over the upper space, considering different RMS height (σ). The backscattering location is marked by the white star symbol.

mechanisms. It is understood that when the RMS height increases, the long scale undulation is stronger, and thus the two-scale multiplicative scattering process is enhanced. However, it is not clear how the counted α keeps gets larger as the RMS height increases from 4 mm to 12 mm (Table 9.8), simply from a two-scale multiplicative scattering process. Actually, it is most likely that both of the two effects should be considered in analyzing and modeling the very high-resolution speckle from exponentially correlated rough surface.

When the RMS height is 2 mm, the slope of the rough surface is very small (0.042). Although the predicted equivalent number of scatterers within resolution cell size is small, those short-scale scatters remain a random Gaussian process without notable long-scale modulation. Therefore, with a large number of realizations (or return cells) the speckle remains the fully developed Rayleigh description. When the RMS height is 12mm, the slope of the rough surface gets to 0.25, the process of long scale modulation of scattering from short-scale roughness should be notable. However, the equivalent scatterer number is too large (thousands) in this case, it is very likely that such a large amount of scatters overwhelms the modulation effect, so that the very high-resolution speckle is close to the fully developed

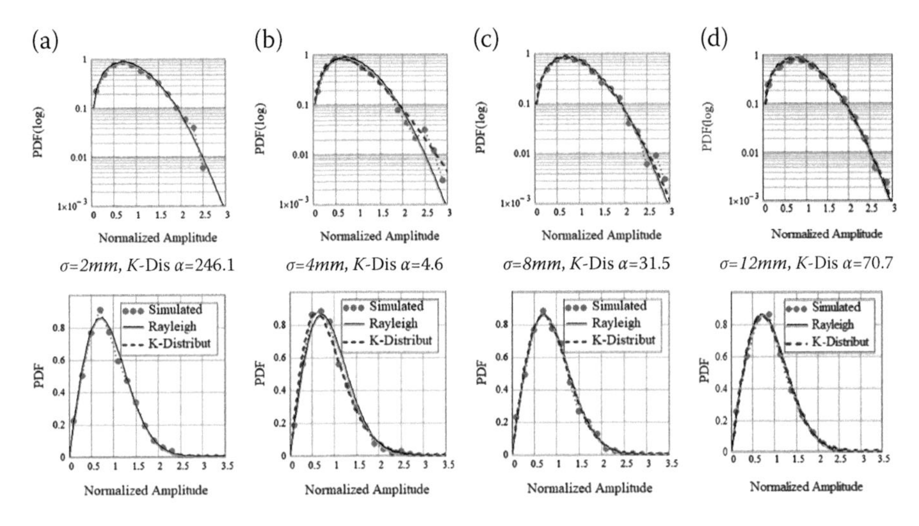

FIGURE 9.16 PDF curves of FDTD-simulated backscattering amplitude at $\theta_i = 30°$, considering exponentially correlated surface with different RMS heights σ, (a): $\sigma = 2$ mm; (b): $\sigma = 4$ mm; (c): $\sigma = 8$ mm; (d): $\sigma = 12$ mm. VV, compared with referencing Rayleigh distribution and K-distribution, 1600 realizations. (**a**) $\sigma = 2$ mm, K-Dis $\alpha = 246.1$ (**b**) $\sigma = 4$ mm, K-Dis $\alpha = 4.6$ (**c**) $\sigma = 8$ mm, K-Dis $\alpha = 31.5$ (**d**) $\sigma = 12$ mm, K-Dis $\alpha = 70.7$.

TABLE 9.8
Comparisons of the Average Intensity Values, and Amplitude Speckle Values from Simulated Backscattering Results by 1600 Realizations, at the Footprint Size of 30 mm (3.2λ, 0.63*l*)

RMS Height σ	2 mm	4 mm	8 mm	12 mm
AIEM σ^0 (dB)	−10.4	−6.8	−11.8	−18.2
Numerical σ^0 (dB)	−10.0	−7.3	−12.0	−17.7
Amplitude γ_a	0.525	0.56	0.535	0.528
α for K-dis	246.1	4.6	31.5	70.7
ENS PredictionN	2.4	43.4	719.9	3668
ENS Predicted α_e	0.24	4.34	72.0	366.8
α_s	102.5	0.106	0.04	0.019

speckle again. It is worthy to look back to the case of $\sigma = 4$ mm. One can find that in this case, the scatterer number is moderate (tens) and not large enough; then the notable two-scale modulation effect drives the speckle PDF away from the Rayleigh distribution to the K-distribution with a moderate α.

From the above discussions on the very high-resolution speckle properties from exponentially correlated rough surfaces representing ground roughness by means of experimental measurements and numerical simulations, we can summarize that both the equivalent scatterer number theory and scattering scale description of two-scale multiplicative process is informative and perspective in analyzing the experimental and numerical results of very high-resolution SAR images. However, by neither of them one can comprehensively explain the observed trends in the results. It is evident from the results that if the equivalent number of scatterers is not sufficiently large, the two-scale multiplicative scattering process may drive the VHR speckle PDF away from the fully developed Rayleigh description. Also, the dominance of the sufficient large equivalent number of scatterers is also observed, and that leads the VHR speckle properties close to the fully developed model even when the resolution cell size is smaller than the correlation length. It is confirmed that the equivalent number of scatterers may be useful in setting a bar for the very high-resolution speckle modeling on exponentially correlated rough surfaces, below which the multiplicative scattering process should be considered, and above that bar the Rayleigh model can be sufficient for modeling speckles.

REFERENCES

1. Chen, K. S., Tsang, L., Chen, K. L., Liao, T. H., and Lee, J. S., Polarimetric simulations of SAR at L-Band over bare soil using scattering matrices of random rough surfaces from numerical three-dimensional solutions of Maxwell equations, *IEEE Transactions on Geosciences and Remote Sensing*, 52(11), 7048–7058, 2014.
2. Ulaby, F. T., Moore, R. K., and Fung, A. K., *Microwave Remote Sensing: Active and Passive*, Addison-Wesley, Reading, MA, 1981.
3. Tsang, L., Kong, J. A., and Ding, K. H., *Scattering of Electromagnetic Waves: Theories and Applications*, John Wiley & Sons, New York, 2000.
4. Lee, J. S., and Pottier, E., *Polarimetric Radar Imaging: From Basics to Applications*, CRC Press: Boca Raton, FL, 2009.
5. Sarabandi, K., and Oh, Y., Effect of antenna footprint on the statistics of radar backscattering from random surfaces. *In Proceedings of the IEEE IGARSS*, Firenze, Italy, 10–14, pp. 927–929, July 1995.
6. Allain, S., Ferro-Famil, L., Pottier, E., and Fortuny, J., Influence of resolution cell size for surface parameter retrieval from polarimetric SAR data. *Proceedings of the IEEE IGARSS*, Toulouse, France, 21–25, pp. 440–442, July 2003.
7. Park, S. E., Ferro-Famil, L., Allain, S., and Pottier, E., Surface roughness and microwave surface scattering of high-resolution imaging radar, *IEEE Geoscience and Remote Sensing Letters*, 12(4), 756–760, 2015.
8. Nesti, G., Fortuny, J., and Sieber, A. J., Comparison of backscattered signal statistics as derived from indoor scatterometric and SAR experiments. *IEEE Transactions on Geoscience and Remote Sensing*, 34(5), 1074–1083, 1996.
9. Di Martino, G., Iodice, A., Riccio, D., and Ruello, G., Equivalent number of scatterers for SAR speckle modeling, *IEEE Transactions on Geoscience and Remote Sensing*, 52(51), 2555–2564, 2014.
10. Ward, K. D., Tough, R. J. A., Watts, S., *Sea Clutter: Scattering, the K Distribution and Radar Performance,* IET, London, UK, 2006.
11. Valenzuela, G. R., Theories for the interaction of electromagnetic and oceanic waves—A review. *Boudary-Layer Meteorology*, 13, 61–85, 1978.
12. Whitt, M. W., Ulaby, F. T., Polatin, P., and Liepa, V. V., A general polarimetric radar calibration technique, *IEEE Transactions on Antennas and Propagation*, 39(1), 62–67, 1991.

13. Freeman, A., Van Zyl, J. J., Klein, J. D., Zebker, H. A., and Shen, Y., Calibration of Stokes and scattering matrix format polarimetric SAR data, *IEEE Transactions on Geoscience and Remote Sensing*, 30(3), 531–539, May 1992.
14. Van Zyl, J. J., Calibration of polarimetric radar images using only image parameters and trihedral corner reflector responses, *IEEE Transactions on Geoscience and Remote Sensing*, 28(3), 337–348, May 1990.
15. Sarabandi, K., Ulaby, F. T., and Dobson, M. C., AIRSAR and POLARSCAT cross-calibration using point and distributed targets, *IEEE Transactions on Geoscience and Remote Sensing*, 2(4), 18–21, August 1993.
16. Touzi, R., and Shimada, M., Polarimetric PALSAR calibration, *IEEE Transactions on Geoscience and Remote Sensing*, 47(12), 3951–3959, December 2009.
17. Rabus, B., Wehn, H., and Nolan, M. The importance of soil moisture and soil structure for insar phase and backscatter, as determined by FDTD modeling, *IEEE Transactions on Geoscience and Remote Sensing*, 48(5), 2421–2429, 2010.
18. Giannakis, I., Giannopoulos, A., and Warren, C., A realistic FDTD numerical modeling framework of ground penetrating radar for landmine detection, *IEEE Journal of Selected Topics in Applied Earth Observervations and Remote Sensing*, 9(1), 37–51, 2016.
19. Bai, M., Jin, M., Ou, N., and Miao, J., On scattering from an array of absorptive material coated cones by the PWS approach, *IEEE Transactions on Antennas Propagation*, 61(6), 3216–3224, 2013.
20. Jakeman, E., and Pusey, P. N., A model for non-Rayleigh sea echo, *IEEE Transactions on Antennas Propagation*, 24(6), 806– 814, 1976.
21. Nouguier, F., Guérin, C.-A., and Chapron, B., Scattering from nonlinear gravity waves: The 'Choppy Wave' model, *IEEE Transactions on Geoscience and Remote Sensing*, 48(12), 4184–4192, 2010.
22. Pinel, N., Chapron, B., Bourlier, C., de Beaucoudrey, N., Garello, R., and Ghaleb, A., Statistical analysis of real aperture radar field backscattered from sea surfaces under moderate winds by Monte Carlo simulations, *IEEE Transactions on Geoscience and Remote Sensing*, 52(1), 6459–6470, 2014.
23. Durden, S. L., and Vesecky, J. F., A physical radar cross-section model for a wind-driven sea with swell. *IEEE Journal of Oceanic Engineering*, 10(4), 445–451, 1985.
24. Durden, S. L. and Vesecky, J. F., A numerical study of the separation wavenumber in the two-scale scattering approximation (ocean surface radar backscatter). *IEEE Transactions on Geoscience and Remote Sensing*, 28(2), 271–272, 1990.
25. Fung, A. K., *Backscattering from Multiscale Rough Surfaces with Application to Wind Scatterometry,* Artech House, Norwood, MA, 2015.

10 Advanced Topic: A Moon-Based Imaging of Earth's Surface

Earth observation from remote sensing satellites orbiting in a low Earth orbit provides a continuous stream of data that can enable a better understanding of the Earth with respect to climate change [1]. Recently, the concept of observing the Earth from the Moon-based platform was proposed [2–4]. The Moon, as the Earth's only natural satellite, is stable in periodic motion, making an onboard sensor unique in observing large-scale phenomena that are related to the Earth's environmental change [5,6]. Synthetic aperture radar (SAR), an active sensor, provides effective monitoring of the Earth with all-time observation capabilities [7,8]. A SAR placed in the lunar platform was proposed [3], in which the configurations and performance of The Moon-Based SAR system was thoroughly investigated. Also, the concept of the Moon-based Interferometric SAR (InSAR) were analyzed by Renga and Moccia [4]. Later, the performance and potential applications of the Moon-Based SAR were characterized by Moccia and Renga [5], and the scientific and technical issues in the application of lunar-based repeat track and along-track interferometry in [6]. Following this stream of development, an L-band Moon-Based SAR for monitoring large-scale phenomena related to global environmental changes was discussed [9]. These studies are focused on the performance analysis and potential applications with some assumptions, such as a regular spherical Earth, an orbicular circular lunar orbit, a fixed earth's rotational velocity, and a stationary Moon. By so doing, the SAR onboard Moon can be viewed as an inverse SAR (ISAR) or an equivalent sliding spotlight SAR [6,10].

10.1 RADAR MOON-EARTH GEOMETRY

The imaging of the SAR system depends on the relative motion between the earth target and Moon-Based SAR. In the relative motion, it is noted that the Moon revolves around the Earth with an average rotational velocity of 1023 m/s and a sidereal month of 27.32 days. The lunar orbit is elliptical, with an average semi-major axis of 384,748 km and an average eccentricity of 0.0549, which is close to a circular cycle. The lunar perigee is approximately 363,300 km, while the apogee can be up to 405,500 km. The angle between the lunar orbit plane and the Earth's equatorial plane varies from 18.3° to 28.6°, with a period of 18.6 years [11]. Under a right-handed geocentric inertial reference frame (to be defined in the following section), with the Z-axis towards the North Pole and the X-axis pointing to the true equinox of the date, the radar Moon-Earth geometry is shown in Figure 10.1, where the lunar ascension and declination are designated by (a_m, δ_m) and the longitude and latitude of the ground target by (a_g, δ_g). R_c is the slant range at the beam center crossing time and ψ is the grazing angle which determines the spatial coverage of the Moon-based SAR at a specified time. When the ascending node of the lunar orbit points to the true equinox of the date, the lunar declination reaches its maximum value under the same reference frame.

10.1.1 Time and Space Coordinates

To define the Moon-based SAR's imaging geometry as shown in Figure 10.1, it is critical to obtain the accurate position vectors of the ground target and SAR system under a spatial reference system in a specified time coordinate [12,13]. The Barycentric Dynamical Time (TBD) is used for providing the coordinate time scale [14,15]. Then, the Moon-based SAR and ground target are mapped to the same reference frame through a series of coordinate conversions.

To uniquely describe the locations of the SAR system on the Moon's surface and target on the Earth's surface, we introduce four reference frames (see Figure 10.2):

Moon-Centered Moon-Fixed Coordinate System (MCMF): The MCMF is used for providing the location of the SAR system on the lunar surface. It is a right-handed reference frame where the *x-axis* extends from the Moon's center to the mean Earth direction. The *z-axis* points to the mean rotational pole of the Moon.

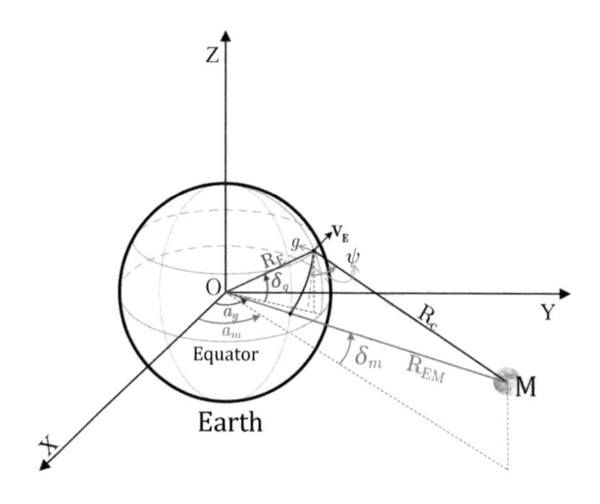

FIGURE 10.1 An unified right-handed geocentric inertial reference coordinate system for Moon-Based synthetic aperture radar (SAR), where point *g* represents the ground target, R_{EM} is the distance between the Earth and the Moon-Based SAR, which can be approximately regarded as the Earth and the Moon, and R_E is the earth radius at point *g* (*sketch not to scale*).

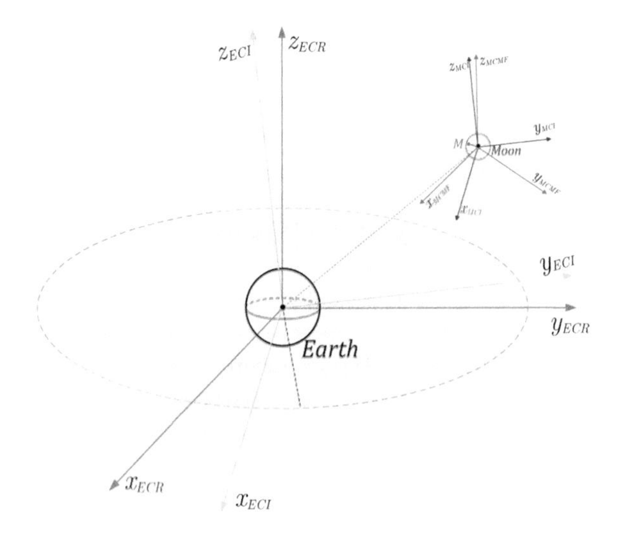

FIGURE 10.2 The definitions of the reference frames, where '*M*' represents the site of the Moon-based SAR on the lunar surface. For the sake of clarity, this diagram is not to scale.

Moon-Centered Inertial Coordinate System (MCI): The origin of the MCI is defined as the Moon's center with the *z-axis* along the direction of the celestial north and the *x-axis* toward the vernal equinox. The *y-axis* obeys a right-hand rule.

Earth-Centered Inertial Coordinate System (ECI): This is a right-handed reference frame and is served to describe the location of the Moon along its orbit. In the ECI, its origin is located at the Earth's center, with the *x-axis* pointing to the vernal equinox and the *z-axis* pointing to the celestial north.

Earth-Centered Rotational Coordinate System (ECR): The ECR is a right-handed reference frame; it is also known as the conventional terrestrial system. The origin of the ECR is Earth's center. The *x-axis* is directed toward the Greenwich Meridian. The *z-axis* points to the North Pole of the Earth.

The position vector of the SAR system on the Moon's surface in the ECI can be determined by [5]

$$\mathbf{R}_{\mathrm{SAR}}^{\mathrm{ECI}} = \mathbf{M}_{\mathrm{MCI}}^{\mathrm{MCMF}} \mathbf{R}_{\mathrm{SAR}}^{\mathrm{MCMF}} + \mathbf{D}_{\mathrm{EM}}^{\mathrm{ECI}}, \tag{10.1}$$

where $\mathbf{R}_{\mathrm{SAR}}^{\mathrm{MCMF}}$ and $\mathbf{D}_{\mathrm{EM}}^{\mathrm{ECI}}$ are the position vector of the SAR system in the MCMF and that of the Moon's center in the ECI, respectively. Generally, $\mathbf{D}_{\mathrm{EM}}^{\mathrm{ECI}}$ can be obtained from the lunar ephemeris. $\mathbf{M}_{\mathrm{MCI}}^{\mathrm{MCMF}}$, the coordinate transformation matrix from the MCMF to MCI, is defined as [16]

$$\mathbf{M}_{\mathrm{MCI}}^{\mathrm{MCMF}} = \mathbf{R}_z(-\phi_m)\mathbf{R}_x(-\theta_m)\mathbf{R}_z(-\psi_m), \tag{10.2}$$

where ϕ_m, θ_m, and ψ_m are three auxiliary angles assigned to describe the libration of the Moon [17]. $\mathbf{R}_x$, $\mathbf{R}_y$, and $\mathbf{R}_z$ are rotation matrices about the x-, y-, and z-axes, respectively.

Then, the position vector of the Moon-based SAR in the ECR is acquired through the following transformation [18,19]

$$\mathbf{R}_{\text{SAR}}^{\text{ECR}} = \mathbf{U}_{\text{ECR}}^{\text{ECI}} \mathbf{R}_{\text{SAR}}^{\text{ECI}}, \tag{10.3}$$

with the transformation matrix expressed as

$$\mathbf{U}_{\text{ECR}}^{\text{ECI}} = \Pi\Theta\mathbf{NP}. \tag{10.4}$$

where matrices $\mathbf{P}$, $\mathbf{N}$, Θ, and Π represent coordinate changes due to the precession, nutation, earth rotation, and polar motion, respectively. The transformations from the ECI to ECR are given next.

10.1.2 Transformations from the ECI to ECR

The transformation from the ECI to ECR is achieved by Equation 10.4, where the matrix $\mathbf{P}$ represents the coordinate change because of the precession and is given by [12]:

$$\mathbf{P} = \mathbf{R}_z(-z)\mathbf{R}_y(\vartheta)\mathbf{R}_z(-\xi), \tag{10.5}$$

where ξ, ϑ, and z are all the auxiliary angles that describe precession in units of seconds. They are expressed as

$$\begin{cases} \xi = 2306''.2182T + 0''.30188T^2 + 0''.017998T^3, \\ \vartheta = 2004''.3109T + 0''.42665T^2 + 0''.014183T^3, \\ z = 2306''.2182T + 1''.09468T^2 + 0''.018203T^3. \end{cases} \tag{10.6}$$

where T is the Barycentric Dynamical Time (TDB) in Julian centuries from J2000.0 $\mathbf{R}_x$, $\mathbf{R}_y$, and $\mathbf{R}_z$, are rotation matrices about the x-, y-, and z-axes, respectively, explicitly given by

$$\begin{aligned} \mathbf{R}_x(\theta) &= \begin{bmatrix} 1 & 0 & 0 \\ 0 & \cos\theta & \sin\theta \\ 0 & -\sin\theta & \cos\theta \end{bmatrix}, \\ \mathbf{R}_y(\theta) &= \begin{bmatrix} \cos\theta & 0 & -\sin\theta \\ 0 & 1 & 0 \\ \sin\theta & 0 & \cos\theta \end{bmatrix}, \\ \mathbf{R}_z(\theta) &= \begin{bmatrix} \cos\theta & \sin\theta & 0 \\ -\sin\theta & \cos\theta & 0 \\ 0 & 0 & 1 \end{bmatrix}. \end{aligned} \tag{10.7}$$

The transformation matrix from the mean equator and equinox of the Earth to the true equator and equinox of the Earth can be written as

$$\mathbf{N} = \mathbf{R}_x(-\varepsilon - \Delta\varepsilon)\mathbf{R}_z(-\Delta\psi)\mathbf{R}_x(\varepsilon), \tag{10.8}$$

where ε is the mean obliquity of the ecliptic, it can be expressed as follows

$$\varepsilon = 84381''.448 - 46''.815T - 0''.00059T^2 + 0''.001813T^3, \tag{10.9}$$

$\Delta\varepsilon$ and $\Delta\psi$ in Equation 10.8 are the nutation in longitude and obliquity, respectively, and are, keeping the consistency with the DE430, given by

$$\begin{cases} \Delta\psi = -17''.206262 \sin \Omega, \\ \Delta\varepsilon = 9''.205348 \cos \Omega. \end{cases} \tag{10.10}$$

where Ω is the mean longitude of the ascending node of the lunar orbit measured on the ecliptic plane from the mean equinox of date, it is

$$\begin{aligned} \Omega = {} & 125°02'40''.280 \\ & -1934°08'10''.549T + 7''.445T^2 + 0''.008T^3. \end{aligned} \tag{10.11}$$

The matrix Θ is related to the Earth's self-rotation, which takes the form

$$\Theta = \mathbf{R}_z(\text{GAST}), \tag{10.12}$$

with

$$\text{GAST} = \text{GMST} + \Delta\psi \cos \varepsilon. \tag{10.13}$$

where GAST represents the Greenwich Apparent Sidereal Time, GMST is the Greenwich Mean Sidereal Time, which can be obtained by

$$\begin{aligned} \text{GMST} = {} & 24110^s.54841 + 8640184^s.812866 \cdot T_0 \\ & + 1.002737909350795 \cdot UT1 \\ & + 0^s.093104 \cdot T_u^2 - 0^s.0000062 \cdot T_u^3, \end{aligned} \tag{10.14}$$

where T_0 and T_u are given by

$$\begin{cases} T_0 = [JD(0^h UT1) - 2451545]/36525, \\ T_u = [JD(UT1) - 2451545]/36525. \end{cases} \tag{10.15}$$

The time unit of T_0 and T_u are Universal Time1 (UT1), which is another time scale [12].

Finally, the matrix related to the polar motion of the Earth is given by

$$\mathbf{\Pi} = \mathbf{R}_y(\Phi_y)\mathbf{R}_x(\Phi_x), \tag{10.16}$$

where Φ_x and Φ_y are computed with an estimated linear correction in the DE430, as

$$\begin{cases} \Phi_x = \Phi_{x0} + 100T \cdot d\Phi_x/dt, \\ \Phi_y = \Phi_{y0} + 100T \cdot d\Phi_y/dt. \end{cases} \tag{10.17}$$

where Φ_{x0} and Φ_{y0} are the rotations of the x-axis and y-axis at J2000.0, respectively. $d\Phi_x/dt$ and $d\Phi_x/dt$ are the linear corrections rates of Φ_x and Φ_y, respectively. They are

$$\begin{cases} \Phi_{x0} = 0''.005675420332289347, \\ \Phi_{y0} = -0''.01702265691498953, \\ d\Phi_{x0}/dt = 0''.0002768991557448355\ yr^{-1}, \\ d\Phi_{y0}/dt = -0''.001211859121655924\ yr^{-1}. \end{cases} \tag{10.18}$$

10.2 SPATIOTEMPORAL COVERAGE

Once the coordinate transformation is accomplished, the distance from the Earth to the Moon-based SAR (see Figure 10.1), nadir point's latitude and longitude can be readily obtained, respectively, by

$$R_{SAR} = \sqrt{x_m^2 + y_m^2 + z_m^2}, \tag{10.19}$$

$$\delta_m = \tan^{-1}[z_m/(x_m^2 + y_m^2)^{0.5}], \tag{10.20}$$

$$a_m = \begin{cases} \tan^{-1}(y_m/x_m), & x \geq 0 \\ \tan^{-1}(y_m/x_m) + \pi, & x < 0,\ y \leq 0 \\ \tan^{-1}(y_m/x_m) - \pi, & x < 0,\ y > 0 \end{cases} \tag{10.21}$$

where (x_m, y_m, z_m) are the coordinates of the Moon-based SAR in the ECR.

10.2.1 GEOMETRIC PARAMETERS

Since a specific site of the Moon-based SAR is yet to be determined, to proceed with analyzing the coverage on the Earth's surface, the position of the Moon-based SAR is assumed to be the same as that of the Moon because of the extremely long distance from the Earth to the Moon. The relevant parameters for investigating the coverage are listed in Table 10.1.

TABLE 10.1
Parameters of the Moon-Based SAR in the ECR

Symbol	Parameters	Symbol	Parameters
M	Site of Moon-based SAR	B	Nadir point
g	Ground target	R_E	Earth's radius at g
a_g	Earth's longitude at g	a_m	Nadir point's longitude
δ_g	Earth's latitude at g	$\delta_m + \tan^{-1}\left(\frac{R_M}{R_{SAR}}\right)$	Nadir point's latitude
R_{SAR}	Earth–Moon-based SAR distance	R_c	Slant range of the Moon-based SAR
$a_m - \tan^{-1}\left(\frac{R_M}{R_{SAR}}\right)$	Linear velocity of the Earth's rotation at g	ψ	Grazing angle
θ_e	Geocentric angle	θ_{el}	Elevation angle
ϕ	Azimuthal angle	$\delta_m - \tan^{-1}\left(\frac{R_M}{R_{SAR}}\right)$	Longitudinal deviation

The geocentric angle, defined as the angle between the nadir point of the Moon-based SAR and the ground target referring to the Earth's center, is

$$\theta_e = \cos^{-1}(\cos\delta_m \cos\delta_g \cos\alpha + \sin\delta_m \sin\delta_g) \tag{10.22}$$

where α is the longitudinal deviation and is defined as

$$\alpha = a_g - a_m. \tag{10.23}$$

The elevation angle that is the angle between the Moon-based SAR's position vector and slant range vector is specified by

$$\theta_{el} = \cos^{-1}\{[R_{SAR} - R_E(\cos\delta_m \cos\delta_g \cos\alpha + \sin\delta_m \sin\delta_g)]/R_c\}. \tag{10.24}$$

The grazing angle, the angle between the slant range and ground range plane, can be expressed as

$$\psi = \sin^{-1}\{[R_{SAR}(\cos\delta_m \cos\delta_g \cos\alpha + \sin\delta_m \sin\delta_g) - R_E]/R_c\}. \tag{10.25}$$

The azimuthal angle, the angle between the slant range vector in the ground plane and velocity vector of the Earth's self-rotation (eastward), takes the form

$$\phi = \cos^{-1}\{\cos\delta_m sin\alpha[1 - (\cos\delta_g \cos\delta_m \cos\alpha + \sin\delta_g \sin\delta_m)^2]^{-0.5}\}. \tag{10.26}$$

It should be noted that the direction of Earth's self-rotation is used as a reference, represented by an azimuthal angle of 0°.

The slant range R_c of the Moon-based SAR, is given by

$$R_c = \sqrt{R_{SAR}^2 + R_E^2 - 2R_{SAR}R_E(\cos\delta_m \cos\delta_g \cos\alpha + \sin\delta_m \sin\delta_g)}. \quad (10.27)$$

Detailed derivations of Equations 10.22–10.25 are given in [19].

The geocentric and azimuthal angles, together with the grazing angle or elevation angle, determine the spatial coverage of a Moon-based SAR. Here, we choose the geocentric, azimuthal, and grazing angles to examine the Moon-based SAR's spatial coverage.

From Figure 10.3, the grazing angle should be bounded by

$$\psi_{far} \le \psi \le \psi_{near}. \quad (10.28)$$

where ψ_{far} and ψ_{near} are the lower and upper bounds of the grazing angle, respectively.

Substituting Equation 10.25 into Equation 10.22 and imposing Equation 10.28, Equation 10.22 becomes

$$\cos\theta_e(\delta_m, \delta_g, \alpha) = \sin\psi \cdot R_c/R_{SAR} + R_E/R_{SAR}. \quad (10.29)$$

As far as the Moon-based SAR's spatial coverage is concerned, the slant range in Equation 10.27 can be further approximated to

$$R_{ca} \approx R_{SAR}\{1 + [0.5R_E^2 - R_{SAR}R_E \cos\theta_e(\delta_m, \delta_g, \alpha)]/R_{SAR}^2\}. \quad (10.30)$$

To confirm the validity of above equation in the context Moon-based SAR's spatial coverage, a relative error is defined as

$$\delta R = (R_c - R_{ca})/R_{SAR}. \quad (10.31)$$

Figure 10.4 presents the relative errors at perigee and apogee of the lunar orbit in the $\alpha - \delta_g$ domain. For the sake of simplicity but without loss of generality, the latitude of the Moon-based SAR's nadir point is set to 0°, and the grazing angle is ranged from 0° to 90°. As detailed in Figure 10.4, the maximum value of the relative error δR is far smaller than 10^{-3}, which appears at the nadir point regardless of the distance from the Earth to the Moon-based SAR. This suggests that the approximation of the slant range in Equation 10.30 has little bearing on the dimensionless quantity, δR. In other words, the error induced by the approximation of the slant range can be reasonably ignored in determining the spatial coverage of a Moon-based SAR. However, we must compensate for this error given the rise in the image focusing. Accordingly, Equation 10.29 can be approximated by taking account of Equation 10.30 to

$$\cos\theta_e(\delta_m, \delta_g, \alpha) \approx \kappa(\psi, R_{SAR}), \quad (10.32)$$

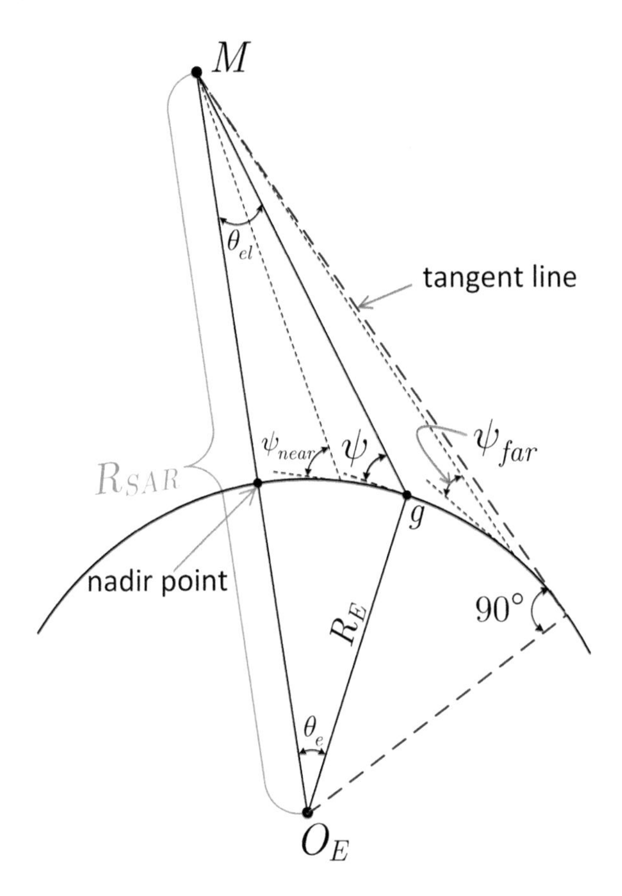

FIGURE 10.3 Illustration of ψ_{far} and ψ_{near}, the lower and upper bounds of the grazing angle, respectively (sketch not to scale).

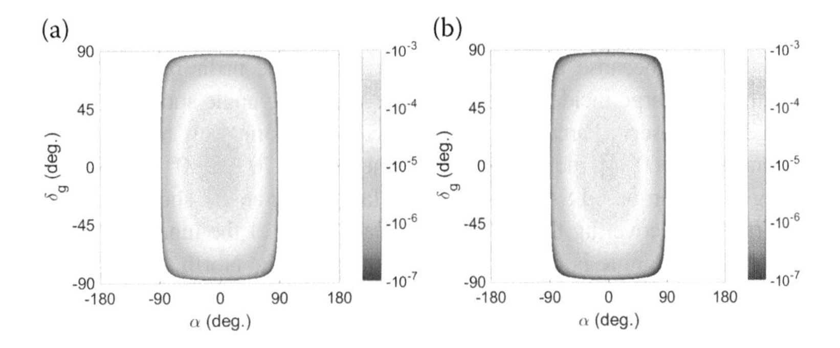

FIGURE 10.4 The relative errors δR in the $\alpha - \delta_g$ domain at (a) perigee of the lunar orbit, (b) apogee of the lunar orbit.

with

$$\kappa(\psi, R_{SAR}) = \frac{\sin\psi + 0.5R_E/R_{SAR}}{\sin\psi \times R_E/R_{SAR} + 1} + 0.5R_E/R_{SAR}. \tag{10.33}$$

Note that the ratio of Earth's radius to Earth–Moon-based SAR distance is spatiotemporally varying as follows

$$R_E/R_{SAR} \approx R_E^A/R_{SAR}^A \times (1 - \Delta R_{SAR}^T/R_{SAR}^A + \Delta R_E^S/R_E^A). \tag{10.34}$$

where ΔR_E^S is the spatially varying component of the Earth's radius at an arbitrary position on the Earth's surface for the average Earth's radius, R_E^A; ΔR_{SAR}^T is the temporally varying part of the Earth–Moon-based SAR distance at any time referring to the average distance from the Earth to the Moon-based SAR, R_{SAR}^A. It can be calculated that $\Delta R_{SAR}^T/R_{SAR}^A$ is approximately 30 times larger than $\Delta R_E^S/R_E^A$ during one period of the lunar revolution. Thus, the spatially varying component of the Earth's radius contributes little to Equation 10.34 (i.e., the ratio of Earth's radius to Earth–Moon-based SAR distance). Hence, it is reasonably to accept a constant Earth's radius (6371.0 km) here.

10.2.2 Effective Range

Generally, a single antenna can observe a limited portion of the Earth's surface in a given time frame. However, multi-antennas or antenna array can achieve the maximum extent of the spatial coverage by adjusting the pointing direction of each antenna. For example, the half-power beam width of a single antenna is [37]

$$\beta \approx k_{xy}\lambda/\ell, \tag{10.35}$$

where λ is the wavelength, ℓ is the aperture length, k_{xy} is the aperture-illumination taper factor for the antenna. Generally, a value of k_{xy} =1.5 is appropriate for an antenna with high beam efficiency. The spatial coverage of the Moon-based SAR is subject to the imaging geometry because the SAR system cannot have nadir imaging and it may perform poorly at large incident angles. Thus, the grazing angle should be limited to a valid range: The observed areas of the satellite SAR are commonly restricted within the range from 20° to 60° of incident angles, namely from 70° to 30° of grazing angles. Regarding the Moon-based SAR, the near and far grazing angles should be limited to a practical range for covering the globe within each cycle of the lunar revolution.

The far grazing angle is determined such that both the North and South Poles of the Earth are observable during one lunar revolution. By letting $|\delta_g| = 90°$, the following relationship is readily obtained

$$\psi_{far} \leq \sin^{-1}\left\{\frac{\sin|\delta_m| - R_E/R_{SAR}}{(1 + R_E^2/R_{SAR}^2 - 2\sin|\delta_m|\cdot R_E/R_{SAR})^{0.5}}\right\}. \tag{10.36}$$

Once the far grazing angle exceeds its upper bound, neither the North Pole nor the South Pole is covered by the Moon-based SAR wherever the nadir point is located. It is noted that even when the far grazing angle is bounded, the North Pole (or the South Pole) is not always observed by the Moon-based SAR. The observation of the North Pole (or the South Pole) is still subjected to the nadir point's latitude, as will be evidenced in the next sub-section.

Similarly, the near grazing angle should satisfy the following relationship to cover the equatorial regions:

$$\psi_{near} \geq \sin^{-1}\left\{\frac{R_E/R_{SAR} - \cos\delta_m}{\cos\delta_m \cdot R_E/R_{SAR} - 0.5 \cdot R_E^2/R_{SAR}^2 - 1}\right\}. \quad (10.37)$$

Note that only when the near grazing angle is larger than the lower bound can the equatorial regions of the Earth be observed during each cycle of the lunar revolution.

The inclination of the lunar orbit to the Earth's equatorial plane stands at between 18.3° and 28.6° [20]; in other words, the maximum scale of the nadir point's latitude varies from 18.3° to 28.6°. In Figure 10.5(a), we plot the bounds of the far grazing angle at apogee and perigee of the lunar orbit versus nadir point's latitude. Then, bounds of the near grazing angle at apogee and perigee of the lunar orbit for the magnitude of the nadir point's latitude are plotted in Figure 10.5(b). The range of latitude for the nadir point of the Moon-based SAR is 18.3° to 28.6°.

Even though the spatial coverage is bounded by the near and far grazing angles, the image quality of the Moon-based SAR is still an issue. The Moon-based SAR system usually works in the squint mode when observing the bounded regions, which would lead to pixel skewing on the ground and raise a significant challenge to the image quality. Under the skewing effect, satisfactory imaging is only available when the

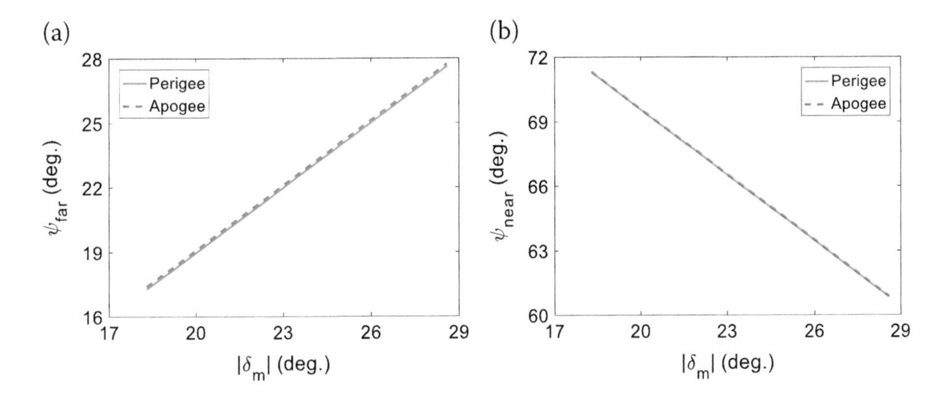

FIGURE 10.5 With the Moon located at the apogee and the perigee of the lunar orbit, (a) shows the upper bound of far grazing angle for covering the Earth's Poles (Neither the North Pole nor the South Pole is observed when the far grazing angle is beyond the lines), and (b) shows the lower bound of near grazing angle for covering the equatorial regions; the equatorial regions will not be observed when the near grazing angle is below the lines.

iso-range and iso-Doppler profiles are relatively far from being parallel. Generally, the quasi-parallel regions of the iso-range and iso-Doppler contours occur at where the azimuthal angles are close to 0° or 180°. Therefore, it is practical to examine the spatial coverage of the Moon-based SAR by limiting the azimuthal angle, namely

$$|\cos \phi| \leq \cos \phi_b, \tag{10.38}$$

where ϕ_b is the bound of the azimuthal angle; here we should set it to 30° to ensure the good image quality.

In Figures 10.6(a–d) we plot the iso-range and iso-Doppler profiles in the $\alpha - \delta_g$ domain within the boundary set by Equation 10.38 with the Moon-based SAR locating at different positions listed in Table 10.2. The beam center crossing time is set to 00:00:00 without loss of generality.

From Figure 10.6, the iso-Doppler and iso-range contours cut across each other wherever the Moon-based SAR is located. This suggests that the image quality is acceptable even when the Moon-based SAR system operates in a squint mode. Thus, the skewing effect exerts a little impact on the spatial coverage of the Moon-based SAR within the scope from Equation 10.38.

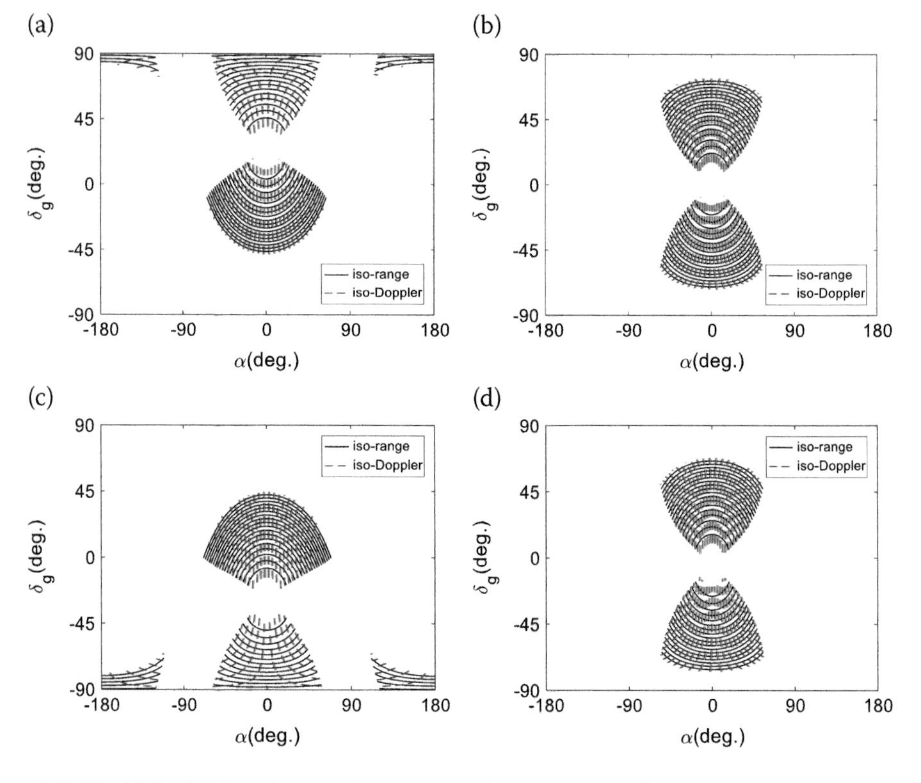

FIGURE 10.6 Iso-Doppler and iso-range profiles in the $\alpha - \delta_g$ domain at 00:00:00 on (a) March 20, 2024, (b) March 25, 2024, (c) April 01, 2024, and (d) April 07, 2024.

TABLE 10.2
Coordinates of the Moon-Based SAR in the ECR at Different Times

Coordinates Time	X(km)	Y(km)	Z(km)
Mar. 20, 2024, 00:00:00	221041.4734	−289559.7324	167171.4715
Mar. 25, 2024, 00:00:00	405558.5205	−8874.6121	4569.0584
Apr. 01, 2024, 00:00:00	74979.4046	330070.1283	−183398.9317
Apr. 07, 2024, 00:00:00	−332302.4149	133331.8257	−30492.2954

*The latitudes of the nadir point on Mar.20 and Apr.01 are close to the maximum and minimum values, while latitudes of the nadir point on Mar.25 and Apr.07 are approaching 0° of the latitude.

Finally, we can obtain the bounds of the geocentric angle that determines the Moon-based SAR's spatial coverage as

$$\theta_e^{low} \leq \theta_e(\delta_m, \delta_g, \alpha) \leq \theta_e^{up}, \tag{10.39}$$

where θ_e^{up} is the upper bound of geocentric angle, defined as

$$\theta_e^{up} = \cos^{-1}[\kappa(\psi_{far}, R_{SAR})], \tag{10.40}$$

θ_e^{low}, the lower bound of geocentric angle is

$$\theta_e^{low} = \begin{cases} \cos^{-1}[\kappa(\psi_{near}, R_{SAR})], & |\alpha| \leq \alpha_{th} \\ \cos^{-1}[\cos\phi_b(\cos^2\phi_b - \cos^2\delta_m sin^2\alpha)^{0.5}], & |\alpha| \geq \alpha_{th} \end{cases} \tag{10.41}$$

where the threshold value α_{th} is given by

$$\alpha_{th} = \sin^{-1}\{\cos\phi_b \cdot \cos^{-1}\delta_m \cdot [1 - \kappa^2(\psi_{near}, R_{SAR})]^{0.5}\} \tag{10.42}$$

Note that the upper bound of geocentric angle, related to the far grazing angle, determines the exterior boundary of the spatial coverage. In comparison, the lower bound of the geocentric angle is correlated to the near grazing angle and bound of the azimuthal angle. It decides the blind region that cannot be observed by the Moon-based SAR within the scope from the upper bound of the geocentric angle.

10.2.3 Moon-based SAR's Spatial Coverage

Now, the spatial coverage of the Moon-based SAR can be examined in terms of the coverage area and ground coverage. To begin with, the coverage area of the Moon-based SAR within the bound of Equation 10.39 is detailed below.

Once the spatial coverage is limited to the range in Equation 10.39, the Moon-based SAR's coverage area on the Earth's surface can be approximated by [19]

$$\begin{aligned} S_c = 4R_E^2\{&\cos^{-1}[\Re(\phi_b)\cdot\kappa(\psi_{far}, R_{SAR})\cdot\zeta(\psi_{far}, R_{SAR})] \\ &- \cos^{-1}[\Re(\phi_b)\cdot\kappa(\psi_{far}, R_{SAR})\cdot\zeta(\psi_{near}, R_{SAR})] \\ &+ \sin^{-1}[\zeta(\psi_{far}, R_{SAR})] - \sin^{-1}[\zeta(\psi_{near}, R_{SAR})]\}, \end{aligned} \tag{10.43}$$

with

$$\begin{cases} \Re(\phi_b) = \cos\phi_b - \sin\phi_b, \\ \zeta(\psi_{far}, R_{SAR}) = [1 + \kappa^2(\psi_{far}, R_{SAR})]^{-0.5}, \\ \zeta(\psi_{near}, R_{SAR}) = [1 + \kappa^2(\psi_{near}, R_{SAR})]^{-0.5}. \end{cases} \tag{10.44}$$

Accordingly, the ratio of the Moon-based SAR's coverage area to the global area can be written as

$$\begin{aligned} r_c = 4R_E^2\{&\cos^{-1}[\Re(\phi_b)\cdot\kappa(\psi_{far}, R_{SAR})\cdot\zeta(\psi_{far}, R_{SAR})] \\ &- \cos^{-1}[\Re(\phi_b)\cdot\kappa(\psi_{far}, R_{SAR})\cdot\zeta(\psi_{near}, R_{SAR})] \\ &+ \sin^{-1}[\zeta(\psi_{far}, R_{SAR})] - \sin^{-1}[\zeta(\psi_{near}, R_{SAR})]\}/\pi \times 100\%. \end{aligned} \tag{10.45}$$

Detailed derivations of Equations 10.43 and 10.45 can be found in [19].

It follows that the coverage area on the Earth's surface is related to the ratio of the Earth's radius to the Earth–Moon-based SAR distance, the near and far grazing angles, and bound of the azimuthal angle. As the bounds of the grazing and azimuthal angles are ascertained in the preceding analysis, the coverage area of the Moon-based SAR depends only on the ratio of the Earth's radius to the distance from the Earth to the Moon-based SAR.

Given the imaging geometry of the Moon-based SAR, the maximum ground coverage's latitude occurs at 0° of the longitudinal deviation. Then, the maximum and minimum latitudes of the ground coverage can be obtained from Equations 10.39 and 10.40, as given by

$$\delta_g = \pm\cos^{-1}[\kappa(\psi_{far}, R_{EM})] + \delta_m. \tag{10.46}$$

By noting that the peak values of the Earth's latitude are ±90°, a threshold for the nadir point's latitude is defined

$$\delta_{mth} = 90° - \cos^{-1}[\kappa(\psi_{far}, R_{EM})]. \tag{10.47}$$

When the grazing angle of the Moon-based SAR is bounded by the far grazing angle, the coverages of the North and South Poles of the Earth are still pertinent to

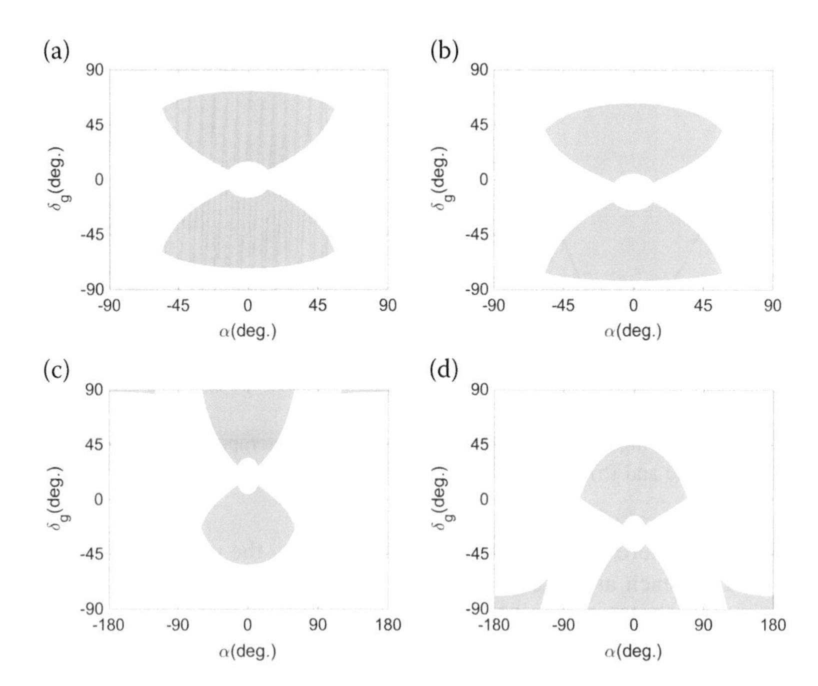

FIGURE 10.7 The Moon-based SAR's ground coverage with an Earth–Moon-based SAR distance of 385,000 km at the nadir point's latitude of (a) 0°, (b) −10°, (c) 18°, and (d) −28°.

the latitude of the nadir point, threshold value δ_{mth}, and opposite value of the threshold, $-\delta_{mth}$.

In Figure 10.7, we plot the ground coverage of the Moon-based SAR at various latitudes of the nadir point (0°, −10°, 18°, and −28°) in the $\alpha - \delta_g$ domain. The Earth–Moon-based SAR distance is set to 385,000 km, the average distance from the Earth to the Moon-based SAR; it can be calculated that the threshold δ_{mth} for this distance is to 15.916°. As illustrated in Figure 10.7, there exist blind regions within the ground coverage bounded by the far grazing angle due to the effects of the near grazing angle and bound of the azimuthal angle. Besides, the longitudinal ground coverage is rotationally symmetric about the axis of $\alpha = 0°$, whereas the latitudinal ground coverage is asymmetric except the nadir point being at the Earth's equator. Moreover, when the nadir point's latitude is smaller than the threshold of δ_{mth}, neither of the Earth's Poles is covered by the Moon-based SAR. Meanwhile, the maximum size of the longitudinal deviation is always smaller than 90°. Regarding the Moon-based SAR's nadir point whose latitude is larger than δ_{mth}, such as in Figure 10.7(c), the North Pole is covered by the Moon-based SAR. However, the high latitudes in the southern hemisphere of the Earth cannot be observed. On the contrary, if the nadir point's latitude is smaller than $-\delta_{mth}$, the South Pole can be covered by the Moon-based SAR, but the high latitudes in the

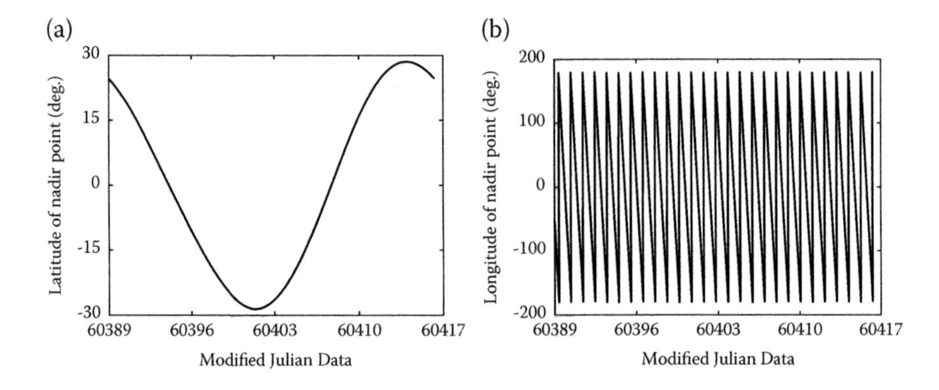

FIGURE 10.8 During one period of the lunar revolution, temporal variations in the nadir point's (a) latitude and (b) longitude.

northern hemisphere are not. Under both conditions, the maximum longitudinal deviation can reach as large as 180°.

10.2.4 Temporal Variations in the Spatial Coverage

The spatial coverage of the Moon-based SAR is determined by the Earth–Moon-based SAR distance, bounds of grazing and azimuthal angles, and position of the nadir point at a specified time. The nadir point, which temporally varies, effectively induces the temporal variation in the spatial coverage. In what follows, the geographic coordinates of the nadir point, in light of the temporal variations are discussed.

The nadir point's latitude and longitude are provided by Equations 10.20 and 10.21, both of which are a function of time and can be numerically obtained when the lunar ephemeris is applied. Now, the temporal variations in the latitude and longitude of the nadir point during one cycle of the lunar revolution are plotted in Figures 10.8(a) and (b), respectively.

We see that the latitude of the nadir point is mainly determined by the lunar revolution. By comparison, the nadir point's longitude is primarily decided by the Earth's self-rotation. Furthermore, the nadir point's longitude differs from that at the initial time after one period of the lunar revolution. This phenomenon is attributed to the Earth's self-rotation and lunar revolution around the Earth. The temporal variation of the nadir point's geographical coordinate is a continuous process. However, the temporal variation of the spatial coverage is not exactly equivalent to that of the nadir point. A time interval related to synthetic aperture, namely the synthetic aperture time, still needs to be considered in the imaging formation of the coverage region. A synthetic aperture time of 200 s can realize an azimuth resolution around 10 m, which is suitable for the Earth observation from the Moon-based SAR. Nevertheless, such a synthetic aperture time requires a large antenna with an aperture length in azimuth larger than 1000 m, which poses a significant challenge to build in practice. The sub-aperture imaging technique provides a feasible method to implement the desired synthetic aperture time without utilizing the large antenna [21]. Thus, the synthetic aperture time of the Moon-based SAR may be set to 200 s.

Figure 10.9 displays the track of the Moon-based SAR's nadir point within one week, from which we observe the daily change of the nadir point's latitude is relatively small (around 3°). The temporal variation of the longitude of the nadir point within one day can exceed 340°. Hence, the temporally varying nadir point, together with the extensive spatial coverage, leads the revisit time to be less than one day for most of the regions covered.

10.2.5 Numerical Illustration of Spatiotemporal Coverage

Based on the lunar ephemeris from the DE430, numerical simulations are performed to visualize the spatiotemporal coverage of the Moon-based SAR. We begin by discussing the hourly variations in the spatial coverage of the Moon-based SAR within one day.

10.2.5.1 Hourly Variations

Assume that the initial visit time is 00:00:00 on March 20, 2024, when the nadir point's latitude is relatively large. The hourly variations in the Moon-based SAR's spatial coverage (eight-hour time intervals) within one day are plotted in Figure 10.10, where we see that most of the northern hemisphere of the Earth is measured by the Moon-based SAR, while the southern hemisphere of the Earth is less observed. Besides, it is understood that the spatial coverage moves from east to west and the revisit time of the covered region is commonly less than one day, implying that Earth's self-rotation is a dominant factor in determining the temporal variation of the spatial coverage. Comparing Figure 10.10(d) with (a), we see that after a one-day lapse, the spatial coverage has moved to the east. This phenomenon is caused by the lunar revolution around the Earth. Moreover, it can be observed a large proportion of the Earth surfaces are viewed (e.g., South America is almost entirely covered by the Moon-based SAR at 00:00:00). Nevertheless, there are still blind regions within the coverage bounded by the far grazing angle. This suggests that the effects of the near grazing angle and bound of the azimuthal angle on the spatial coverage are significant.

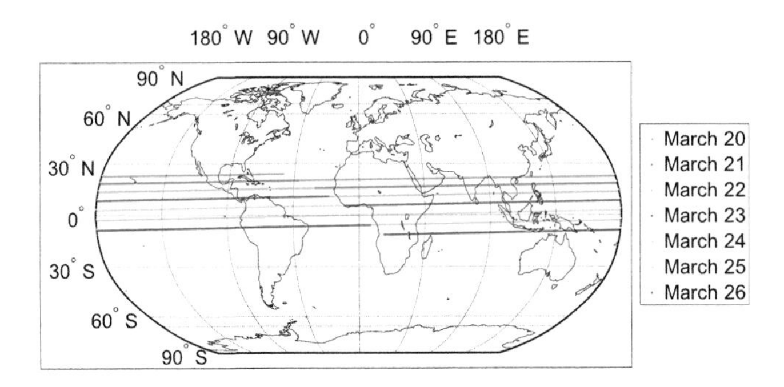

FIGURE 10.9 The track of the Moon-based SAR's nadir point from March 20, 2024 to March 26, 2024. The synthetic aperture time is set to 200 s.

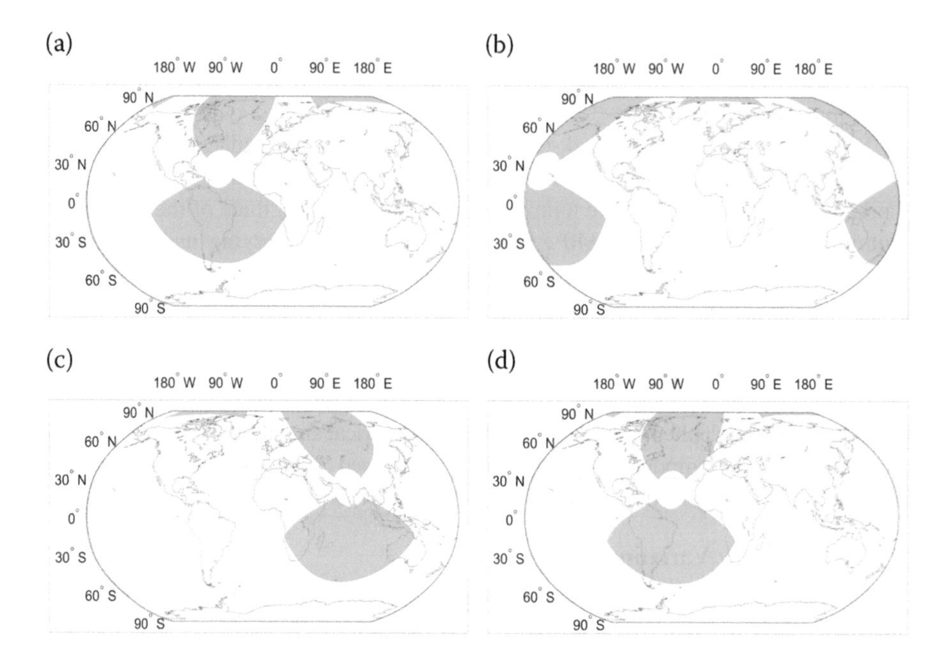

FIGURE 10.10 The Moon-based SAR's spatial coverage at (a) 00:00:00, (b) 08:00:00, (c) 16:00:00, and (d) 24:00:00 on March 20, 2024.

Hence, the spatial coverage is jointly determined by the near and far grazing angles and bound of the azimuthal angle.

For comparison, the hourly variations in the spatial coverage of the Moon-based SAR with a small size of the nadir point's latitude are presented in Figure 10.11. The visit time is reset to 00:00:00 on March 25, 2024, when the latitude of the nadir point approaches to 0°. The spatial coverage of the Moon-based SAR within one day is still significant when the size of the nadir point's latitude is small, but neither the North Pole nor the South Pole of the Earth can be observed. Also, the revisit time of the observed region is less than one day as a result of the extensive spatial coverage and Earth's self-rotation. Moreover, the effects of the lunar revolution result in the eastward motion of the spatial coverage after a one-day lapse, as well. Furthermore, a blind region of considerable extent appears on the region bounded by the far grazing angle, implying the effects of the near grazing angle and the bound of the azimuthal angle are indeed not negligible even when the size of the nadir point's latitude is small.

10.2.5.2 Global Accumulated Visible Time within Different Periods

To exploit the spatiotemporal variation in the Moon-based SAR's ground coverage for a specified period, we analyze the accumulated visible time that is the accumulated time for a specific area covered by the Moon-based SAR within a specified period. The global daily-accumulated visible time of the Moon-based SAR on March 20, March 25, April 01, and April 07 of 2024 are shown in

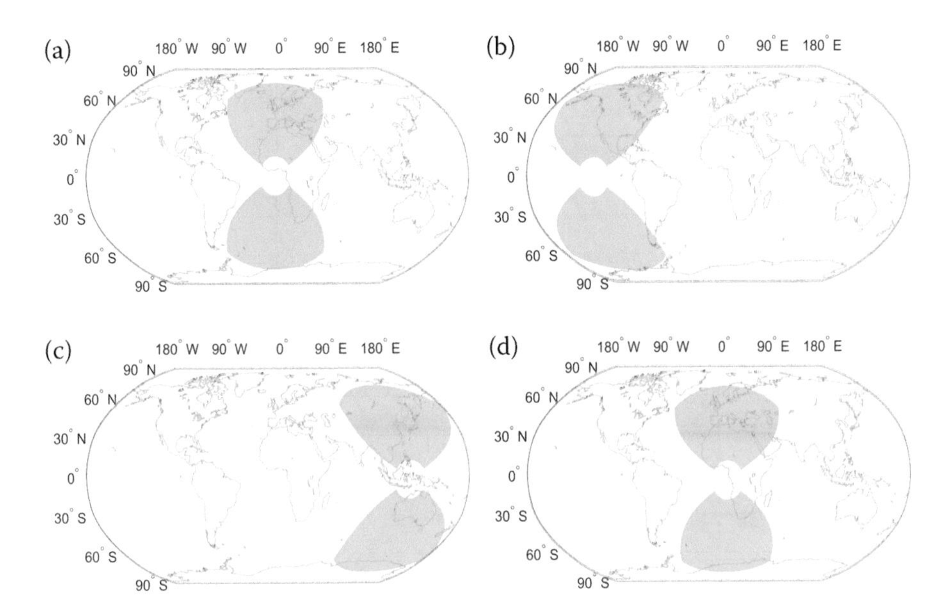

FIGURE 10.11 The Moon-based SAR's spatial coverage at (a) 00:00:00, (b) 08:00:00, (c) 16:00:00, and (d) 24:00:00 on March 25, 2024.

Figures 10.12(a–d), respectively. As is seen, the majority of the world can be covered by the Moon-based SAR within one day, and those regions are observed continuously, ranging from several hours to tens of hours. Besides, there is a conspicuous stripe that lies along the track of the nadir point, indicating that the daily-accumulated visible time is affected by the near grazing angle. Also, the longitudinal accumulated visible time approximates to uniformly distribute. By contrast, the distribution of the latitudinal accumulated visible time depends on the nadir point's latitude. When the nadir point's latitude is in the northern hemisphere of the Earth, the maximum latitudinal accumulated visible time appears at the region near the North Pole whereas the fully Antarctica cannot be observed. The trend reverses when the nadir point lies in the southern hemisphere of the Earth. Once the nadir point's latitude approaches 0°, the latitudinal accumulated visible time trends to be symmetrically distributed. In such a situation, neither of the Earth's poles could be observed by the Moon-based SAR within one day.

In general, there is not distinct regularity in the lunar revolution for one Julian year (365.25 days). Thus, the accumulated visible time over one Julian year cannot certainly reflect the tendency of the spatiotemporal coverage of the Moon-based SAR. As a result, the accumulated visible time of the Moon-based SAR within one Julian year is not presented. However, the inclination angle of the lunar orbit with respect to the Earth's equator periodically varies with a period of 18.6 years [18,22]. Hence, the range of the latitude of the nadir point reaches a maximum or minimum every 18.6 years. Hence, it is of interest to illustrate the accumulated visible time of the Moon-based SAR over 18.6 years. By setting

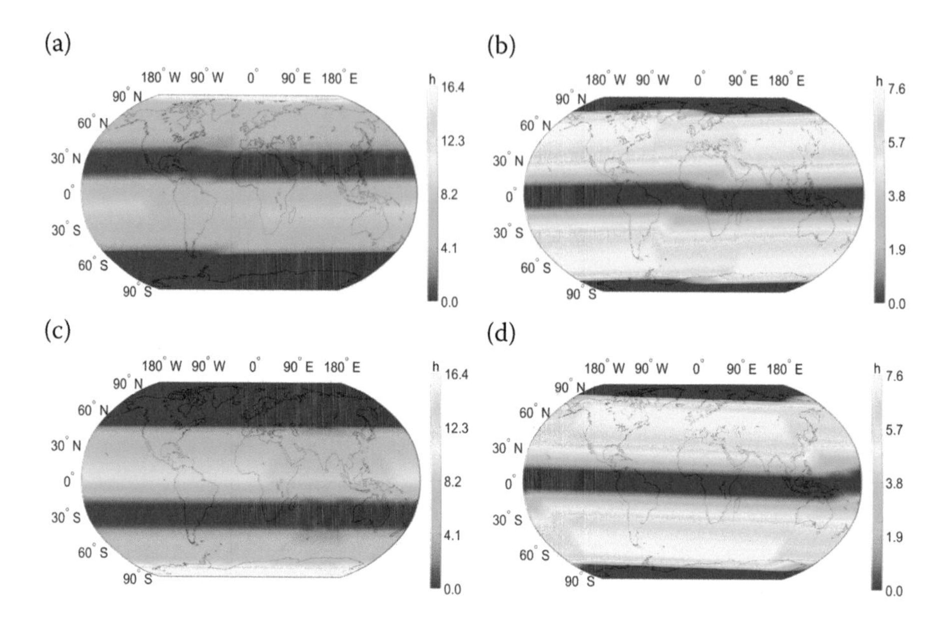

FIGURE 10.12 The global daily accumulated visible times of the Moon-based SAR on (a) March 20, (b) March 25, (c) April 01, and (d) April 07, 2024.

the initial visit time to 00:00:00 on March 20, 2024, the global accumulated visible time of the Moon-based SAR over 18.6 years is presented in Figure 10.13. The global accumulated visible time of the Moon-based SAR over 18.6 years is on the order of several years. Also, the longitudinal accumulated visible time trends uniformly distributed. In comparison, the latitudinal accumulated visible time is symmetrically distributed about the Earth's equator. What is more, the peak values of the latitudinal accumulated visible time appear at ±40° of latitudes, whereas the valleys occur at high latitudes with a size around 80°. Interestingly, there are apparent stripes above and below the Earth's equator in global accumulated visible time within an 18.6-year period, which is given rise by the effect of the near grazing angle. This implies the near grazing angle exerts a considerable impact on the spatiotemporal coverage of the Moon-based SAR with a period of 18.6 years.

Typically, the accumulated visible time over a period of one nodical month or one Julian year is different from that over the next period due to the temporal variation in the lunar orbital inclination. Thus, the accumulated visible time within one nodical month or one Julian year is not suitable for setting a Moon-based SAR. In contrast, the lunar orbital inclination is periodically varying every 18.6 years, thus the accumulated visible time of the Moon-based SAR within a period of 18.6 years is approximately the same as that in the next period of 18.6 years.

For the effective range-limited grazing and azimuthal angles, the far grazing angle plays a dominant role in determining the spatiotemporal coverage of the Moon-based SAR. The effects of the near grazing angle and bound of azimuthal

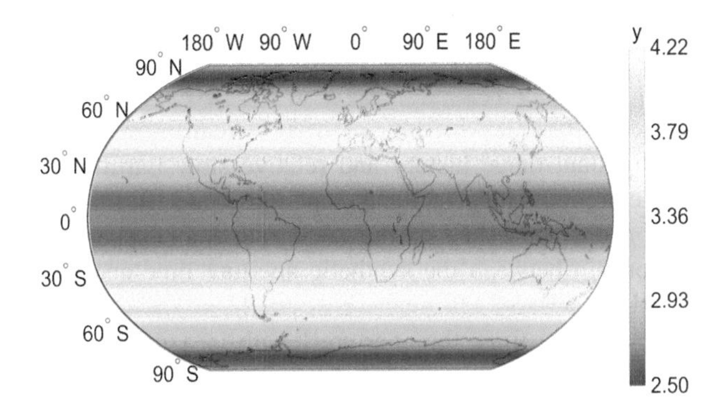

FIGURE 10.13 The global accumulated visible time of the Moon-based SAR over a period of 18.6 years. The symbol 'y' represents the Julian year.

angle are also profoundly significant to affect the spatiotemporal coverage. Regarding the optimal site selection, the site near the lunar North Pole is preferred for observing the Arctic of the Earth, while the site near the lunar South Pole may be a good choice for observing Antarctica. For other regions on the Earth, the site of the Moon-based SAR has no bearing on the spatiotemporal coverage of the Moon-based SAR over a long period. Consequently, the Moon-based SAR can perform long-term, continuous Earth observations on a global scale to enhance our capabilities of understanding the planet.

10.3 PROPAGATION THROUGH IONOSPHERIC LAYERS

By considering the application requirements and system performance of the Moon-based SAR, L-band is chosen as one of the main carrier frequencies of Moon-based SAR [9]. The L-band propagation effect through troposphere is relatively small, though not negligible, compared to ionospheric effect [23]. The lunar ionospheric electron content is two orders of magnitude less than that of the Earth's ionosphere with a far thinner thickness, thus its impact may be ignored [24]. Hence, the ionospheric layer of the Earth is the primary factor that affects the imaging geometric and radiometric qualities of the Moon-based SAR.

Ionosphere exerts profound but measurable effects on the SAR imaging by at least three accounts. First, the Faraday rotation (FR) effect rotates the energy from co-channels to cross-channels, subsequently leading to attenuation of the SAR signal [25]. Second, the phase dispersion of the background ionosphere causes a phase advance in signal, resulting in target image shift, resolution deterioration and image defocusing [26]. Third, it should be mentioned that the ionospheric scintillation gives rise to fluctuations in the amplitude and phase of the SAR signal, further impair the SAR focusing [27]. Besides, as long as estimating the ionospheric effects on SAR imaging is concerned, the ionospheric inhomogeneous nature is a considerable factor [28].

10.3.1 Phase Error due to Temporal-Spatial Varying Background Ionosphere

The synthetic aperture of the Moon-based SAR is realized mainly by the Earth's rotation. The long slant range history and relative slow Earth's rotating velocity bring about an extremely long synthetic aperture time, further resulting in the temporal-variation characteristics of the background ionosphere. For a rigorous study, the relationship between the synthetic aperture time and the azimuthal resolution of the Moon-based SAR is now examined.

The synthetic aperture time, the aperture length in azimuth, can be accurately obtained by

$$T_{sar} = \eta_{end} - \eta_{star} \tag{10.48}$$

where η_{end} and η_{star} are defined as the start time and end time when a point target is illuminated of one full aperture in azimuth, respectively. Both of them can be calculated by:

$$\cos^{-1}\left(\frac{\mathbf{R}_{\mathbf{star}}\cdot\mathbf{R}_{\mathbf{c}}}{|\mathbf{R}_{\mathbf{star}}|\cdot|\mathbf{R}_{\mathbf{c}}|}\right) = \frac{\theta_b}{2}, \tag{10.49}$$

$$\cos^{-1}\left(\frac{\mathbf{R}_{\mathbf{end}}\cdot\mathbf{R}_{\mathbf{c}}}{|\mathbf{R}_{\mathbf{end}}|\cdot|\mathbf{R}_{\mathbf{c}}|}\right) = -\frac{\theta_b}{2}. \tag{10.50}$$

where $\mathbf{R}_{\mathbf{star}}$ is the slant range vector of the Moon-based SAR at time η_{star}, $\mathbf{R}_{\mathbf{end}}$ is the slant range vector at time η_{end}, $\mathbf{R}_{\mathbf{c}}$ is the slant range vector at the beam center crossing time η_c; θ_b is the half-power beam width in radians defined as

$$\theta_b = \lambda/\ell_a \tag{10.51}$$

where λ is the wavelength, ℓ_a is the aperture length in azimuth.

The azimuthal resolution of the Moon-based SAR is determined by

$$\rho_a = \frac{\ell_a}{2}\frac{R_E}{R_{EM}}\frac{\cos\delta_g}{\cos\delta_m}\cos\alpha \tag{10.52}$$

where R_E is the Earth's radius, and R_{EM} is the distance between the Earth and Moon-based SAR (see Figure 10.1); δ_g is the latitude of the Earth's target, δ_m is the declination of the Moon-based SAR. α is the angular difference between the Moon-based SAR's ascension and the longitude of the Earth's target at beam center crossing time. As can be seen, three extra terms: R_E/R_{EM}, $\cos\delta_g/\cos\delta_m$ and $\cos\alpha$ are presented to modify the ideal azimuth resolution [12].

By combining Equations 10.48–10.50, the relationship between the synthetic aperture time and azimuthal resolution is plotted in Figure 10.14, with, as an

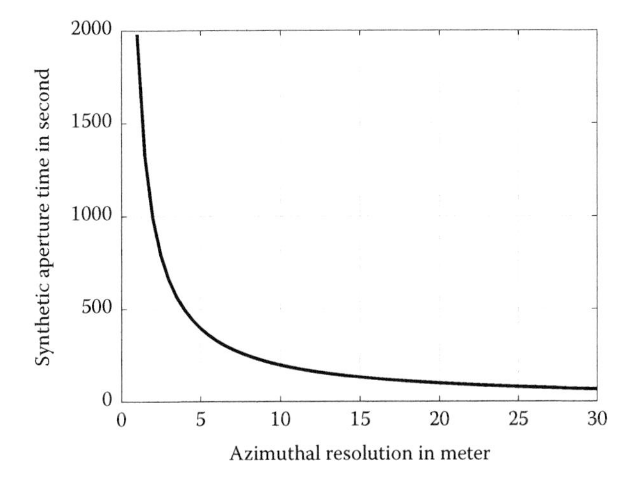

FIGURE 10.14 The relationship between the synthetic aperture time and azimuthal resolution, here the angular difference between the ascension of Moon-based SAR and the longitude of the Earth's target is assumed to be 0°.

example, the declination of the Moon-based SAR set to be 28.5° and latitude of the Earth's target assumed to be 0°.

As can be seen from Figure 10.14, the synthetic aperture time of the Moon-based SAR can be up to hundreds or even thousands of seconds. In contrast, the synthetic aperture time of the LEOSAR is usually limited to 1–2 s. As a result, it seems that the ionospheric freezing assumption for LEOSAR loses its effect in the Moon-based SAR. To catch a better notion, in Figure 10.15, we plot the measured vertical TEC at Guangzhou, China (113.23°E, 23.16°N), with a synthetic aperture time of 1800 s, a typical value of the Moon-based SAR. The TEC data used in the schematic diagram for the temporal variation of background ionosphere is acquired in October 2016 and reported by International Reference Ionosphere 2012 (IRI 2012) [29].

It is clear that the variation of the vertical TEC over a period during the synthetic aperture time exceeds 3 TECU, which seems quite appreciable, and the TEC temporal variation is nonlinear. Consequently, the temporal-varying background ionosphere should be taken into consideration, thus the temporal-varying ionospheric TEC is supposed to express in the form of slow time:

$$TEC_t(\eta) = TEC_{t0} + k_1\eta + k_2\eta^2 + k_3\eta^3 + k_4\eta^4 + \mathcal{O}(\eta^5) \tag{10.53}$$

where TEC_{t0} is the constant component of TEC with respect to slow time, which remains constant with the varying time for a given position; k_i, $i = 1, ..., n$, is the nth derivative of the temporal-varying TEC against the slow time η.

Because the Moon-based SAR is observing the Earth on a global scale, the spatial-varying ionospheric TEC at different positions within the imaging swath

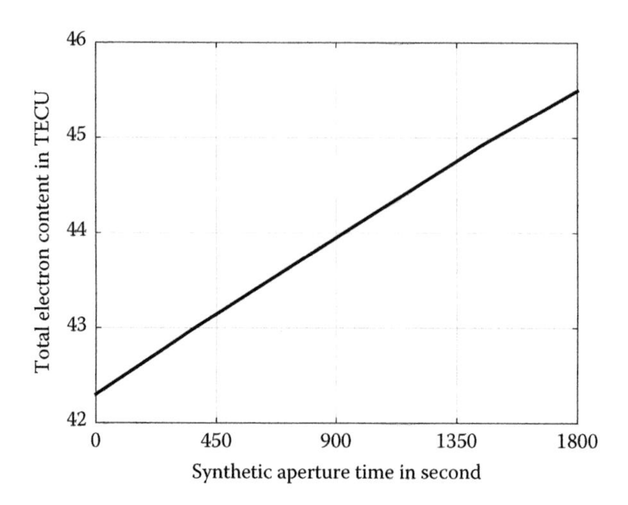

FIGURE 10.15 The measured temporal-varying ionospheric vertical TEC at 12:00–12:30 local time.

should be taken account, as is evident from the strongly varying TEC with relative positions of the ground target illustrated in Figure 10.16.

Assume that the spatial variation of the TEC along the azimuth direction is time-invariant, the spatial-varying TEC can be written as:

$$TEC_s(x) = TEC_{s0} + s_1x + s_2x^2 + s_3x^3 + s_4x^4 + \mathcal{O}(x^5) \tag{10.54}$$

where TEC_{s0} is the constant component of TEC regarding the azimuth distance x at specified time; s_i, $i = 1, ..., n$, are the change rates of spatial-varying TEC over azimuth distance x at various orders; note that x is the relative motion distance between the Moon-based SAR and the ground target taking account of spatial-varying background ionosphere along azimuth direction within an epoch of η, which is defined as:

$$x = V_{iono}\eta \tag{10.55}$$

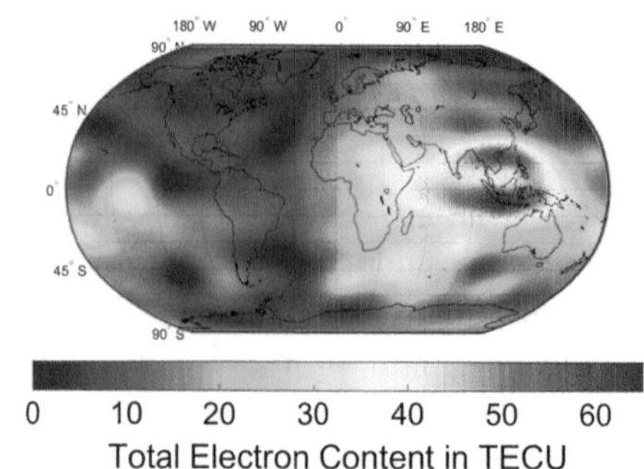

FIGURE 10.16 The global distribution of the vertical TEC in TECU at 08:00 UTM, 22 October 2016. Data was taken from [30].

where V_{iono} is the relative velocity of the ionospheric TEC above the rotating Earth, given by

$$V_{iono} = (R_E + h_{iono})\cos\delta_g(\chi\omega_E) \tag{10.56}$$

where R_E is the Earth's radius, h_{iono} is the ionospheric equivalent height, δ_g is the ground target's latitude, ω_E is the Earth's rotation velocity, χ is the scale factor given rise by lunar revolution as to be detailed below.

It is known that the temporal and spatial variations of the background ionosphere are highly-coupling in practice. Thus, the temporal-spatial varying ionospheric TEC can be readily expressed in terms of slow time η:

$$TEC(\eta) = TEC_0 + t_{g1}\eta + t_{g2}\eta^2 + t_{g3}\eta^3 + t_{g4}\eta^4 + O(\eta^5) \tag{10.57}$$

where TEC_0 is the constant component of temporal-spatial varying ionospheric TEC; t_{gi} is n^{th} derivative of the temporal-spatial varying TEC with respect to slow time defined as: $t_{gi} = k_i + s_i V_{iono}^i$, $i = 1, \ldots, n$.

When the SAR signal propagates through the ionosphere, the phase-path length will be changed under the impact of the background ionosphere [31]. The phase error induced by the change of the phase-path length in the round-trip can be defined as

$$\Delta\varphi_{iono} = \frac{-4\pi A}{c(f_\tau + f_c)} TEC \tag{10.58}$$

where $A = 40.32 m^2/s^3$, c is the propagation velocity of the electromagnetic wave in free space, f_τ is the range frequency and f_c is the carrier frequency.

It turns out that the phase error induced by the temporal-spatial varying background ionosphere should be modified by substituting Equation 10.57 into Equation 10.58:

$$\Delta\varphi_{iono}(\eta) = \frac{-4\pi A}{c(f_\tau + f_c)} TEC(\eta) \tag{10.59}$$

It shows in Equation 10.59 that the spatial-varying ionospheric TEC, together with the temporal-varying ionospheric TEC, further makes the background ionospheric effects even more complicated. It may be drawn that the spatial-variation of background ionosphere within an image scene is approximately time-invariant, implying that the ionospheric TEC in Equation 10.57 potentially can be used to probe the temporal-spatial background ionospheric effects [32].

10.3.2 Slant Range in the Context of Background Ionospheric Effects

To derive an expression of the slant range, a right-handed geocentric inertial reference frame is defined, with Z-axis towards the North Pole, and X-axis pointing to the true equinox of the date (referring to Figure 10.1). Without loss of generality, let's initialize the azimuth time to zero for the shortest distance between the ground target and the Moon-based SAR. The slant range between the Moon-based SAR and the ground target at the time $t = \eta$ takes the following expression:

$$R(\eta) = \{R_E^2 + R_{EM}^2 - 2X_m(\eta)R_E \cos\delta_g \cos a_g(\eta) - 2Y_m(\eta)R_E \cos\delta_g \sin a_g(\eta) - 2Z_m(\eta)R_E \sin\delta_g\}, \tag{10.60}$$

where $a_g(\eta) = \omega_E\eta + a_{g0}$ is the angle that the ground target rotates from the zero azimuth time position to the position at $t = \eta$, ω_E is the angular velocity of the Earth's self-rotation, a_{g0} is the longitude of ground target at zero azimuth time; $(X_m(\eta), Y_m(\eta), Z_m(\eta))$ are coordinates of the Moon-based SAR in considering of the lunar revolution with:

$$X_m(\eta) = R_{EM}\cos(\delta_m + \omega_M\eta\sin\vartheta_S)\cos(a_m + \omega_M\eta\cos\vartheta_S) \tag{10.61}$$

$$Y_m(\eta) = R_{EM}\cos(\delta_m + \omega_M\eta\sin\vartheta_S)\sin(a_m + \omega_M\eta\cos\vartheta_S) \tag{10.62}$$

$$Z_m(\eta) = R_{EM}\sin(\delta_m + \omega_M\eta\sin\vartheta_S) \tag{10.63}$$

In above equations, ω_M is the angular velocity of the lunar revolution; ϑ_S is the inclination of the lunar orbit to of the Earth's equator.

Recalled that the propagation distance of the Moon-based SAR is over hundreds of times longer than that of the LEOSAR, thus the "stop-and-go" assumption used in the LEOSAR is no longer applicable for the Moon-based SAR [16]. Now let's calibrate the error in "stop-and-go" assumption: Suppose that the transmitted signal arrives to the ground target at time T_1, where $T_1 = R(\eta)/c$. In an epoch of propagation delay T_D, the Moon-based SAR has moved forward for a certain distance to the new position, so now the slant range between the ground target and radar is $R(\eta + T_D)$. The backscattering signal is received at a time $T_2 = R(\eta + T_D)/c$. The total time delay of wave propagation is the sum of T_1 and T_2:

$$T_D = T_1 + T_2, \tag{10.64}$$

where T_D is typically between 2.3s and 2.7s [16]. Notice that the time delay given rise by the atmosphere is far smaller than T_D; therefore, it can be reasonably ignored in removing the "stop-and-go" assumption. The Moon-based SAR observation geometry under "non-stop-and-go" assumption shown in Figure 10.17 depicts a scenario whereas the transmitted signal scatters back, the radar moves a distance within an epoch of T_D. Thus, the Moon-based SAR can no longer be regarded as operating in monostatic, but instead in bistatic mode.

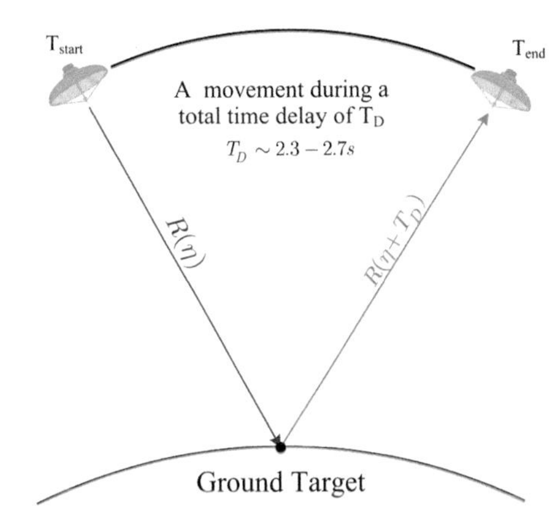

FIGURE 10.17 The equivalent observation geometry of the Moon-based SAR with "non-stop-and-go" assumption. During TD the Moon-based SAR has moved for an equivalent distance where the ground target is regarded as stationary while the Moon-based SAR rotates (Sketch not to scale).

Finally, the slant range for the Moon-based SAR under the equivalent bistatic mode could be written as:

$$R_S(\eta) = [R(\eta) + R(\eta + T_D)]/2 \tag{10.65}$$

It is seen that the 4th order Taylor series range equation is capable of dealing with the curved trajectory of the Moon-based SAR's signal during the long synthetic aperture time, which is most suitable for Moon-based SAR imaging [16]. Consequently, the slant range can be expanded into Taylor series up about $\eta = 0$ to 4th order,

$$R_S(\eta) = R_0 + R_1\eta + R_2\eta^2 + R_3\eta^3 + R_4\eta^4 + \mathcal{O}(\eta^5) \tag{10.66}$$

where R_0 is the shortest slant range. The derivation of expansion coefficients is straightforward but tedious. Complete expressions for R_i are given in [16].

10.3.3 SAR Signal in the Context of Background Ionospheric Effects

We shall establish a Moon-based SAR signal model in considering of background ionospheric effects, based on the curved trajectory, in which the relative positions between the Earth target and the Moon-based SAR are accounted for. The received signal of the Moon-based SAR system from the point of interest can be written as [33]

$$\begin{aligned} s_r(\tau, \eta) = {} & w_r[\tau - 2R_S(\eta)/c]\, w_a(\eta) \exp\{-j4\pi f_c R_S(\eta)/c\} \\ & \cdot \exp\{j\pi K_r[\tau - 2R_S(\eta)/c]^2\} \end{aligned} \tag{10.67}$$

where w_r and w_a are window functions in fast time and slow time domains, respectively; K_r is the chirp rate.

Fourier transform in range direction is first applied to Equation 10.67 by using the principle of stationary phase (POSP):

$$s_r(f_\tau, \eta) = w_r(f_\tau) w_a(\eta) \exp(-j\pi f_\tau^2 / K_r) \cdot \exp[-j4\pi R_S(\eta)(f_\tau + f_c)/c] \tag{10.68}$$

Consequently, the signal under the effects of the temporal-varying background ionosphere can be obtained by including the phase shift Equation 10.59 into the signal Equation 10.68:

$$s_{r.iono}(f_\tau, \eta) = w_r(f_\tau) w_a(\eta) \exp(-j\pi f_\tau^2 / K_r) \exp[-j4\pi R_S(\eta)(f_\tau + f_c)/c + j\Delta\varphi_{iono}(\eta)] \tag{10.69}$$

Taking Fourier transform along azimuth direction in Equation 10.69 by virtue of the POSP and the method of series reversion (MSR) [33], we have

$$s_{r.iono}(f_\tau, f_\eta) = w_r(f_\tau)\, w_a(f_\eta) \exp\{j\Psi(f_\tau, f_\eta)\}, \tag{10.70}$$

where the phase takes the form

$$\begin{aligned}\Psi(f_\tau, f_\eta) = & -\pi \frac{f_\tau^2}{K_r} - \frac{2\pi}{c} a_0 (f_c + f_\tau) - \frac{2\pi}{c} \frac{2A \cdot TEC_0}{(f_c + f_\tau)} \\ & + 2\pi \frac{1}{4P_2} \left(\frac{c}{f_c + f_\tau} \right) \left(f_\eta + P_1 \frac{(f_c + f_\tau)}{c} \right)^2 \\ & + 2\pi \frac{P_3}{8P_2^3} \left(\frac{c}{f_c + f_\tau} \right)^2 \left(f_\eta + P_1 \frac{(f_c + f_\tau)}{c} \right)^3 \\ & + 2\pi \frac{9P_3^2 - 4P_2P_4}{64P_2^5} \left(\frac{c}{f_c + f_\tau} \right)^3 \left(f_\eta + P_1 \frac{(f_c + f_\tau)}{c} \right)^4 \end{aligned} \tag{10.71}$$

where $a_0 = 2R_0$, $P_i = 2A/f_c^2 t_{gi} + 2R_i$, $i = 1, ..., 4$.

From Equation 10.71, we see that the range and azimuth frequency are highly coupled in the phase term. To process the signal and to identify the background ionospheric effects on the Moon-based SAR imaging, Equation 10.71 is further expanded as follows:

$$\Psi(f_\tau, f_\eta) = \Psi_r(f_\tau) + \Psi_a(f_\eta) + \Psi_{rcm}(f_\tau, f_\eta) + \Psi_{src}(f_\tau, f_\eta) + \Psi_{res} \tag{10.72}$$

where $\Psi_r(f_\tau)$ is related to range compression through

$$\begin{aligned}\Psi_r(f_\tau) = 2\pi \Bigg[& \left(-\frac{a_0}{c} + \frac{2ATEC_0}{cf_c^2} + \frac{P_1^2}{4P_2c} + \frac{P_1^3P_3}{8P_2^3c} + \frac{9P_3^2 - 4P_2P_4}{64P_2^5c} P_1^4 \right) f_\tau \\ & + \left(-\frac{1}{2K_r} - \frac{2ATEC_0}{cf_c^3} \right) f_\tau^2 + \frac{2ATEC_0}{cf_c^4} f_\tau^3 - \frac{2ATEC_0}{cf_c^5} f_\tau^4 \Bigg] \end{aligned} \tag{10.73}$$

$\Psi_a(f_\eta)$ is in connection with the azimuth compression and is expressed as

$$\Psi_a(f_\eta) = 2\pi\left[\left(\frac{P_1}{2P_2} + \frac{3P_1^2P_3}{8P_2^3} + \frac{9P_3^2 - 4P_2P_4}{16P_2^5}P_1^3\right)f_\eta + \left(\frac{c}{4P_2f_c} + \frac{3P_1P_3c}{8P_2^3f_c} + \frac{9P_3^2 - 4P_2P_4}{32P_2^5f_c}3P_1^2c\right)f_\eta^2 + \left(\frac{P_3c^2}{8P_2^3f_c^2} + \frac{9P_3^2 - 4P_2P_4}{16P_2^5f_c^2}P_1c^2\right)f_\eta^3 + \frac{9P_3^2 - 4P_2P_4}{64P_2^5f_c^3}c^3f_\eta^4\right] \tag{10.74}$$

$\Psi_{rcm}(f_\tau, f_\eta)$ is the range cell migration term which takes the form of

$$\Psi_{rcm}(f_\tau, f_\eta) = 2\pi\left[-\left(\frac{1}{4P_2f_c^2} + \frac{3P_1P_3}{8P_2^3f_c^2} + \frac{9P_3^2 - 4P_2P_4}{32P_2^5f_c^2}3P_1^2\right)cf_\eta^2 - \left(\frac{9P_3^2 - 4P_2a_4}{8P_2^5f_c^3}P_1 + \frac{P_3}{4P_2^3f_c^3}\right)c^2f_\eta^3 - \frac{9P_3^2 - 4P_2P_4}{64P_2^5f_c^4}3c^3f_\eta^4\right]f_\tau \tag{10.75}$$

where $\Psi_{src}(f_\tau, f_\eta)$ accounts for secondary range compression, explicitly given by

$$\Psi_{src}(f_\tau, f_\eta) = 2\pi\left\{\left[\left(\frac{1}{4P_2f_c^3} + \frac{3P_1P_3}{8P_2^3f_c^3} + \frac{9P_3^2 - 4P_2P_4}{32P_2^5f_c^3}3P_1^2\right)cf_\eta^2 + \left(\frac{3P_3}{8P_2^3f_c^4} + \frac{9P_3^2 - 4P_2P_4}{16P_2^5f_c^4}3P_1\right)c^2f_\eta^3 + \frac{9P_3^2 - 4P_2P_4}{32P_2^5f_c^5}3c^3f_\eta^4\right]f_\tau^2 + \left[\left(-\frac{1}{4P_2f_c^4} - \frac{3P_1P_3}{8P_2^3f_c^4} - \frac{9P_3^2 - 4P_2P_4}{32P_2^5f_c^4}3P_1^2\right)cf_\eta^2 + \left(-\frac{P_3}{2P_2^3f_c^5} - \frac{9P_3^2 - 4P_2P_4}{4P_2^5f_c^5}P_1\right)c^2f_\eta^3 - \frac{9P_3^2 - 4P_2P_4}{32P_2^5f_c^6}5c^3f_\eta^4\right]f_\tau^3\right\} \tag{10.76}$$

and Ψ_{res} is the residual phase term given by

$$\Psi_{res} = 2\pi\left(-\frac{a_0}{c}f_c - \frac{2ATEC_0}{cf_c} + \frac{P_1^2}{4cP_2}f_c + \frac{P_1^3P_3}{8cP_2^3}f_c + \frac{9P_3^2 - 4P_2P_4}{64cP_2^5}P_1^4f_c\right) \tag{10.77}$$

At this point, we have established the signal model and 2-D spectrum, based on the curved trajectory, taking accounts of he temporal-spatial varying background ionospheric effects. It can be seen from Equation 10.73 through Equation 10.77 that all the terms are closely related to TEC. Besides, the imaging performance in the range direction is impacted by the range compression term $\Psi_r(f_\tau)$. The residual phase term Ψ_{res} does not affect the Moon-based SAR imaging. As for the azimuthal imaging, it is impacted by the azimuth compression term $\Psi_a(f_\eta)$, the range cell migration term $\Psi_{rcm}(f_\tau, f_\eta)$ and the secondary range compression term $\Psi_{src}(f_\tau, f_\eta)$ under the effects of background ionosphere.

Numerical results [32] show that the absolute phase errors of the range cell migration term, and second range compression term are negligibly small in comparison with the azimuth compression term. Consequently, the azimuth compression term dominates the focusing of the Moon-based SAR. Also, the range cell migration term, and secondary range compression term, unlike the azimuth compression term, exercise little effects on azimuthal imaging.

10.4 IMAGE DISTORTIONS BY DISPERSIVE EFFECTS

The background ionospheric effects on the Moon-based SAR imaging are examined from the phase terms given in Equations 10.73 and 10.74. The background ionospheric effects can be split into the range and azimuth aspects, and are illustrated by using the simulation parameters given in Table 10.3. For the sake of brevity but without loss of generality, the angular difference between the ground target's longitude and Moon-based SAR's ascension at beam center crossing time is assumed 0° in the following analysis.

10.4.1 Ionospheric Effects on Range Imaging

Through the range compression term given in Equation 10.73, the background ionospheric impacts on range imaging by manifesting the range shift and

TABLE 10.3
Simulation Parameters for the Moon-Based SAR System

Symbol	Parameters	Quantity	Unit
δ_g	Latitude of the ground target	22.5	degree
δ_m	Declination of the Moon-based SAR	28.5	degree
θ_S	Inclination of the lunar orbit to the equator of the Earth	28.6	degree
R_E	Earth radius	6,371	km
R_{EM}	The distance between the Earth and Moon-based SAR	389,408	km
ω_E	Earth's rotation angular velocity	7.292×10^{-6}	rad/s
ω_M	Lunar revolution angular velocity	2.662×10^{-6}	rad/s
f_c	Carrier frequency	1.2	GHz
B	System bandwidth	50	MHz
h_{iono}	Ionospheric equivalent height	400	km

defocusing. It is recognized that the range shift is caused by the linear term due to group delay, while the range defocusing is induced by the quadratic term and cubical term. Besides, the range resolution is deteriorated due to broaden chirped pulse, which is silently imbedded in the quadratic term. Moreover, a ghost image may appear in the range direction as a result of asymmetric distortion of the signal caused by the cubical term [34]. In what follows, the detailed effects of phase shift on the range imaging are analyzed by Equation 10.73.

10.4.1.1 Range shift

According to the linear term in Equation 10.73, the range shift can be expressed as:

$$\Delta L_r = \Delta L_{r0} + \Delta L_{r1} \tag{10.78}$$

$$\Delta L_{r0} = A \cdot TEC_0 / f_c^2 \tag{10.79}$$

$$\Delta L_{r1} = \frac{P_1^2}{8P_2} + \frac{P_1^3 P_3}{16P_2^3} + \frac{9P_3^2 - 4P_2 P_4}{128P_2^5} P_1^4 - \frac{a_1^2}{8a_2} - \frac{a_1^3 a_3}{16a_2^3} - \frac{9a_3^2 - 4a_2 a_4}{128a_2^5} a_1^4 \tag{10.80}$$

As can be seen from Equation 10.78, the range shift consists of two terms: one is the image shift caused by the constant component of ionospheric TEC under the ionospheric freezing assumption, and the other one is the range shift caused by the temporal and spatial gradients of ionospheric TEC related to the temporal-spatial varying part of background ionosphere. The second term of the range shift can be further divided into two parts based on the different contribution parts from temporal-spatial varying background ionosphere: ΔL_{r1t} induced by the temporal-varying part of the background ionosphere and ΔL_{r1s} caused by the spatial-varying part of the background ionosphere. Since we can get each of them by letting another part of the ionospheric TEC gradient be zero, thus they are not expressed separately.

The range shift caused by the constant component of background ionosphere, ΔL_{r0}, is proportional to the constant component of the TEC and is inversely proportional to squared of the carrier frequency. The range shift ΔL_{r1}, induced by the temporal-spatial varying part of background ionosphere, has an obvious dependency on the temporal and spatial gradients of ionospheric TEC and the carrier frequency. Also, the range shift ΔL_{r1} is dependent on the relative position between the ground target and the Moon-based SAR as well.

Figure 10.18(a) and (b) show the temporal-varying background ionosphere contribution part of the range shift, ΔL_{r1t} with different Moon-based SAR's declinations with the Earth's latitude under normal ionosphere and during an ionospheric storm. Similarly, the spatial-varying ionospheric TEC contribution part of range shift, ΔL_{r1s}, under normal ionosphere and ionospheric storm are plotted in Figure 10.18(c) and (d), respectively. The overall range shift induced by the temporal-spatial varying part of the background ionosphere, ΔL_{r1}, are plotted in Figure 10.18(e) and (f), eventually.

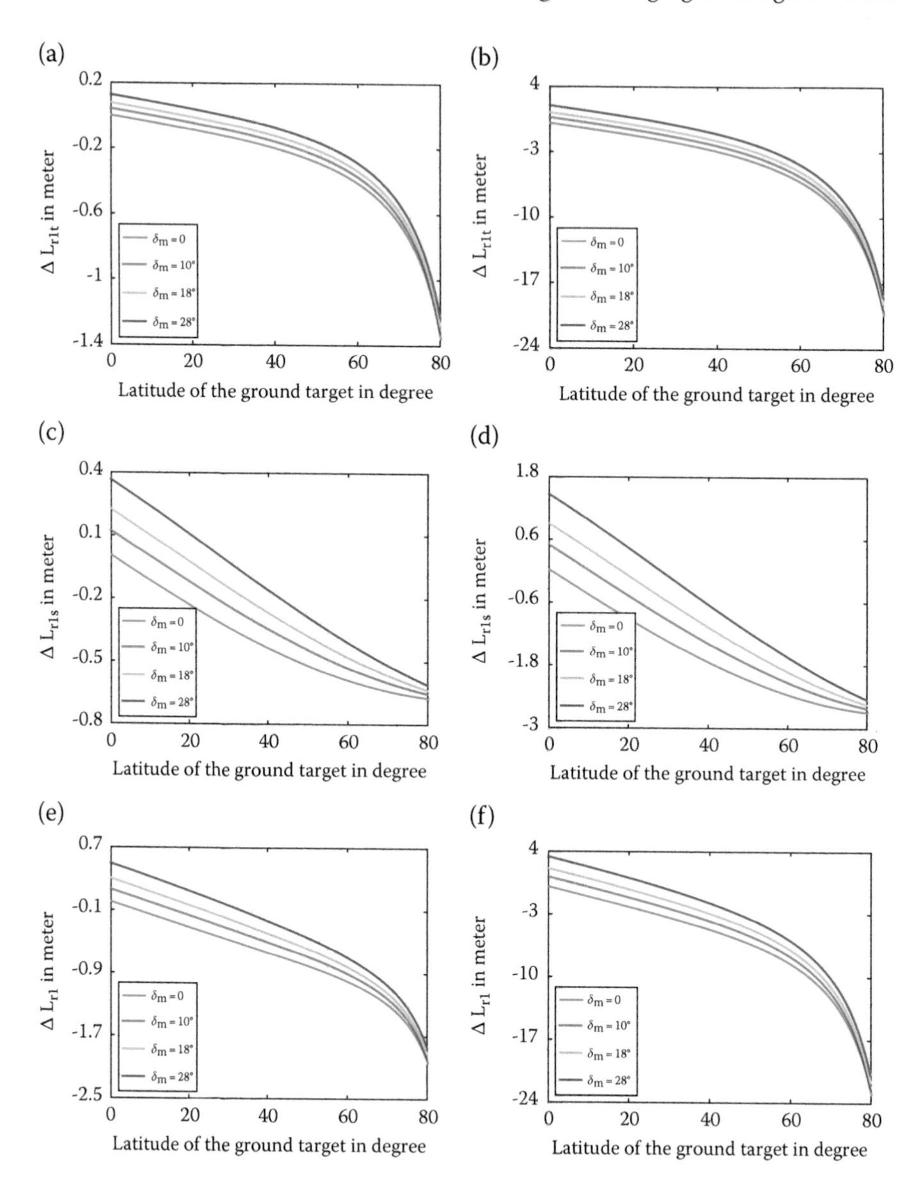

FIGURE 10.18 The range shift caused by temporal-varying background ionosphere under (a) normal ionosphere, (b) ionospheric storm; The range shift caused by spatial-varying background ionosphere under (c) normal ionosphere, (d) ionospheric storm; The range shift caused by temporal-spatial varying background ionosphere under (e) normal ionosphere, (f) ionospheric storm.

From Figure 10.18(a)–(d), we see that both the temporal-varying and spatial-varying background ionospheres bring about range shifts, which are in negative connection with the ground target's latitudes but positively correlated with the Moon-based SAR's declination. Yet their variations with respect to the Earth's latitudes and Moon-based SAR's declinations are different because the range shift caused by the temporal-varying part of the background ionosphere is only related to the change rate of slant range, while ΔL_{r1s}, induced by the spatial-varying part of the background ionosphere, is dependent on both change rate of slant range and ionospheric TEC relative motion velocity that is related to the relative position of the ground target's latitude and Moon-based SAR's declination. Besides, the range shift ΔL_{r1} is the contribution from the TEC temporal gradient and spatial gradient; either variation of them leads to different ΔL_{r1}.

Next, we plot the 1-D range profile of point target response at the of the Earth's latitude of 22.5° under the effect of the range shift for the cases of normal ionosphere and ionospheric storms in Figure 10.19(a) and (b). Then the 1-D range profile of point target response with the same TEC status as Figure 10.19(a) and (b) at the latitude of 62.5° are drawn in Figure 10.19(c) and (d), respectively.

As illustrated from Figure 10.19, the range shift caused by the temporal-spatial varying background ionosphere is on the order of tens of meters under normal

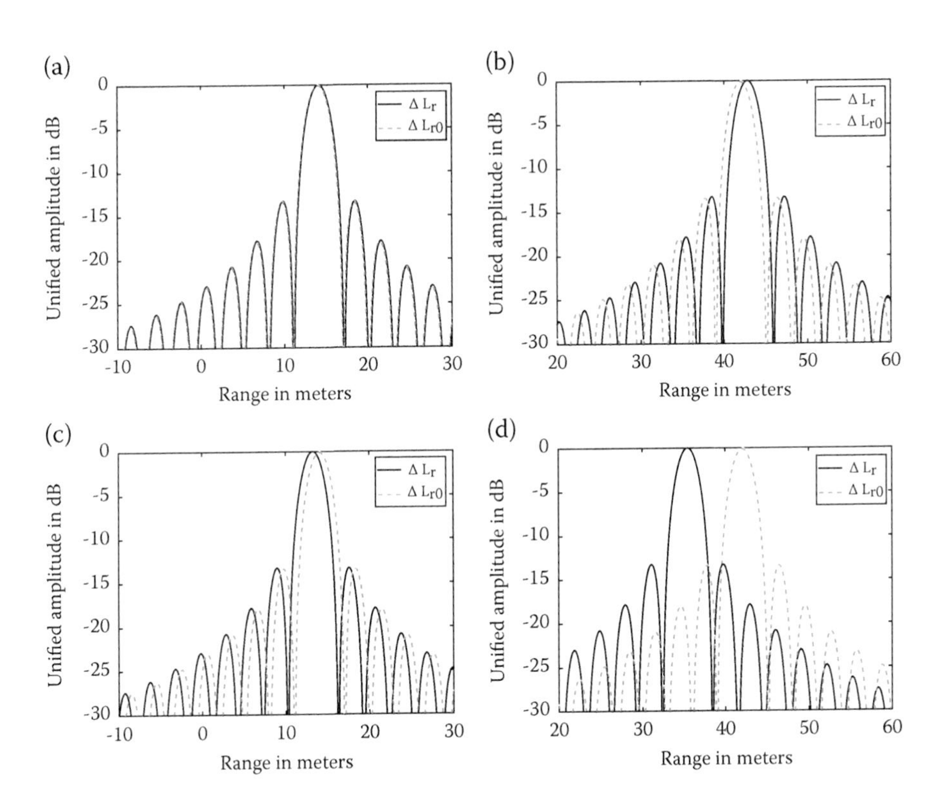

FIGURE 10.19 The 1-D range profile of point target response at the latitude of 22.5° under the effect of range shift under: (a) normal ionosphere $\Delta L_r = 14.12$ m, $\Delta L_{r0} = 14.00$ m; (b) ionospheric storm $\Delta L_r = 42.83$ m, $\Delta L_{r0} = 42.00$ m. The 1-D range profile of point target response at the latitude of 62.5° under the effect of range shift with: (c) normal ionosphere $\Delta L_r = 13.24$ m, $\Delta L_{r0} = 14.00$ m; (d) ionospheric storm $\Delta L_r = 35.36$ m, $\Delta L_{r0} = 42.00$ m.

ionosphere while it can be up to dozens of meters during an ionospheric storm. Moreover, the range shift ΔL_{r0} caused by the constant component of ionospheric TEC under the ionospheric freezing assumption plays a dominating role in determining the whole range shift. The range shift induced by the temporal-spatial varying background ionosphere is relatively small but not negligible under normal ionosphere while it is potential enough during an ionospheric storm.

If we project the slant range shift onto ground range, then the ground range shift is:

$$\Delta L_{gr} \approx \frac{\Delta L_r}{\cos\theta_i \sin\theta_i} \tag{10.81}$$

where the incident angle is given by

$$\theta_i = \sin^{-1}\left\{\frac{R_{EM}\sin|\delta_g - \delta_m|}{|\mathbf{R}_c|}\right\} \tag{10.82}$$

The incident angle in the above equation is related to the latitude of the ground target and declination of the Moon-based SAR, and is spatial-varying as well. As a result, the spatial-varying incident angle, together with the temporal-spatial varying range shift, brings about the temporal-spatial variation of the ground range shift, further complicating the geometric distortion.

Figure 10.20(a) displays the ground range shift under the normal ionosphere as a function of the ground target's latitude with different declinations with the simulation parameters given in Table 10.4 taken from [32]. The ground range shift under the ionospheric storm is drawn in Figure 10.20(b) by using the same simulation parameters as Figure 10.20(a).

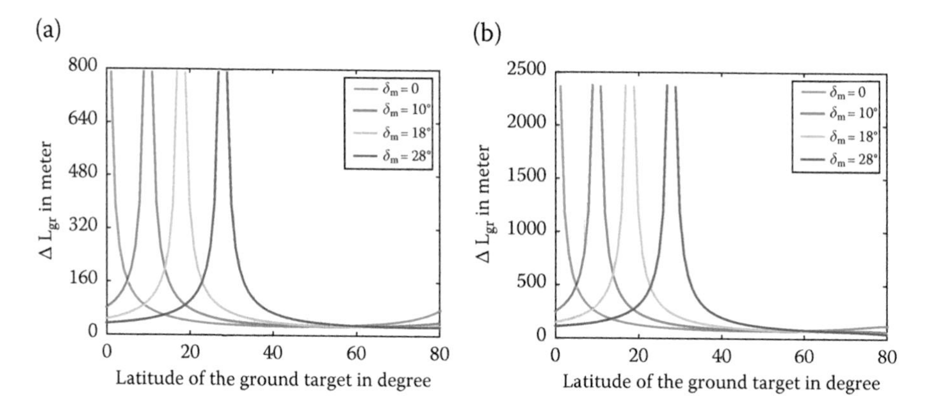

FIGURE 10.20 The ground range shift caused by temporal-spatial varying background ionosphere as a function of the ground target's latitude with different Moon-based SAR's declinations under (a) normal ionosphere, (b) ionospheric storm. Note that the point of $\delta_g - \delta_m = 0$ is the nadir point.

TABLE 10.4
Temporal and Spatial Gradient of Temporal-Spatial Varying TEC [32]

Temporal Derivative	Quantity	Unit	Spatial Derivative	Quantity	Unit
s_1	5.28×10^{-3}	TECU/s	k_1	-6.55×10^{-5}	TECU/m
s_2	-3.56×10^{-7}	TECU/s^2	k_2	7.49×10^{-14}	TECU/m^2
s_3	-7.81×10^{-11}	TECU/s^3	k_3	-2.43×10^{-20}	TECU/m^3
s_4	-5.24×10^{-15}	TECU/s^4	k_4	2.59×10^{-26}	TECU/m^4

It can be observed in Figure 10.20 that the ground range shift is on the order of dozens of meters under normal ionosphere while the maximum value of the ground range shift can be up to thousands of meters during an ionospheric storm. When the latitude of the ground target is within 60°, the ground range shift depends on the difference between the ground target's latitude and Moon-based SAR's declination. However, if the ground target's latitude is larger than 60°, the ground range shift becomes a positive correlation from a negative correlation with the difference between the ground target's latitude and Moon-based SAR's declination. The cause of this phenomenon is attributed to that the range shift caused by the temporal-varying background ionosphere begins to show an obvious impact on the whole ground range shift.

The 1-D range profile in the presence of the ground range shift at different ground target's latitude and Moon-based SAR's declinations are simulated and plots in Figure 10.21. The simulation parameters are given in Table 10.3 except for the latitudes of ground target and declination of Moon-based SAR.

From Figure 10.21, the ground range shift, ranging from dozens of meters up to hundreds of meters even under a normal ionospheric condition, can easily pose a threat to correctly pinpoint the ground target. As can be seen from Figure 10.21(a)–(d) that the ground range shift decreases with the increasing of the difference between ground target's latitude and Moon-based SAR's declination. Results, as displayed in Figure 10.21(e)–(f), show a contrary tendency with the ground target's latitude of 79.6° in contrast with Figure 10.21(a)–(d).

10.4.1.2 Range Defocusing

Now examining the effects of the quadratic and cubical terms in Equation 10.73 is in order. As can be seen from Equation 10.73, both quadratic and cubical terms are merely correlated with the constant component of the TEC, indicating the range focusing is not disturbed by the temporal-spatial varying part of the background ionosphere.

According to Equation 10.73, the quadratic term inherently causes the filter mismatched, main lobe broaden, and side-lobes risen. These factors altogether contribute to deteriorate the range resolution. The cubical term may result in asymmetric distortion of the received signal, and further, bring about a ghost image in range direction.

A quadratic phase error (QPE) is defined in time domain to measure the pulse broadening [35]:

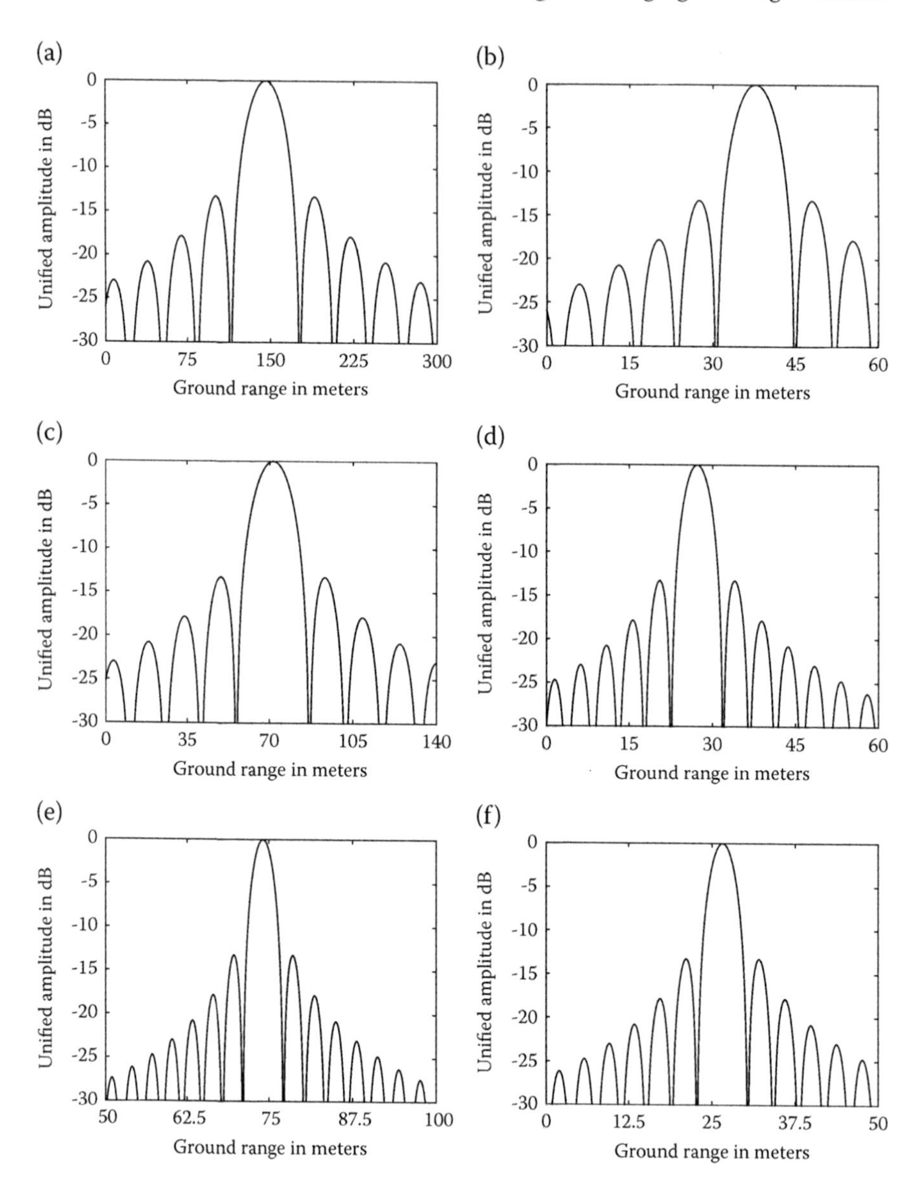

FIGURE 10.21 The 1-D range profile of point target response under the effect of range shift with: (a) $\delta_g = 22.5°$, $\delta_m = 28°$, $\Delta L_{gr} = 145.48\,\text{m}$; (b) $\delta_g = 3.4°$, $\delta_m = 28°$, $\Delta L_{gr} = 37.71\,\text{m}$; (c) $\delta_g = 39.2°$, $\delta_m = 28°$, $\Delta L_{gr} = 71.24\,\text{m}$; (d) $\delta_g = 39.2°$, $\delta_m = 0°$, $\Delta L_{gr} = 27.20\,\text{m}$; (e) $\delta_g = 79.6°$, $\delta_m = 0°$, $\Delta L_{gr} = 74.06\,\text{m}$; (f) $\delta_g = 79.6°$, $\delta_m = 28°$, $\Delta L_{gr} = 25.26\,\text{m}$. The ionospheric TEC status is set to $TEC_0 = 50$ TECU, $k_1 = 0.2$ TECU/min and $s_1 = 0.02$ TECU/km.

$$QPE = \frac{\pi AB^2}{cf_c^3}TEC_0 \tag{10.83}$$

Equation 10.83 states that the QPE is proportional to the constant component of TEC and the squared of the system bandwidth for a specific carrier frequency. From SAR imaging theory, if QPE exceeds π/4, the filter mismatch and thus the range defocusing occur. The QPE as a function of TEC_0 with different system bandwidths from 10 MHz to 200 MHz is plotted in Figure 10.22. It appears from Figure 10.22 that for the Moon-based SAR with a system bandwidth of 50 MHz, the range focusing is little affected by the QPE unless there is an ionospheric storm. But for a larger bandwidth, say 100 MHz, the range imaging severely degrades even if the TEC is just around 30 TECU.

In Figure 10.23, we plot the 1-D range profile of point target to see the effects of the QPE (no range shift). Range resolution, both idea and real, and measures of image quality in terms of peak to sidelobe ratio (PSLR) and integrated sidelobe ratio (ISLR) are listed in Table 10.5. We see that at 50 TECU the range resolution is slightly degraded for a bandwidth of 50 MHz and 100 MH; image quality in terms of PSLR and ISLR worsens for a larger bandwidth. For higher TEC at 150 TECU, range resolution at a bandwidth of 50 MHz is not affected much though PSLR and ISLR are higher– undesirable. At bandwidth of 100 MHz and 150 TECU, the range resolution is seriously degraded and PSLR and ISLR are unacceptable. In this regard, larger bandwidth makes it more difficult to maintain good image quality under the ionospheric exposure, especially in the case of ionospheric storm.

Similarly, the cubical phase error (CPE) can be defined as

$$CPE = \frac{\pi A B^3}{2 c f_c^4} TEC_0 \tag{10.84}$$

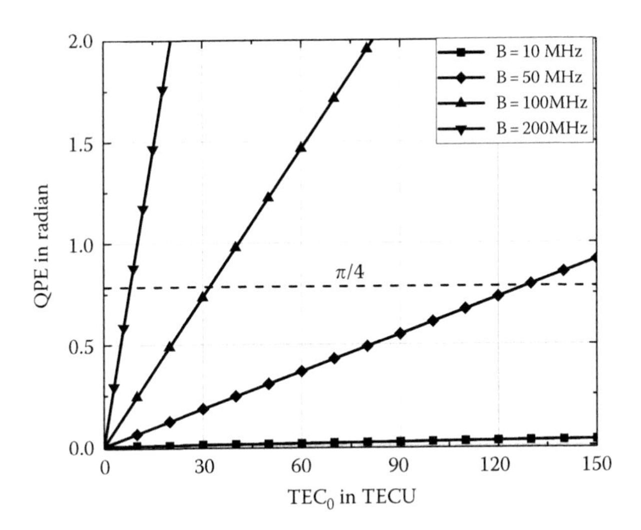

FIGURE 10.22 The quadratic phase error (QPE) as a function of TEC_0 with different bandwidths.

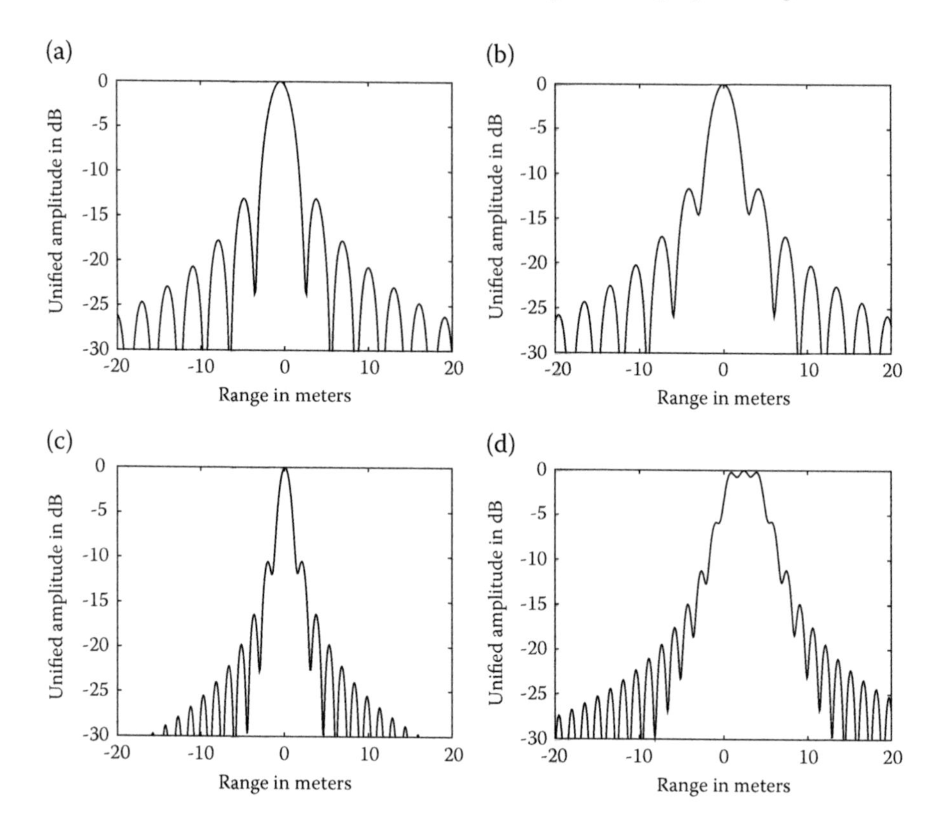

FIGURE 10.23 The 1-D range profile of point target response under the effect of the QPE with (a) B = 50 MHz, TEC_0 = 50TECU; (b) B = 50 MHz, TEC_0 = 150TECU; (c): B = 100 MHz, TEC_0 = 50TECU; (d) B = 100 MHz, TEC_0 = 150TECU.

TABLE 10.5

Numeric Measures of Image Quality of a Point Target Response in Range Direction due to QPE

	TEC_0 = 50 TECU		TEC_0 = 150 TECU	
B = 50MHz	Ideal resolution (m)	3.00	Ideal resolution (m)	3.00
	Real resolution (m)	3.10	Real resolution (m)	3.04
	PSLR (dB)	–13.07	PSLR (dB)	–11.63
	ISLR (dB)	–14.70	ISLR (dB)	–10.88
B = 100MHz	Ideal resolution (m)	1.50	Ideal resolution (m)	1.50
	Real resolution (m)	1.55	Real resolution (m)	5.29
	PSLR (dB)	–10.50	PSLR (dB)	–5.76
	ISLR (dB)	–10.00	ISLR (dB)	–14.79

Equation 10.84 indicates that the CPE is proportional to the constant component of TEC and the cube of system bandwidth. In addition, the range CPE also strongly depends on the carrier frequency. When the CPE is larger than $\pi/8$, the range imaging will be defocused. The CPE as a function of TEC_0 with bandwidths from 10 MHz to 200 MHz is plotted in Figure 10.24. It reveals that the impact of the CPE has a lesser extent than that of the QPE. It can be observed that Moon-based SAR with a bandwidth smaller than 100 MHz is barely affected by the CPE. However, when the bandwidth of 200 MHz is chosen, the CPE begins to affect the range imaging with a TEC around 50 TECU. The range profiles affected by CPE with different TECs are plotted in Figure 10.25 (ignoring the range shift) and the corresponding evaluations are given in Table 10.6. The simulation results manifest that the CPE defocus range image with a system bandwidth of 200 MHz and TEC larger than 50 TECU. It is found that a ghost image appears in range direction due to the asymmetric distortion of the received signal on account of cubical phase error when the range imaging is impacted by the CPE. As checked in Table 10.6, the image quality is deteriorated by the CPE when range defocusing appears while the range resolution varies little.

Now consider the coupled effects of the QPE and CPE on imaging focus. To see this we simulate the 1-D range profile in Figure 10.26 (ignoring the range shift), and numeric quantities of resolution and image quality measures are listed in Table 10.7. When the effects of QPE and CPE on Moon-based SAR imaging are coupled—image quality is seriously deteriorated and the range resolution is degraded dramatically with the increasing of system bandwidth and TEC. Interestingly, the PSLR of a point target response due to the effects of the range QPE and CPE strongly depends on the TEC_0 and system bandwidth.

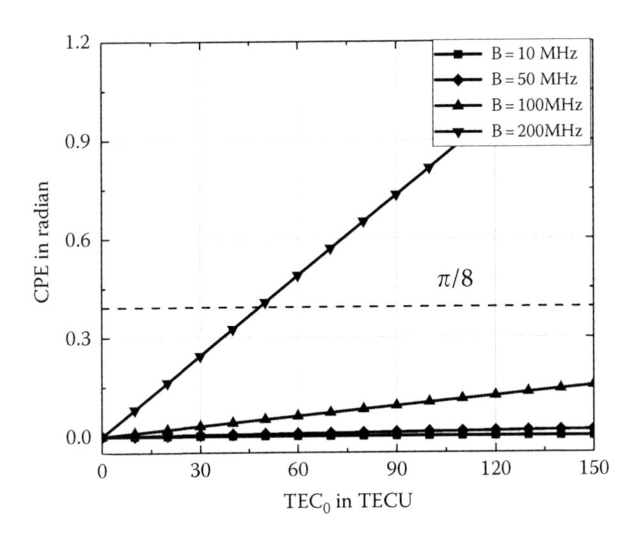

FIGURE 10.24 The cubical phase error (CPE) as a function of TEC_0 with different bandwidths.

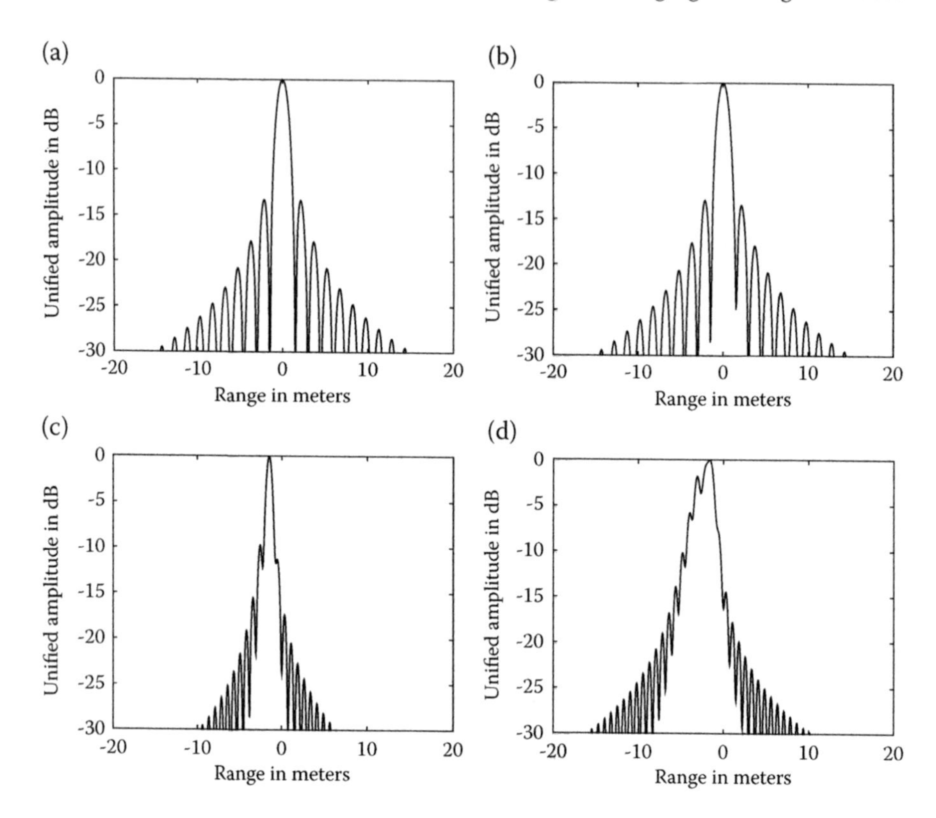

FIGURE 10.25 The 1-D range profile of point target response under the effect of the CPE with (a) B = 100 MHz, TEC0 = 50TECU; (b) B = 100 MHz, TEC0 = 150TECU; (c) B = 200 MHz, TEC0 = 150TECU; (d) B = 200 MHz, TEC0 = 150TECU.

TABLE 10.6
Numeric Measures of Image Quality of a Point Target Response in Range Direction due to CPE

	TEC = 50 TECU		TEC = 150 TECU	
B = 100MHz	Ideal resolution (m)	1.50	Ideal resolution (m)	1.50
	Real resolution (m)	1.55	Real resolution (m)	1.52
	PSLR (dB)	–13.16	PSLR (dB)	–12.91
	ISLR (dB)	–15.08	ISLR (dB)	–13.49
B = 200MHz	Ideal resolution (m)	0.75	Ideal resolution (m)	0.75
	Real resolution (m)	0.78	Real resolution (m)	2.11
	PSLR (dB)	–9.74	PSLR (dB)	–1.77
	ISLR (dB)	–9.91	ISLR (dB)	–6.54

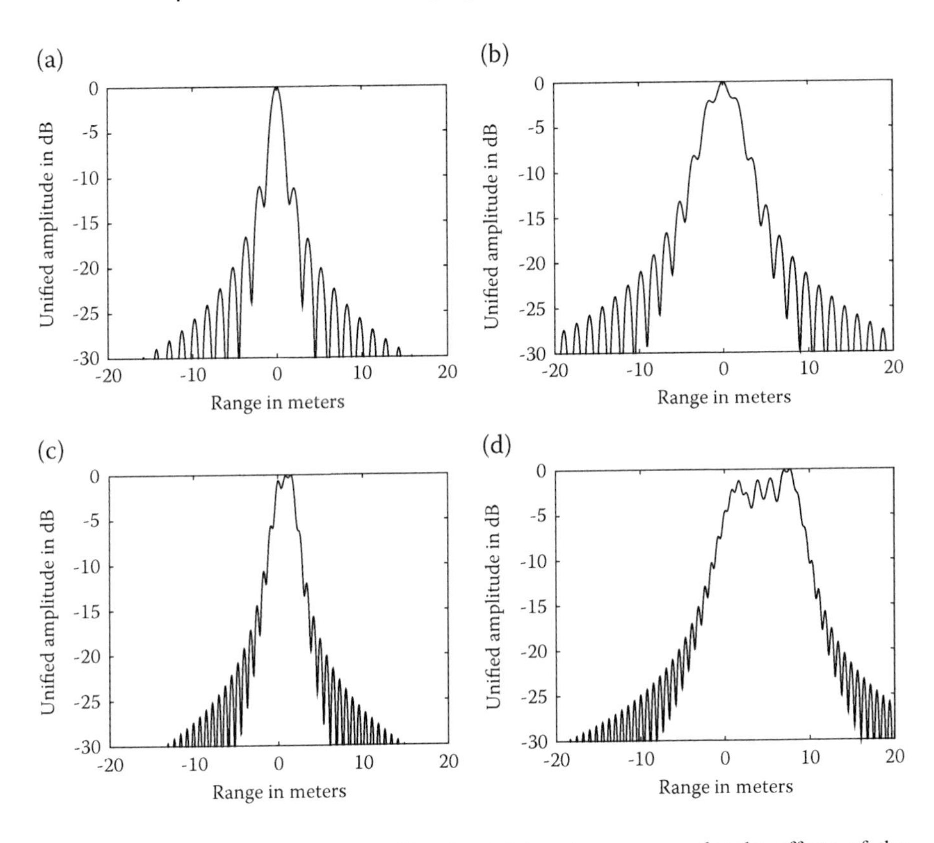

FIGURE 10.26 The 1-D range profile of point target response under the effects of the QPE and CPE with (a) B = 100 MHz, TEC_0 = 50TECU; (b) B = 100 MHz, TEC_0 = 150TECU; (c) B = 200 MHz, TEC_0 = 150TECU; (d) B = 200 MHz, TEC_0 = 150TECU.

TABLE 10.7
Numeric Measures of Image Quality of a Point Target Response in Range Direction due to QPE and CPE

	TEC = 50 TECU		TEC = 150 TECU	
B = 100MHz	Ideal resolution (m)	1.50	Ideal resolution (m)	1.50
	Real resolution (m)	1.55	Real resolution (m)	4.51
	PSLR (dB)	−11.02	PSLR (dB)	−1.81
	ISLR (dB)	−10.00	ISLR (dB)	−14.63
B = 200MHz	Ideal resolution (m)	0.75	Ideal resolution (m)	0.75
	Real resolution (m)	2.61	Real resolution (m)	7.48
	PSLR (dB)	−0.11	PSLR (dB)	−0.16
	ISLR (dB)	−9.67	ISLR (dB)	−5.36

Figure 10.27 shows the PSLR as a function of the TEC_0 with different bandwidths, from which we observe that PSLR increases with the increase of the TEC_0. Apart from that, there is a positive correlation between the PSLR and system bandwidths. It is desirable to compensate for the QPE and CPE when a meter scale of range resolution is required. Notice that at a range resolution of decametric level (B~10 MHz), both the QPE and CPE are not an issue.

10.4.2 Ionospheric Effects on Azimuth Imaging

The azimuth compression term given in Equation 10.74 is just related to the change rate of the TEC over slow time at various orders (temporal gradient and spatial gradient of TEC) but has no connection with the constant component of TEC. The TEC derivatives with respect to slow time result in changes in the Doppler frequency and Doppler FM rate and thus give rise to the distortion of the azimuth imaging. The linear term of Equation 10.74 brings about image shift in azimuth direction due to the Doppler shift whereas the quadratic term and cubical term may lead to the azimuth defocusing if it is possible.

10.4.2.1 Azimuth Shift

The azimuth shift aroused by the Doppler shift can be expressed as

$$\Delta L_a = V_{EM}\cdot\left(\frac{P_1}{2P_2} + \frac{3P_1^2P_3}{8P_2^3} + \frac{9P_3^2 - 4P_2P_4}{16P_2^5}P_1^3 - \frac{R_1}{2R_2} - \frac{3R_1^2R_3}{8R_2^3} - \frac{9R_3^2 - 4R_2R_4}{16R_2^5}R_1^3\right) \tag{10.85}$$

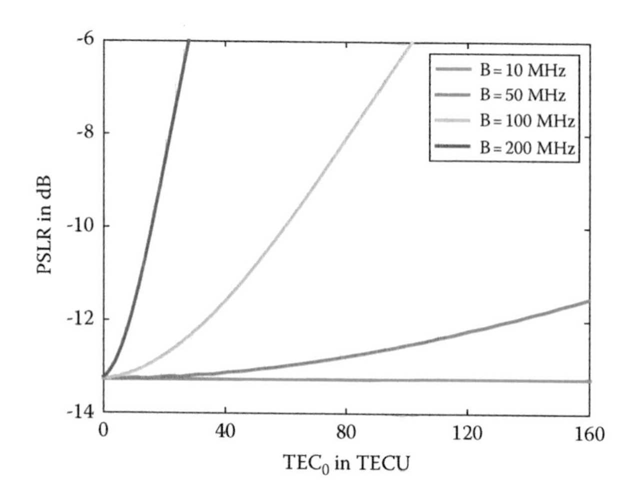

FIGURE 10.27 The PSLR as a function of the TEC_0 with different bandwidths.

where $P_i = 2A/f_c^2 t_{gi} + 2R_i$, $i = 1, \ldots, 4$, R_i, $i = 1, \ldots, 4$ is the derivative of the slant range against the slow time at i^{th} order, their specified expressions are given in [16]; V_{EM} is the relative velocity of the ground target in the context of lunar revolution on the rotating Earth, it is defined by

$$V_{EM} = R_E \cos \delta_g (\chi \omega_E) \tag{10.86}$$

where the scale factor χ accounts the lunar revolution, taking the form

$$\begin{aligned} \chi = \{1 + \omega_M^2/\omega_E^2 - 2\omega_M \cos \vartheta_S \\ /\omega_E - 2 \tan \delta_m \tan \alpha \sin \vartheta_S \omega_M (\omega_E + \omega_M \cos \vartheta_S)/\omega_E^2\}^{0.5} \end{aligned} \tag{10.87}$$

A closer look at Equation 10.85 indicates that the azimuth shift is related to the TEC temporal gradient, TEC spatial gradient, and the slant range expansion coefficients at various orders. Additionally, the azimuth shift is negatively correlated to the carrier frequency. Similar to the range shift caused by the temporal-spatial varying part of the background ionosphere, the azimuth shift is comprised of the azimuth shift ΔL_{as} caused by the spatial-varying part of the background ionosphere and ΔL_{at}induced by the temporal-varying part of the background ionosphere.

Figure 10.28 presents the azimuth shift caused by the temporal-varying, spatial-varying and temporal-spatial varying parts of the background ionosphere, from which we observe that the azimuth shift caused by the effects of the temporal-spatial varying background ionosphere is on the order of tens of meters and can be up to hundreds of meters under the ionospheric storm. In addition, the azimuth shift caused by the temporal-varying part of the background ionosphere is negatively connected with the Earth's latitude. Yet the spatial-varying background ionosphere induced azimuth shift is in positive correlation with the Earth's latitude. As for the Moon-based SAR's declination, there is a positive correlation with the azimuth shifts caused by temporal-spatial varying background ionosphere.

Figure 10.28 presents the azimuth shift caused by the temporal-varying, spatial-varying and temporal-spatial varying parts of the background ionosphere, from which we observe that the azimuth shift caused by the effects of the temporal-spatial varying background ionosphere is on the order of tens of meters and can be up to hundreds of meters under the ionospheric storm. In addition, the azimuth shift caused by the temporal-varying part of the background ionosphere is negatively connected with the Earth's latitude. Yet the spatial-varying background ionosphere induced azimuth shift, is in positive correlation with the Earth's latitude. As for the Moon-based SAR's declination, there is a positive correlation with the azimuth shifts caused by temporal-spatial varying background ionosphere.

Now that in Figure 10.29 we plot the 1-D azimuth profile of point target response under normal and storm. Ionospheres. For all cases, the exposure time is 200 s. For comparison 22.5° [see Figures 10.29(a), (b)] and 62.5° [see Figures 10.29(c), (d)] of Earth's latitudes are shown. The azimuth shift extends from tens to hundreds of meters and varies with the relative position of the ground

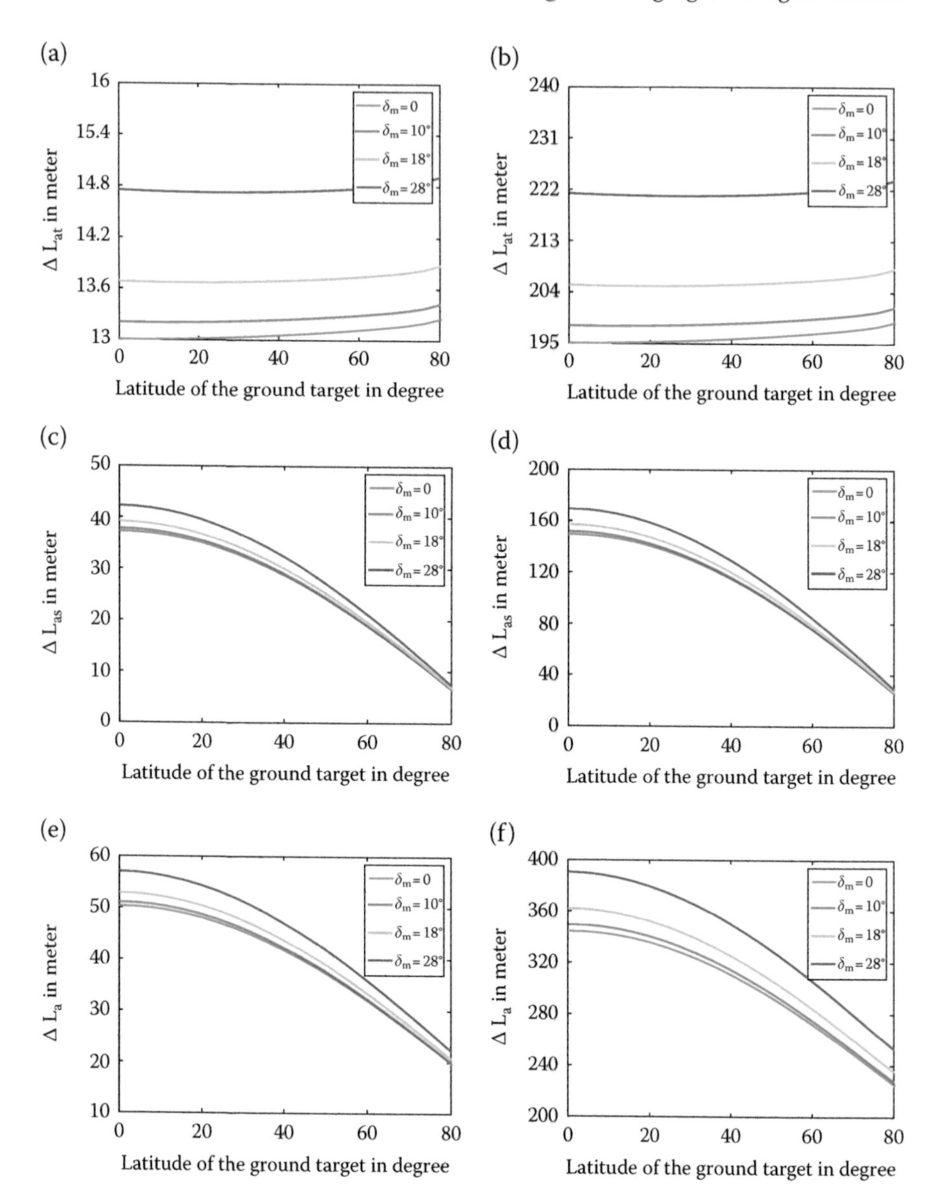

FIGURE 10.28 Azimuth shift caused by temporal-varying background ionosphere under (a) normal ionosphere, (b) ionospheric storm; the azimuth shift caused by spatial-varying background ionosphere under (c) normal ionosphere, (d) ionospheric storm; the azimuth shift caused by temporal-spatial varying background ionosphere under (e) normal ionosphere, (f) ionospheric storm.

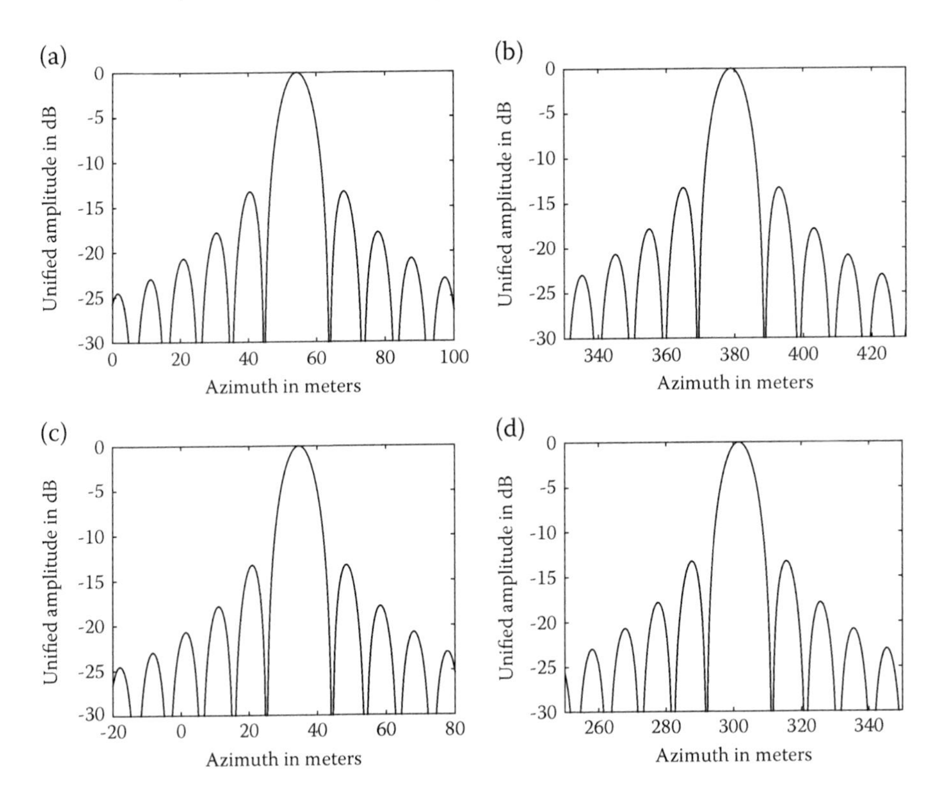

FIGURE 10.29 1-D azimuthal profile of point target response at 22.5°of latitude of under (a) normal ionosphere $\Delta L_a = 54.01$ m, (b) ionospheric storm $\Delta L_a = 378.80$ m; and at 62.5° of the latitude under (c) normal ionosphere $\Delta L_a = 34.52$ m, (d) ionospheric storm $\Delta. L_a = 301.43$ m.

target and the Moon-based SAR. It can be observed that both the ground targets' latitude and Moon-based SAR's declination conceivably induce the azimuth shift. The spatial-varying TEC temporal gradient variations further give rise to an irregular pattern of the azimuth shift, making its prediction and compensation even more difficult if not impossible.

10.4.2.2 Azimuth Defocusing

The temporal-spatial varying background ionosphere causes both azimuth shift and azimuth defocusing. The quadratic and cubical terms in Equation 10.74 broaden the main lobe, raise the side-lobe, and cause azimuthal defocusing. Here the azimuth quadratic phase error (QPE_a) is given by

$$QPE_a = \frac{\pi c}{2f_c} \Psi_Q f_{dr}^2 T_{sar}^2 \tag{10.88}$$

where f_{dr} is the Doppler frequency modulation rate, T_{sar} is the synthetic aperture time, with phase Ψ_Q given by

$$\Psi_Q = \frac{c}{4P_2f_c} + \frac{3P_1P_3c}{8P_2^3f_c} + \frac{9P_3^2 - 4P_2P_4}{32P_2^5f_c}3P_1^2c - \frac{c}{4a_2f_c} - \frac{3a_1a_3c}{8a_2^3f_c} - \frac{9a_3^2 - 4a_2a_4}{32a_2^5f_c}3a_1^2c \tag{10.89}$$

Similarly, the azimuth cubical phase error (CPE_a) is defined as

$$CPE_a = \frac{\pi c^2}{4f_c^2} \cdot \Psi_C \cdot f_{dr}^3 \cdot T_{sar}^2 \tag{10.90}$$

where Ψ_C is expressed as

$$\Psi_C = \frac{P_3c^2}{8P_2^3f_c^2} + \frac{9P_3^2 - 4P_2P_4}{16P_2^5f_c^2}P_1c^2 - \frac{a_3c^2}{8a_2^3f_c^2} - \frac{9a_3^2 - 4a_2a_4}{16a_2^5f_c^2}a_1c^2 \tag{10.91}$$

The typical thresholds of azimuth QPE and CPE are set to $\pi/4$ and $\pi/8$, respectively. As can be checked from Equations 10.88 and 10.90, the azimuth QPE and CPE are closely related to the time change rate of ionospheric TEC and the synthetic aperture time, or equivalently the synthetic aperture length. Furthermore, the TEC spatial gradient can be expressed in the form of a change rate of the TEC over time. In Figure 10.30 we plot both phase errors as a function of change rate of TEC over time by using the simulation parameters given in Table 10.3, with a synthetic aperture time of 600 s, which corresponds to an azimuthal resolution around 3 meters. It is seen that both the azimuth QPE and CPE increase with the TEC temporal, but neither of them exceeds their respective thresholds even under a TEC

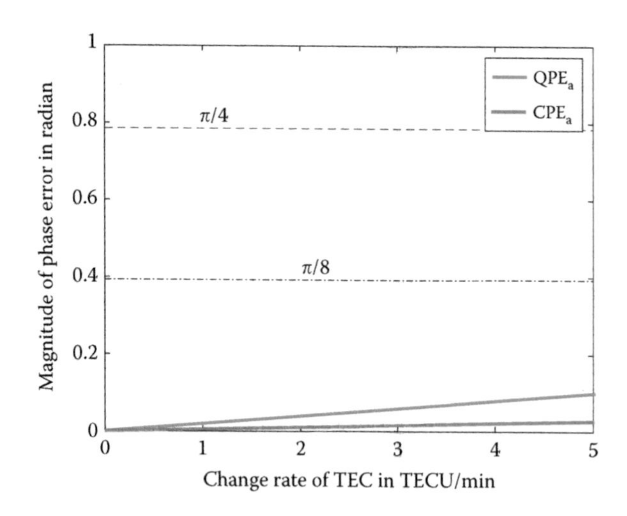

FIGURE 10.30 Azimuth QPE and CPE as a function of TEC temporal gradient with a synthetic aperture time of 600 s.

gradient of 5 TECU/s, thus the azimuth imaging of the Moon-based SAR is not disturbed by azimuth QPE or CPE in such a situation.

We now turn our attention to the magnitudes of the azimuth QPE and CPE at different Moon-based SAR's declinations as a function of the Earth's latitude. The synthetic aperture time is set to be 600 s and the ionospheric parameters are given: $k_I = 3$ *TECU/min* and $s_I = 0.08$ *TECU/km*. As detailed in Figure 10.31, the azimuth QPE has a dependency on the relative positions of the ground target's latitude and Moon-based SAR's declination where the azimuth CPE is independent of the relative positions. It can be identified that both the azimuth

QPE and CPE are smaller than their thresholds, suggesting that the azimuth focusing is not affected by neither of them wherever the Moon-based SAR and ground target locate.

For a more rigorous check, we simulate the 1-D azimuthal profile of point target response at the Earth's latitude of 22.5°and 67.5° under the influence of the azimuthal QPE and CPE in Figure 10.32, with synthetic aperture time of 600 s. The ionospheric cases are the same as Figure 10.31 while the simulation parameters are given in Table 10.3. The magnitude of azimuthal QPE and CPE and numeric measures of image quality are summarized in Table 10.8. It is found that there is no sign of defocusing in azimuth imaging at low and high latitudes, even though the synthetic aperture time is up to 600 s. At this point, it may be argued that the background ionospheric effects hardly defocus the azimuth imaging of the Moon-based SAR with an azimuthal resolution larger than 3 meters.

10.5 IMAGE SIMULATIONS AND ERROR ANALYSIS

Reminding that the background ionosphere is temporal-spatial varying which embodies in the diversity of ionospheric TEC and temporal and spatial gradients of TEC at different locations within the same imaging scene. To gain a more complete

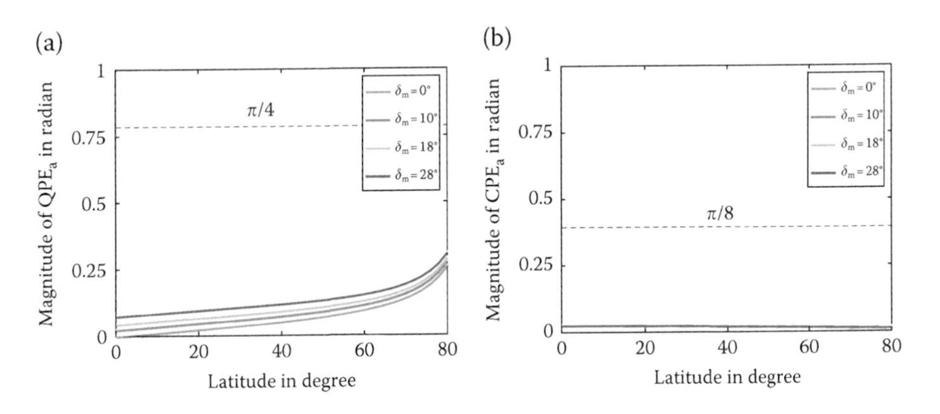

FIGURE 10.31 Azimuth phase errors as a function of Earth's latitude with a synthetic aperture time of 600 s: (a) azimuth QPE, (b) azimuth CPE.

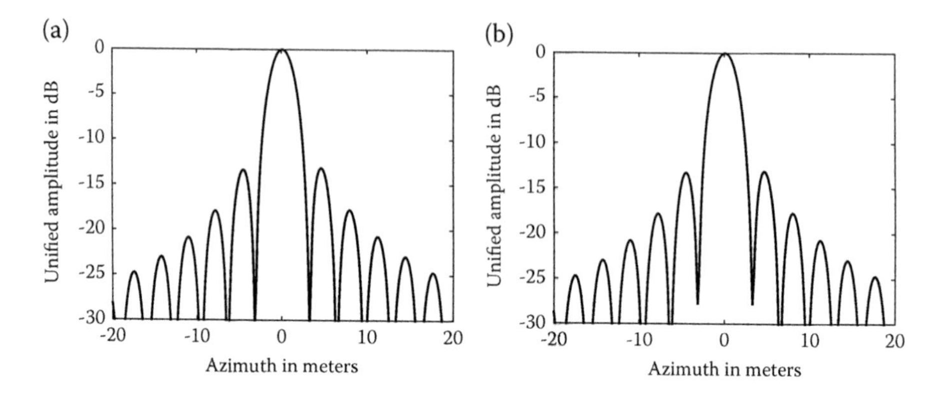

FIGURE 10.32 1-D azimuthal profile of point target response under the effects of azimuth QPE and CPE with a synthetic aperture time of 600s at the Earth's latitude of (a) 22.5° (b) 67.5°.

TABLE 10.8
Numeric Measures of Image Quality of a Point Target Response in Azimuth Direction due to Azimuth QPE and CPE

$\delta_g = 22.5°$, $\delta_m = 28.5°$		$\delta_g = 67.5°$, $\delta_m = 28.5°$	
QPE_a (rad)	−0.1031	QPE_a (rad)	−0.1826
CPE_a (rad)	0.0281	CPE_a (rad)	0.0208
Ideal resolution (m)	3.20	Ideal resolution (m)	3.20
Real resolution (m)	3.24	Real resolution (m)	3.25
PSLR (dB)	−13.14	PSLR (dB)	−13.12
PSLR (dB)	−13.68	PSLR (dB)	−13.59
QPE_a (rad)	−0.1031	QPE_a (rad)	−0.1826
CPE_a (rad)	0.0281	CPE_a (rad)	0.0208

picture, we perform image simulations to examine the image quality in range and azimuth directions, using point targets and extended target.

10.5.1 Imaging of Point Targets

Three point targets placed in the image scene, forming a triangular setting with two along azimuth direction and another in range direction. Their relative positions and system parameters are given in Table 10.9. The simulation results are shown in Figure 10.33, and numeric measures of image shift and image quality are listed in Table 10.10.

It can be perceived from Table 10.10 that the image shift and image defocusing are spatially dependent and are determined by the locally varying ionospheric TEC. Thus the temporal-spatial varying background ionosphere produces a certain amount of image shift

TABLE 10.9
Simulation Parameters of Three Point Targets at Different Positions

Parameter	Point 1	Point 2	Point 3
Position (km)	(–50,0)	(50,0)	(0,50)
TEC_0 (TECU)	20	50	150
TEC temporal gradient (TECU/s)	0.001	–0.003	0.05
TEC spatial gradient (TECU/km)	–0.005	0.03	0.07
System bandwidth (MHz)	50 (~ground range resolution of 28 m)		
synthetic aperture time (s)	70 (~azimuth resolution of 27.4 m)		
Latitude of ground target		22.5°	
Declination of Moon-based SAR		28.5°	

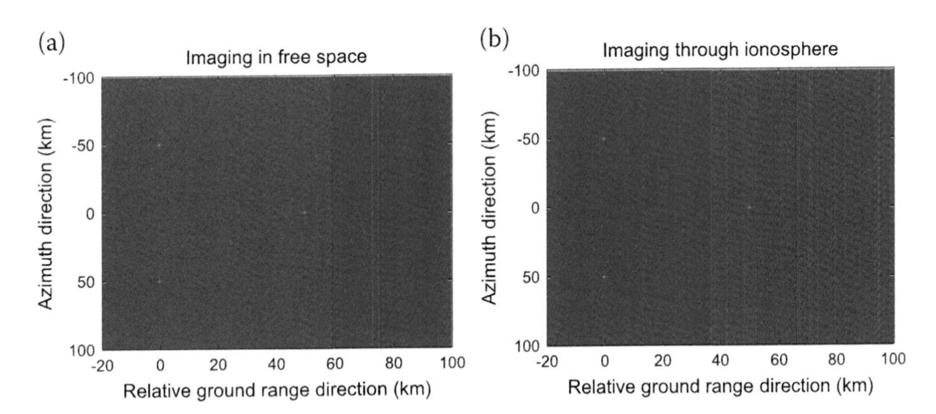

FIGURE 10.33 The multi-points' imaging (a) in free space; (b) through ionosphere (The ionosphere is exaggerated for the purpose of illustration.).

and image distortion on the single-point response. However, results in Figure 10.33 reveal that the three point targets, which form in triangular, in the image varies little simply because the image shift of each three point targets is negligibly small within the image scene. As we have already known from previous analysis, the temporal and spatial variations of background ionosphere are slow and continuous processes and they are not so extreme in practice as we assumed in the simulation. Consequently, the overall image distortion caused by the spatial-variation image shift may be ignored.

10.5.2 Imaging of Extended Target

The examination of the background ionospheric effects on the imaging of the extended area is in order. The total electron content and the change rate of TEC over time and space are assumed constant since the coverage of the earth's surface is relatively small. The digital elevation model (DEM) is shown in Figure 10.34(a), and the corresponding simulated SAR image in free space is shown in

TABLE 10.10
Numeric Measures of Image Quality of Point Targets Response

		Point 1	Point 2	Point 3
Range	Image shift (m)	58.89	133.43	404.94
	Ideal resolution (m)	3.00	3.00	3.00
	Real resolution (m)	3.10	3.04	3.04
	PSLR (dB)	−13.23	−13.05	−11.72
	ISLR (dB)	−13.71	−13.36	−11.14
Azimuth	Image shift (m)	−5.36	45.49	359.19
	Ideal resolution (m)	27.41	27.41	27.41
	Real resolution (m)	28.12	28.12	28.12
	PSLR (dB)	−13.15	−13.15	−13.15
	ISLR (dB)	−13.70	−13.70	−13.70

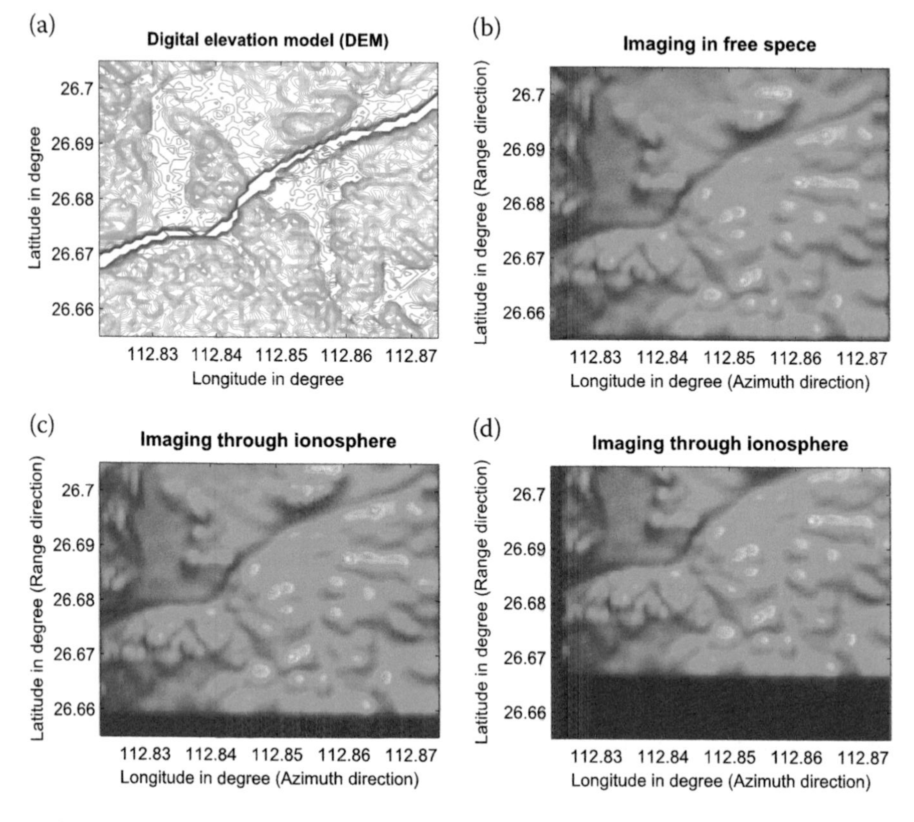

FIGURE 10.34 (a) Digital elevation model of the earth's surface; (b) Imaging in free space. Imaging through ionosphere with (c) under normal ionosphere, (d) during ionospheric storm.

Figure 10.34(b), while imaging through ionosphere with different ionospheric conditions are plotted in Figure 10.34(c) and (d), where the ground range resolution and azimuth resolution are set to 112.3 and 100.0 meters, respectively. It is clear that the background ionospheric effects show a distinct influence on a hundred meters level Moon-based SAR imaging. Though the image focusing seems not much disturbed by the background ionospheric effects, the image shift is significant, particularly at high TEC, implying that geometric correction and calibration are critical in the context of the Moon-based SAR imaging of Earth.

The background ionospheric effects on the geometric quality and fidelity of Moon-based SAR system are investigated in this paper. In so doing, a signal model considering background ionospheric effects for the Moon-based SAR based on a curved trajectory is established. Simulation results show that the range shift is on the order of hundreds of meters while the magnitude of the azimuth shift under normal solar activity condition is an order of ten meters. The image focusing by a Moon-based SAR with a scale of decameter is barely affected by the background ionospheric effects. It is found that background ionospheric effects seem not to defocus the azimuthal imaging in Moon-based SAR with an azimuthal resolution larger than 3 meters. When a geomagnetic storm occurs, which is closely related to the solar cycle, the azimuth shift can be even up to hundreds of meters, which is unacceptable for most applications.

The background ionospheric effects defocus the range imaging with a metric scale resolution. Further delving deeper into the ionospheric scintillation effects on Moon-based SAR imaging should be attempted in the future study. It is remarked that since the image quality is affected through phase distortion due to the ionospheric variations, the Moon-based SAR would potentially be an alternative to map the high-density spatiotemporal ionospheric disturbance.

REFERENCES

1. Campbell, J. B., and Wynne, R. H., *Introduction to Remote Sensing,* Guilford Press, New York, 2011.
2. Johnson, J., Lucey, P., Stone, T., and Staid, M., Visible/near-infrared remote sensing of Earth from the moon. In NASA Advisory Council Workshop on Science Associated with the Lunar Exploration Architecture White Papers; 2007. Available online: http://www.lpi.usra.edu/meetings/LEA/whitepapers/Johnson_etal_v02.pdf (accessed on 19 July 2017).
3. Renga, A., Configurations and Performance of Moon-based SAR Systems for very high-resolution Earth remote sensing. *In Proceedings of the AIAA Pegasus Aerospace Conference*, Naples, Italy, 12–13 April 2007.
4. Renga, A., and Moccia, A., Preliminary analysis of a Moon-based interferometric SAR system for very high-resolution Earth remote sensing. *In Proceedings of the 9th ILEWG International Conference on Exploration and Utilisation of the Moon, Sorrento*, Italy, 22–26 October 2007.
5. Moccia, A., and Renga, A., Synthetic aperture radar for earth observation from a lunar base: Performance and potential applications. *IEEE Transactions on Aerospace Electronic Systems,* 2010, 46(3), 1034–1051.
6. Fornaro, G., Franceschetti, G., Lombardini, F., Mori, A., and Calamia, M., Potentials and limitations of moon-borne SAR imaging, *IEEE Transactions on Geoscience and Remote Sensing,* 48(7), 3009–3019, 2010.

7. Purkis, S. J., and Klemas, V. V., *Remote Sensing and Global Environmental Change,* John Wiley & Sons, Chichester, West Sussex, UK, 2011.
8. Elachi, C., *Spaceborne Radar Remote Sensing: Applications and Techniques*, IEEE Press, New York, 1988.
9. Guo, H. D., Ding, Y. X., Liu, G., Zhang, D. W., Fu, W. X., and Zhang, L., Conceptual study of lunar-based SAR for global change monitoring. *Science China Earth Sciences* 57, 1771–1779. 2014
10. Guo, H. D., Liu, G., and Ding, Y. X., Moon-based earth observation: scientific concept and potential applications, *International Journal of Digital Earth*, 11(6), 546–557, 2018.
11. Ouyang, Z. Y., *Introduction to Lunar Science,* China Astron Publishing House, Beijing, China, 2005.
12. Chen, K. S., *Principles of Synthetic Aperture Radar Imaging: A System Simulation Approach*, CRC Press, Boca Raton, FL, 2016.
13. Xu, Z., and Chen, K. S., Temporal-spatial varying background ionospheric effects on the moon-based synthetic aperture radar imaging: a theoretical analysis, *IEEE Access*, 6, 66767–66786, June, 2018.
14. Folkner, W. M., Williams, J. G., Boggs, D. H., Park, R. S., and Kuchynka, P., The Planetary and Lunar Ephemerides DE430 and DE431, *Interplanetary Network Progress Report*, 42–196, 1–81, February 2014.
15. Xu, Z., and Chen, K. S., Effects of the Earth's curvature and lunar revolution on the imaging performance of the moon-based synthetic aperture radar, *IEEE Transactions on Geoscience and Remote Sensing*, 57(8), 5868–5882, August 2019.
16. Xu, Z., and Chen, K. S., On signal modeling of moon-based sar imaging of earth, *Remote Sensing,* 10(3), 486, March 2018. doi:10.3390/rs10030486.
17. Ye, H., Guo, H., Liu, G., and Ren, Y., Observation scope and spatial coverage analysis for earth observation from a moon-based platform, *International Journal of Remote Sensing*, 39(18), 1–25, October 2017. doi:10.1080/01431161.2017.1395976.
18. Montenbruck, O., and Gill, E., *Satellite Orbits: Models, Methods and Applications,* Springer, Berlin, 2012.
19. Xu, Z., Chen, K. S., Liu, G., and Guo, H. D., Spatiotemporal coverage of a moon-based synthetic aperture radar: theoretical analyses and numerical simulations. *IEEE Transactions on Geoscience and Remote Sensing*, 19, 1–16, May. doi:10.1109/TGRS.2020.2990433
20. Gutzwiller, M. C., Moon-Earth-Sun: The oldest three-body problem, *Reviews of Modern Physics*, 70(2), 589–639, April 1998.
21. Moreira, A., Real-time Synthetic Aperture Radar (SAR) processing with a new sub-aperture approach, *IEEE Transactions on Geoscience and Remote Sensing*, 30(4), 714–722, July 1992.
22. Gutzwiller, M. C., Moon-Earth-Sun: The Oldest Three-Body Problem, *Reviews of Modern Physics*, 70(2), 589–639, April 1998.
23. Quegan, S., and J. Lamont, Ionospheric and tropospheric effects on synthetic aperture radar performance. *International Journal of Remote Sensing*, 7(4), 525–539, April 1986.
24. Stern, S. A., The lunar atmosphere: History, status, current problems, and context. *Review of Geophysics*, 37(4), 453–492, November 1999.
25. Wright, P. A., Quegan, S., Wheadon, N. S., and Hall, C. D. Faraday rotation effects on L-band spaceborne SAR data, *IEEE Transactions on Geoscience and Remote Sensing*, 41(12), 2735–2744, December 2003.
26. Hu, C., Tian, Y., Yang, X., Zeng, T., Long, T., and Dong, X., Background ionosphere effects on geosynchronous sar focusing: Theoretical analysis and verification based on the BeiDou Navigation Satellite System (BDS). *IEEE Journal of Selected Topics in Applied Earth Observations and Remote Sensing*, 9, 1143–1162, March 2016.
27. Ishimaru, A., Kuga, Y., Liu, J., Kim, Y., and Freeman, T., Ionospheric effects on synthetic aperture radar at 100 MHz to 2 GHz. *Radio Science*, 34(1), 257–268, October 1999.

28. Liu, J., Kuga, Y., Ishimaru, A., and Pi, X., Ionospheric effects on SAR imaging: A numerical study, *IEEE Transactions on Geoscience and Remote Sensing*, 41(51), 939–947, 2003.
29. Heliospheric Physics Lab of NASA/GSFC, "International Reference Ionosphere (IRI 2012)," May, 23, 2017 [Online]. Available: http://iri.gsfc.nasa.gov/.
30. GIPP in Academy of Opto-electronics of Sciences, "Global ionospheric VTEC map produced based on real time GNSS data," July 27, 2014 [Online]. Available: http://www.gipp.org.cn/
31. Lawrence, R. S., Little, C. G., and Chivers, H. J. A., A survey of ionospheric effects upon earth-space radio propagation. *Proceedings of the IEEE*, 52(1), 4–27, January 1964.
32. Xu, Z., and Chen, K. S., Temporal-spatial varying background ionospheric effects on the moon-based synthetic aperture radar imaging: A theoretical analysis, *IEEE Access*, 6, 66767–66786, July 2018. doi:10.1109/ACCESS.2018.2853163.
33. Cumming, I. G., and Wong, F. H., *Digital Signal Processing of Synthetic Aperture Radar Data: Algorithms and Implementation,* Artech House, Boston, MA, 2005.
34. Wang, C., Zhang, M., Xu, Z. W., Chen, C., and Guo, L. X., Cubic phase distortion and irregular degradation on SAR imaging due to the ionosphere, *IEEE Transactions on Geosciences and Remote Sensing*, 53(6), 3442–3451, 2015.
35. Jehle, M., Frey, O., Small, D., and Meier, E., Measurement of ionospheric TEC in spaceborne SAR data, *IEEE Transactions on Geoscience and Remote Sensing*, 48(6), 2460–2468, June 2010.

Index

A

B

P

Q

R

9780367627997